The Smaller Mammals of KwaZulu-Natal

The Smaller Mammals
of KwaZulu-Natal

Peter Taylor

University of Natal Press
Pietermaritzburg
1998

ISBN 0 86980 942 3

Cover Design: Christeen Grant

The following mammals are illustrated on the front cover (clockwise from top):
Mops condylurus Angola free-tailed bat; *Cercopithecus mitis* Samango monkey
(Oryx Enterprises, Rosetta, KZN); *Lepus saxatilis* Scrub hare; *Elephantulus myurus*
Rock elephant shrew; *Galago moholi* Lesser bushbaby (Oryx Enterprises, Rosetta, KZN).
Back cover: *Otomys sloggetti* Sloggett's rat (Oryx Enterprises, Rosetta, KZN).

Acknowledgement to Sponsors
Richards Bay Minerals and the Durban Natural Science Museum
Friends Society are thanked for financial contributions to the
publication costs of the book.

Typeset by the University of Natal Press
Printed by Kohler Carton and Print
Box 955, Pinetown 3600, South Africa

CONTENTS

Foreword *by Dr George Hughes*   vi
Acknowledgements   vii

Colour plates appear between pages 56 and 57

Introduction   1
Species Accounts   7

Gazetteer of Collecting Localities   119
Bibliography   136
Index of Scientific Names   140
Index of Common Names   141

FOREWORD

This comprehensive record of the smaller mammals found in KwaZulu-Natal will be of benefit to both the scientific community and members of the public who have an interest in the lesser known hairy creatures. The inclusion of colour plates and the style in which the book is written, makes it user-friendly. Apart from the importance as a scientific record, the book will also go a long way towards creating an awareness of the smaller mammals in the region.

The geographical composition of the province provides a very wide range of habitats which has resulted in a high level of biodiversity. Less than 10 per cent of the terrestrial area of the region is formally protected in a network of nature reserves and parks. The area, as a whole, is generally highly modified, and nearly 25 per cent of the South African human population is to be found in KwaZulu-Natal.

Fortunately, unlike many larger mammals, the smaller ones do not require extensive areas to survive, and a number of species are quite at home in urban gardens and buildings. Others have adapted to agricultural areas and have become more abundant. However, some species are rare (e.g. the pangolin), while others (e.g. the four-toed elephant shrew) require specialised habitat, such as coastal forests which are under threat.

These smaller mammals are often at the lower end of the food chain, and birds such as Barn Owls and Black-shouldered Kites are dependent on them as a food source. Some are useful in gardens to control certain invertebrates, while others are regarded as agricultural pests. All-in-all, they are an extremely interesting group, and an integral part of KwaZulu-Natal's ecological systems.

Without these smaller mammals, our greater systems would not function as they should.

Dr G. R. Hughes
Acting Chief Executive Officer
KwaZulu-Natal Nature Conservation Service

ACKNOWLEDGEMENTS

My employer, the Director of the Durban Natural Science Museum, Dr Q. Brett Hendey, provided me with unlimited support and the freedom to pursue this project wholeheartedly over a long period of time.

The Acting Chief Executive Officer and staff of KwaZulu-Natal Nature Conservation Service, and particularly David Rowe-Rowe and Ortwin Bourquin, are thanked for permission to collect mammals, logistic support in the field, and access to data on distribution and occurrences of mammals in protected areas. David Rowe-Rowe's cheerful encouragement, and editorial and scientific advice from the outset of the project was a continual source of inspiration.

Drs Brian Stuckenberg and Jason Londt, previous and present directors of the Natal Museum, are thanked for the long-term loan and use of the Natal Museum's mammal collection.

This book relied on the efforts of a number of people in computerising and checking museum specimen information, and obtaining latitude and longitude data for a great many localities: Peter and Marian Neal, Robert Grove-Annesley, Ernest Seamark, Eve Johnson and Greg Davies.

The mammal collection of the Natal Museum and parts of the Durban Natural Science Museum collection were curated by Kate Richardson.

Reuben Ngubane, Robert Grove-Annesley, Peter Neal, Neville Pillay, Teresa Kearney, Michael Roberts, A. R. Ali, Michelly Rall and Kate Richardson assisted with mammal collecting trips.

The curators of a number of museums are thanked for contributing museum specimen data on KwaZulu-Natal mammals: Gary Bronner and Chris Chimimba (Transvaal Museum), Lloyd Wingate (Kaffrarian Museum), Denise Drinkrow (South African Museum), Paula Jenkins (The Natural History Museum, London), Duane Schlitter (Carnegie Museum) and M. Carleton (Smithsonian Institution).

Colin Sapsford, Teresa Kearney, Ernest Seamark, Fiona Mackenzie, Kate Richardson and Fiona and Naomi Radford provided information on bat specimens and capture-release data. Members of the Durban Bat Interest Group enthusiastically contributed towards much of the information gathered on bats between 1994 and 1997.

David Rowe-Rowe, Gary Bronner, Peter Henzi, Mike Lawes, Mike Perrin, Ara Monadjem and Adrian Armstrong are thanked for proofreading various drafts of the manuscipt and providing constructive comments for its improvement.

This book benefitted from the enthusiastic expertise and technical competence of the staff of the University of Natal Press, Sally Hines, and Kohler Carton and Print.

The photographers (credits with the colour plates) are thanked for donating photographs for free, or for a nominal fee to help offset production costs.

Finally, my wife, Frances, and four children – Ashleigh, Robyn, Lauren, and Benjamin – are thanked heartily for enduring my absences during numerous long nights and weekends.

I dedicate this book to the memory of my mentor, the late J. A. J. ('Waldo') Meester, a fine mammalogist with a fine heart.

To a Mouse

On Turning Her up in Her Nest with the Plough,
November, 1785.

Wee, sleeket, cowrin, tim'rous beastie,
O, what a panic's in thy breastie!
Thou need na start awa sae hasty,
      Wi' bickerin brattle!
I wad be laith to rin an' chase thee,
      Wi' murderin' pattle!

I'm truly sorry man's dominion,
Has broken nature's social union,
An' justifies that ill opinion,
      Which makes thee startle
At me, they poor, earth-born companion,
      An' fellow-mortal!

But Mousie, thou art no thy lane,
In proving foresight may be vain;
The best-laid schemes o' mice an' men
      Gang aft agley,
An' lea'e us nought but grief an' pain,
      For promis'd joy!

Still thou art blest, compar'd wi'me!
The present only toucheth thee:
But och! I backward cast my e'e,
      On prospects drear!
An' forward, tho' I canna see,
      I guess an' fear!

ROBERT BURNS

INTRODUCTION

This text covers the lesser known 'smaller mammals' of the province of KwaZulu-Natal, South Africa. The term smaller mammals, as used in this book, includes rodents, insectivores (shrews and golden moles), bats, elephant shrews, hares, primates, dassies, the aardvark and the pangolin, i.e. all mammals excluding the larger hoofed mammals (or ungulates) and carnivores (Order Carnivora) of KwaZulu-Natal.

Perhaps because they are seldom seen, the public is generally uninformed or even fearful of bats, mice, shrews and other small mammals. Few realise that they are actually more diverse in terms of numbers of species than the larger mammals – 103 smaller mammal species are found in KwaZulu-Natal, compared to 65 larger mammal species. Because of its climatic and topographic variability, KwaZulu-Natal is faunally extremely rich, containing *over one half* of all the smaller mammal species of *the entire southern African subregion.*

Even fewer people appreciate the (mostly positive) impact that small mammals have on our lives, and on the well being of the planet's ecosystems. A few rodent species can cause extensive damage to crops and forestry plantations in South Africa. Rodent ecologist, Dr Ken Willan, estimated in 1992 that rodents cost at least R600 million nationally in local losses to agriculture, or some four per cent of the country's annual earnings from agriculture. Nevertheless, rodents comprise a vitally important prey source for many predators such as serval, caracal and many birds of prey. Methods of rodent control should avoid the use of environmentally harmful chemicals: trapping and the use of natural predators (e.g. introducing raptor perches) are recommended. The same principle applies to the control of garden pests such as the insectivorous golden moles and the herbivorous rodent moles.

Bats have been unfairly misrepresented over the centuries as hair-clinging, blind harbingers of evil. In recent years, scientists have demonstrated the value of bats as major predators of night-flying insects, and potential controllers of many agricultural insect pests. Fruit bats play a vital role in seed dispersal, and account for up to 95 per cent of forest regeneration following clearing in tropical regions. At least 300 plants are known to be dependent on fruit bats for their pollination services, many of them economically important agricultural species which provide more than 450 economically important products valued at hundreds of millions of $US annually. During the mid 1990s local bat interest groups have become established in the Durban and Pretoria areas of South Africa, with the aim of educating the public about bats and bat conservation.

While the primary aim of this text is to disseminate new scientific data on the distribution and biology of the smaller mammals of KwaZulu-Natal, identification keys, colour photographs and summary notes on habits have been added, in the hope of 'selling' KwaZulu-Natal's fascinating mini wildlife to a wider sector of the educated lay public.

Scientific background

By ratifying the global Convention on Biology Diversity in 1995, South Africa has committed herself to mapping and conserving our nation's biodiversity heritage, our fauna and flora. Species-based conservation efforts depend on accurate species distribution maps based on scientifically valid records (e.g. museum specimens or reliable sightings). For most of the smaller mammals, distribution maps are based on museum records. Distribution maps for the present text were based on over 15 000 museum specimen records compiled from the collections of eight natural science museums around the world.

Accurate regional distribution maps are available for Botswana (Smithers 1971), Zimbabwe (Smithers and Wilson 1979), Mozambique (Smithers and Tello 1976), Lesotho (Lynch 1994) and the following provinces of South Africa: greater Transvaal (Rautenbach 1982), Free State (Lynch 1983), northern Eastern Cape (Lynch 1989), Western Cape south of the Orange River (Swanepoel, unpublished data), and Northern Cape and parts of Northwest Province (B. H. Erasmus, unpublished data). Accurate distribution maps of carnivores are available for the former Cape Province (Stuart 1981), as are distribution maps for ungulates (Rowe-Rowe 1994) and carnivores (Rowe-Rowe 1992) in KwaZulu-Natal. De Graaff (1981) produced distribution maps of rodents (plotted by 1/16th-degree grid squares) for the whole of southern Africa, although actual records of occurrence were not published.

No single publication on KwaZulu-Natal mammals containing accurate distribution maps of all species has hitherto been available. Previous publications on KwaZulu-Natal mammals have concentrated on specific groups such as carnivores (Pringle 1977; Rowe-Rowe 1978; Rowe-Rowe 1992), ungulates (Rowe-Rowe 1994), larger mammals (Mentis 1974), primates, hares, elephant shrews and ant-eating mammals (Pringle 1974).

The first synthesis of available data on the smaller mammals of KwaZulu-Natal (the orders Insectivora, Chiroptera, Primates, Pholidota, Lagomorpha, Rodentia and Hyracoidea) was contained in an internal report of the former Natal Parks Board produced by O. Bourquin in 1988. Distribution maps were based mainly on museum records in the collections of the Transvaal Museum, Natal Museum, Durban Natural Science Museum and Kaffrarian Museum, as well as on sight and literature records.

Between 1989 and 1997, some 5 000 mammal specimens from KwaZulu-Natal were added to the mammal collection of the Durban Natural Science Museum. Of this total, 900 specimens of 48 species were collected during intensive small mammal surveys of 22 localities in KwaZulu-Natal. More recently (1994–1997), many important new records of bats have been collected through the efforts of the Durban Bat Interest Group. Data extracted from eight museum collections have been used to compile the distribution maps in this book: Carnegie Museum of Natural History, Durban Natural Science Museum, Kaffrarian Museum, Natal Museum, the Natural History Museum in London, South African Museum, Smithsonian Institution, and Transvaal Museum. Eight new species have been added to Bourquin's (1988) list of KwaZulu-Natal small mammals.

This book covers all the orders of mammals represented in KwaZulu-Natal apart from the orders Carnivora, Perissodactyla, Artiodactyla and Proboscidea, which were addressed by Rowe-Rowe (1992, 1994). The three books together form a complete series on the mammals of KwaZulu-Natal. The same (1/64th-degree square, 7.5′ x 7.5′) map scale used by Rowe-Rowe (1992, 1994) has been adopted in the present book for the sake of consistency and comparability. While the primary focus of this book (as indicated by the title) is the smaller mammals of KwaZulu-Natal, some larger forms such as the aardvark, pangolin and larger primates are included for completeness.

Apart from the Carnivora (32 species), Perissodactyla, Artiodactyla and Proboscidea (33 species), 103 indigenous, extant terrestrial mammal species of nine orders are represented in KwaZulu-Natal: Insectivora (shrews: 13 species; golden moles: five species), Chiroptera (bats: 38 species), Primates (bushbabies, monkeys and baboons: five species), Hyracoidea (dassies: two species), Tubulidentata (aardvark: one species), Pholidota (pangolin: one species), Rodentia (rodents: 33 species), Lagomorpha (hares: three species) and Macroscelidia (elephant shrews: two species).

In addition to producing high resolution maps for each species, a further aim of this book is to include new information on the taxonomy and biology of each species. New information is presented under several subheadings within each species account, as described in the section below.

How to use this book

This text is primarily a synthesis of new scientific information on 103 small mammal species found in KwaZulu-Natal, obtained primarily from museum specimens. Information is organised under subheadings for each species account as explained below. Identification keys and photographs of some species are included to assist the interested layperson or beginner. In many cases, e.g. some shrews, bats and rodents, identification of species remains extremely difficult, even for experienced mammalogists, and anyone requiring a definite identification of a small mammal should contact the Durban Natural Science Museum or nearest natural science museum.

Taxonomic status

Scientific and common names and taxonomic arrangements follow Meester et al. (1986), except where recent studies have revealed taxonomic changes (e.g. *Amblysomus* (golden moles), *Mastomys* (multimammate mice), *Aethomys* (red veld rats) and *Myosorex* (forest shrews)). The current taxonomic status of each species and subspecies is presented, with views on material from KwaZulu-Natal. In the case of cryptic species (e.g. *Mastomys natalensis* and *M. coucha*), valid subspecies (e.g. *Amblysomus hottentotus* and *Paraxerus palliatus*), or elevation of subspecies to specific rank (e.g. *Myosorex sclateri*), where material could be diagnosed, separate distribution maps were plotted. Where available, karyotypes of chromosomally polymorphic species from KwaZulu-Natal are presented.

Distribution

The distribution of each species in KwaZulu-Natal is briefly described. Distribution maps for each species were plotted by 1/64th-degree grid squares (7.50 x 7.50). For most of the small mammals, distribution maps were based entirely on museum records, or, in the case of bats, verified sight records of roosts (Durban Bat Interest Group, unpublished data) or specimens identified and sent for rabies testing during a rabies scare (Sapsford, unpublished data). In the case of the larger, more visible mammals such as hares, porcupines, dassies, primates, pangolin and aardvark, museum records were supplemented with records from KwaZulu-Natal Nature Conservation Service surveys, questionnaires, or sight records. This information was kindly supplied by D. Rowe-Rowe, formerly of the KwaZulu-Natal Nature Conservation Service. Distributional records for tree dassies (*Dendrohyrax arboreus*) were obtained mostly from a recent study by N. Langley which employed eliciting call responses with tape recordings of tree dassie calls.

Additional localities obtained from the literature are indicated by open squares on distribution maps (as opposed to closed squares for specimen records or verified sight records). Prehistorical distribution of small mammals in KwaZulu-Natal was discussed with reference to Avery's (1991) extensive study of micromammalian ocurrence at upper Pleistocene and Holocene owl pellet deposits from throughout KwaZulu-Natal. In a few instances historical changes in distribution could be inferred from dates associated with specimens in museum collections.

Protected areas

Former Natal Parks Board and KwaZulu Department of Nature Conservation reserve records are given for those species so far recorded in reserves (data provided by D. Rowe-Rowe). This gives some indication of the level of protection, an important consideration in determining conservation status and management of species.

Conservation status

South African Red Data Book status (Smithers 1986) is given wherever applicable. The conservation status of each species in KwaZulu-Natal, based on available information on habitat specificity, distribution, commonness, level of protection and degree of endemism, is described.

Habitat
Detailed habitat data were recorded for all small mammals collected during small mammal surveys undertaken by the Durban Natural Science Museum. For those species which were not personally collected by the author, habitat occurrence was determined from museum specimen label data as well as from the literature.

Habits
Information on habits and diet was summarised briefly from the literature. Personal observations on behaviour are appended where they provide additional or conflicting information.

Breeding
Reproductive data obtained from most small mammals collected in KwaZulu-Natal between 1989 and 1997 were used to estimate litter size and the timing and duration of the reproductive season. This was supplemented by available literature on KwaZulu-Natal mammals.

Linear measurements and mass
Standard measurements were obtained for most of the small mammals collected in KwaZulu-Natal between 1989 and 1997. Such data, although of limited scope for taxonomic purposes, can be used for assessing sexual dimorphism and geographic variation. In particular, body mass is a useful ecological variable. Certain mensural variables (e.g. forearm length in many bats) have diagnostic value. Measurements of length and mass also provide the lay person with a standard reference of size. The following standard measurements were recorded in millimetres (except for mass, which is recorded in grammes):

TL	Total length from nose to tip of tail
HB	Head and body length, excluding the tail
T	Tail length
Hf*su/cu*	Hind foot length, excluding (*sine unguis*) and including (*cum unguis*) the claw. Unless stated otherwise Hf is reported without claw (*su*).
E	Ear length measured from notch to furthest margin
FA	Forearm length (only measured for bats)
Mass	Total body mass

Records of occurrence
In addition to a map illustrating the distribution of each species in KwaZulu-Natal, a list is provided of localities where specimens were collected (specimen records) and where species were sighted or recorded from the literature (additional records). Abbreviations denoting the museum of origin of specimen records are given for the following museums:

BM:	The Natural History Museum, London
CM:	Carnegie Museum of Natural History, Pittsburgh
DM:	Durban Natural Science Museum, Durban
KM:	Kaffrarian Museum, King William's Town
NM:	Natal Museum, Pietermaritzburg
NMB:	National Museum, Bloemfontein
SI:	Smithsonian Institution, Washington DC
SAM:	South African Museum, Cape Town
TM:	Transvaal Museum, Pretoria

Where no museum is stipulated in literature records, the literature reference is given. Both here and in the general body of the text, locality names employed the following abbreviations:

NR	Nature Reserve
GR	Game Reserve
PGR	Private Game Reserve
NP	National Park.

Study Area
Location, size and topography
The province of KwaZulu-Natal, South Africa, lies between 26°45′ and 31°10′S; 28°45′ and 32°50′E and covers 91 800 km² (approximately 450 x 200 km) (Figure 1). The Drakensberg forms a natural western boundary of the province. KwaZulu-Natal is bounded to the south by the former Transkei (Eastern Cape), to the north by the Mpumalanga Province, Swaziland and Mozambique, and to the east by the Indian Ocean. Altitude ranges from sea level in the east to over 3 400 m on the Drakensberg in the west. The drainage for the province comprises several major east-flowing rivers rising either in the Drakensberg or the Lebombo Mountains (Figure 1).

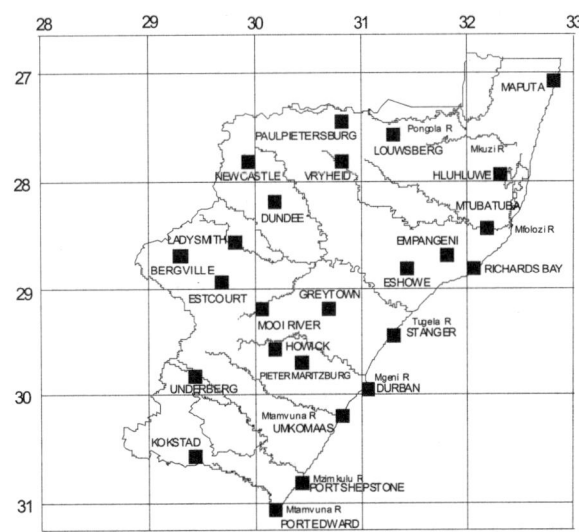

Figure 1: ***Major towns and rivers of KwaZulu-Natal***

Climate and vegetation
Based on previous classifications of vegetation types by Acocks (1975), Camp (1995) and Phillips (1973), and following Rowe-Rowe and Taylor (1996), nine 'bioregions' were recognised (Table 1; Figure 2), namely: **Coast lowlands** (evergreen grassland, and tropical forest and thicket);

Table 1: Bioregions of KwaZulu-Natal and summary of their climatic characteristics (After Rowe-Rowe and Taylor 1996)

Region	Altitude (m)	Annual rainfall (mm)	Temperatures °C* Winter Min	Winter Max	Summer Min	Summer Max
Coast lowland	0–450	800–1 200	10	24	21	32
Coast hinterland	450–900	750–1 300	7	22	17	28
Lowveld	150–1 000	550–900	12	24	22	32
Valley bushveld	0–900	620–720	4	17	22	30
Mistbelt	900–1 400	750–1 500	3	19	16	27
Moist upland	900–1 400	700–1 000	2	21	15	27
Drier upland	900–1 000	720–760	3	24	15	31
Highland	1 400–1 800	700–1 250	1	17	13	25
Montane	1 800–3 500	1 200–1 800	0	16	13	23[1]
			-7	10	6	18[2]

*   Temperatures are mean daily minimum and mean daily maximum for the coldest month in winter and warmest month in summer

[1]   Measured at 1 800 m

[2]   Measured at 3 000 m

Coast hinterland (grassland and semi-deciduous woody vegetation); **Mistbelt** of the midlands (grassveld and Afro-montane forest); **Moist upland** (tall grassveld and open savanna); **Drier upland** (tall grassveld and open savanna, but drier than the previous region); **Highland** (grassland with short, dense cover, and patches of Afro-montane forest); **Montane** (temperate grassland and fynbos); **Lowveld** (low-lying semi-deciduous and evergreen wooded areas of the north-east interior); and **Valley bushveld** (thicket and scrub, mainly *Acacia spp.,* of the lower reaches of the major river valleys). Climatic characteristics of the different bioregions are given in Table 1.

Land use in KwaZulu-Natal

Protected areas in KwaZulu-Natal occupy 9.6 per cent of the surface area (6.3 per cent under former Natal Parks Board control and 3.3 per cent under the former KwaZulu Department of Nature Conservation). Privately-owned farms occupy about 54 per cent and subsistence farming areas in KwaZulu account for about 35 per cent. The distribution of protected areas in KwaZulu-Natal is shown in Figure 3.

Over much of KwaZulu-Natal more than 20 per cent of private land is cultivated. In the coastal belt, midlands, and parts of western KwaZulu-Natal, over 50 per cent of land is cultivated. While certain farming practices – e.g. maize, sheep and beef cattle – are fairly widely distributed over KwaZulu-Natal, other practices are specific to regions. For example, sugar cane is restricted to the coastal belt,

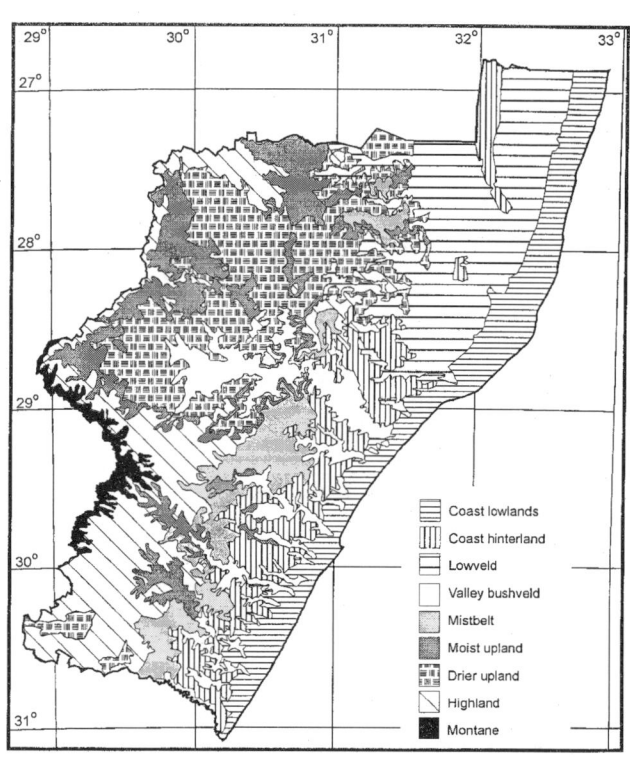

Figure 2:   Distribution of bioregions in KwaZulu-Natal. Division of KwaZulu-Natal based on Acocks (1975), Camp (1995) and Phillips (1973) (After Rowe-Rowe and Taylor 1996)

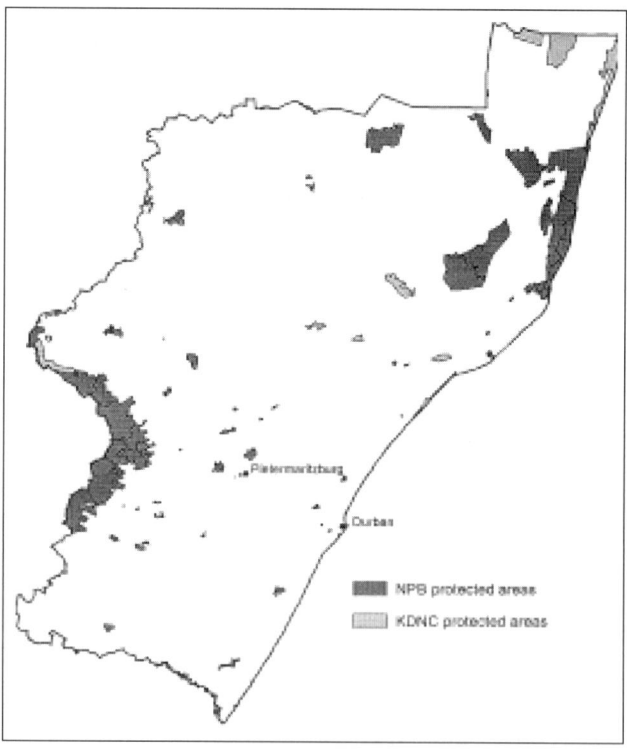

Figure 3:   Distribution of formally protected areas of KwaZulu-Natal (After Rowe-Rowe and Taylor 1996)

while pasture grasses, dairy cattle, gum, wattle and pine plantations are mostly concentrated in the midlands. Mixed farming is practised on a wide scale in KwaZulu-Natal. Further information on agricultural practices in KwaZulu-Natal can be obtained from Fotheringham (1981).

Material and Methods
Data collection
Bourquin's (1988) report on the smaller mammals of KwaZulu-Natal incorporated data from four sources: (1) requests to South African museums (Natal Museum, Durban Museum, Kaffrarian Museum, Transvaal Museum), (2) personal checking of some museum databases, (3) collection of specimens and personal sight records, and (4) published and unpublished literature on KwaZulu-Natal mammals. Information was gathered up to 31 March 1988.

Subsequent to Bourquin's report, an additional 5 000 mammals collected in KwaZulu-Natal have been added to the mammal collection of the Durban Natural Science Museum. In addition to these new records, some 5 000 KwaZulu-Natal mammal records were obtained from the Transvaal Museum, 2 000 from the Natal Museum, 1 000 from the Smithsonian Institution, 600 records from the Kaffrarian Museum, 200 from the South African Museum in Cape Town, 100 from the Natural History Museum in London, and 50 from the Carnegie Museum of Natural History in Pittsburgh. Unpublished data on bats were obtained from C. Sapford, E. Seamark and the Durban Bat Interest Group. Sight records for the larger mammals were obtained with permission from the files of the KwaZulu-Natal Nature Conservation Service. A total of eight new small-mammal species have been added to Bourquin's list of mammals for KwaZulu-Natal.

Of the small mammals accessioned in the Durban Natural Science Museum since July 1989 (when the post of mammalogist was created at this museum), some 900 specimens of 48 species were collected during small-mammal surveys of 22 localities, mostly KwaZulu-Natal Nature Conservation Service reserves. Small-mammal surveys of 3–10 days duration employed a standard method. Traplines

of between 20 and 50 traps each (usually 50) were placed in each habitat represented, and detailed habitat descriptions were made for each trapline. Between 100 and 300 traps in total were set per night. The most commonly used rodent traps were 'Museum Special' kill-traps, 'Titian' folding aluminium live-traps ('Sherman'-type) and 'Willan' PVC construction live-traps (Willan 1979). Between one and 20 'mole' traps were set each night in golden and rodent mole burrows. Mole traps included 'Macabee' traps, two sizes of 'scissors' traps (available from Collingwoods in Durban and Joubert of Cape Town) and 'Hickman' tunnel traps.

On most nights, one or two mistnets (12 x 2.7 m; 6 x 2.1 m) were erected in likely places near water, in open forest clearings, picnic sites, etc. Caves, thatched shelters and other likely daytime roosting sites for bats were systematically searched. Searches were also made for owl pellets which were collected and later analysed. Termitaria were excavated to search for dwarf shrews (*Suncus spp.*)

Every specimen collected was euthenased with ether, measured, and its reproductive condition and habitat details recorded. Study skins were prepared in the field, except in the case of certain chromosomally polytypic species (e.g. *Mastomys spp.* (multimammate mice), *Otomys spp.* (vlei rats), *Graphiurus murinus* (dormice), *Saccostomus campestris* (pouched mice) and *Dasymys incomtus* (water rat)) where live animals were transported back to the laboratory for karyotype analysis. In some cases, liver and kidney tissues were collected into liquid nitrogen.

In addition to collecting trips and specimen donations to the Durban Natural Science Museum, new data were obtained by means of the curation of the Natal Museum and early Durban Museum collections. Both collections contained a substantial number of incorrectly identified or unidentified specimens. This task was mostly undertaken during 1992 and 1993 by one person, Mrs Kate Richardson, Research Associate of the Durban Natural Science Museum, with the author's collaboration. As a result of this exercise, a number of erroneous distribution records in Bourquin (1988) were corrected.

SPECIES ACCOUNTS

<table>
<tr><td>Order   INSECTIVORA</td></tr>
</table>

1. Visible eyes, ears and tail. Fore-claws not
   modified for burrowing . . . .
   Family SORICIDAE, p. 7
－ No visible eyes, ears or tail. Fore-claws
   modified for burrowing . . . .
   Family CHRYSOCHLORIDAE, p. 7

<table>
<tr><td>Family   SORICIDAE   Shrews
Subfamily CROCIDURINAE</td></tr>
</table>

KEY TO GENERA (Meester et al. 1986):

1. Skull with braincase sharply angled
   laterally in squamosal region; paired
   dorsal foramina in frontal region; second
   upper unicuspid (I^3) clearly smaller than
   first (I^2) and third (C); fourth unicuspid
   (P^3) present, minute . . . . *Myosorex*, p. 7
－ Skull with braincase smoothly curved
   or only slightly angled in squamosal
   region, no paired dorsal foramina in
   frontal region; second upper unicuspid
   (I^3) not clearly smaller than first (I^2)
   and third (C); fourth unicuspid (P^3)
   absent or present and conspicuous 2

2. Fourth upper unicuspid absent . . . .
   Crocidura, p. 11
－ Fourth upper unicuspid present,
   conspicuous . . . . *Suncus*, p. 18

<table>
<tr><td>Genus   Myosorex   Gray 1837</td></tr>
</table>

KEY TO SPECIES (Modified from Meester et al. 1986,
Kearney 1993):

1. Colour duller and paler, more greyish-
   brown or brown; ventral colour greyish-
   fawn; hind feet off-white; tail bicoloured;
   anterior palate with rear margin of lateral
   fissures overlapping front margin of
   median fissure . . . . *M. varius*, p. 7
－ Colour richer, more reddish-brown or
   blackish-brown; ventral colour
   yellowish-brown; hind feet and tail
   brown to black; tail not bicoloured;
   anterior palate with lateral palatal
   fissures anterior to median fissure 2

2. Smaller in overall size; head and body
   50–105 mm, condylo-incisor skull length
   21–24 mm . . . . *M. cafer*, p. 9
－ Larger in overall size; head and body
   53–118 mm; condylo-incisor skull
   length 24–27 mm . . . . *M. sclateri*, p. 10

Myosorex varius (Smuts 1832)   PLATE 1
Forest shrew
Bosskeerbek

Taxonomic status
No subspecies have been recognised (Meester et al. 1986).
The species is often difficult to separate from *M. cafer*; in
Zimbabwe the morphological features separating the two
species completely break down, and only one species (*M. cafer*) is recognised. However, Kearney (1993) showed that,
in KwaZulu-Natal, *M. varius* can be separated clearly from
both *M. cafer* and *M. sclateri* using multivariate
morphometric methods.

Distribution

Myosorex varius is sympatric with *M. cafer* over much of its distribution within KwaZulu-Natal (Figure 4). However, owing to its broader habitat tolerance, *M. varius* has a wider distribution than *M. cafer* in KwaZulu-Natal, occupying Montane, Mistbelt, Coast hinterland, Coast lowland and Moist upland bioregions, while the latter occupies Mistbelt, Coast lowland and Coast hinterland regions. Both species are absent from bushveld habitats. *Myosorex varius* and *M. cafer* were collected together at the following localities: Clearwater Farm, Karkloof Nature Reserve, Kilgobbin Farm, Lidgetton, Ngome State Forest, Pietermaritzburg, Redlands, Umtamvuna Nature Reserve, Umtwalume River. Rautenbach (1982) collected the two species together at only one locality in greater Transvaal. *Myosorex varius* was collected together with *M. sclateri* at two localities: Dukuduku Forest and Nkandhla. Avery (1991) recorded *M. varius* from late Quaternary (from upper Pleistocene to very late Holocene) archaeological sites at Border Cave (in the Lebombo Mountains), Umhlatuzana (near Durban), Mgede (north-western KwaZulu-Natal) and Collingham (Drakensberg foothills).

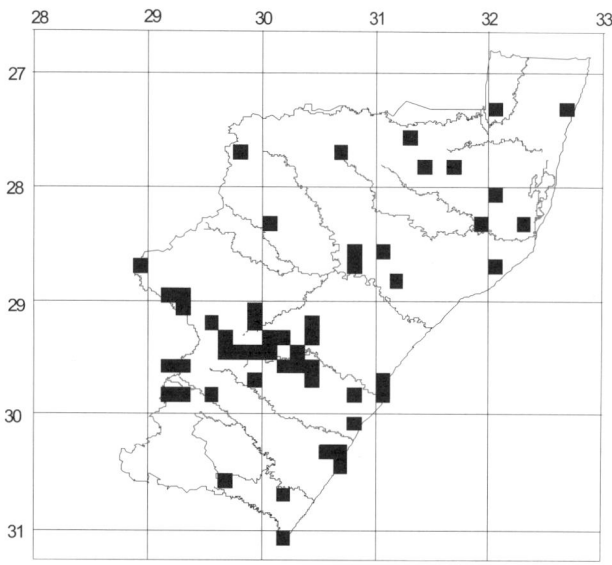

Figure 4

Protected areas

Cathedral Peak, Fort Nottingham, Giant's Castle, Itala, Kamberg, Karkloof, Krantzkloof, Midmar, Mount Currie, Oribi Gorge, Queen Elizabeth Park, Royal Natal, Umgeni Vlei, Umtamvuna, Vernon Crookes.

Conservation status

Common and widespread in KwaZulu-Natal.

Habitat

In KwaZulu-Natal, *M. varius* was typically collected in dense, moist grassland habitats, usually close to streams and dams. Trapping results demonstrated an association between relative abundance of shrews and annual rainfall. High trap success was achieved generally within the Mistbelt and Montane bioregions, moderate success within

Coast hinterland, and very poor success within Drier upland and Moist upland.

Population densities of shrews in the Natal Drakensberg Park have been shown to fluctuate dramatically over time due to temporal variation in rainfall. During the 1991–1992 drought, populations of *M. varius* in the Drakensberg declined from being the most abundant to the second most abundant species of small mammals collected during the late seventies and early eighties (Rowe-Rowe 1977; Rowe-Rowe and Lowry 1982), to very low numbers in 1991 and 1992 (about 0.2/100 trap nights over approximately 4 000 trap nights at four localities; Rowe-Rowe 1995). By December 1993, numbers had recovered slightly to 1.1/100 trap nights over 700 trap nights at two Drakensberg localities (Taylor unpublished data), reaching high densities of 11/100 trap nights by mid-1995 (Rowe-Rowe 1995).

Although predominantly grassland-associated in KwaZulu-Natal, it has been reported from forest habitats in the Drakensberg (Rowe-Rowe 1977; Rowe-Rowe and Meester 1982a), and elsewhere (Rautenbach 1982; habitat observations for specimens in the Transvaal Museum collection). In the Drakensberg, and on grassy plateaux above major river valleys within the Coast hinterland bioregion (e.g. Umtamvuna and Krantzkloof Nature Reserves), it was found in sparser, shorter grassland some distance from water, and often associated with rocky slopes. Kearney (1993) found that, in contradiction to Bergmann's rule, *M. varius* from Drakensberg localities are distinctly smaller than shrews from lower altitude localities in KwaZulu-Natal. Rowe-Rowe and Meester (1985) demonstrated a similar decrease in size in shrews in relation to increasing altitude. This phenomenum may represent an adaptation to reduce absolute energy requirements at lower temperatures.

Habits

The forest shrew is intermittently active throughout the day and night, and in captivity feeds on earthworms and a variety of insects, including locusts, grasshoppers, termites, bagworms, and beetles (Goulden and Meester 1978). Rowe-Rowe (1986) recorded a wide variety of arthropods in the stomachs of *M. varius* from Giant's Castle, including Crustacea, Myriapoda, Arachnida, Diptera (adults and larvae), Coleoptera (adults and larvae), Blattodea, Orthoptera, Hemiptera, Psocoptera, Isoptera. Coleopterans were the dominant food items in both the wet and dry season.

These shrews often use runways constructed by rodents, but are also capable of excavating shallow blind tunnels leading from rocks and other obstacles. They also construct nests of soft grass with one or two entrances, one usually leading to a blind tunnel underneath.

Breeding

Pregnant females were collected during November, December, January and July in KwaZulu-Natal. Lactating females were collected in November and December. Other studies indicate a longer breeding season than suggested by the above data, extending from August to March in the KwaZulu-Natal Drakensberg (Rowe-Rowe and Meester 1982a), from September to March in greater Transvaal (Rautenbach 1982), and from October to February in Lesotho (Lynch 1994). The mean number of foetuses per gravid female was 3.8 (*n*=6; range 3–6), compared with 2.8 (*n*=19; range 1–4) for the KwaZulu-Natal Drakensberg (Rowe-Rowe and Meester 1982a), 3.0 (*n*=3; range 2–4) for the Eastern Cape (Lynch 1989) and 3.8 (*n*=45; range 2–6) for Lesotho (Lynch 1994).

Linear measurements (mm) and mass (g)

	Males					Females				
	$\bar{x}$	s.d.	n	Min	Max	$\bar{x}$	s.d.	n	Min	Max
TL	117.7	9.5	43	99	1441	116.4	8.3	51	101	132
HB	80.9	7.8	43	67	99	78.9	7.1	50	65	99
T	36.9	4.1	43	30	47	37.5	3.5	51	31	46
Hf	13.0	1.4	20	11	16	12.7	1.4	18	9	15
E	8.2	1.6	22	5	10	7.6	1.7	20	5	11
Mass	13.7	3.3	19	10	20	12.7	2.8	10	10	18

See also Rowe-Rowe and Meester (1985) for data on 97 *M. varius* collected in the KwaZulu-Natal Drakensberg.

Records of occurrence

SPECIMEN RECORDS:
DM: Blinkwater NR, Bosch Hoek (SADF base), Bosch Hoek Farm, Cathedral Peak NR, Clearwater Farm, Durban, Fort Nottingham, Garden Castle NR, Highmoor, Hillcrest, Hilton (12 Forest Lane), Hlatikulu Vlei, Karkloof NR, Krantzkloof NR, Mfongosi, Mgeni Vlei, Ngome, Oribi Gorge NR, Royal Natal NP, Umbilo, Umhlanga Rocks, Umtamvuna NR, Virginia Farm, Winterskloof (13 Crompton Road);
KM: Underberg District;
NM: Carter's Nursery, Castle View, Cathedral Peak, Champagne Castle, Donkerhoek Farm, Gwaliweni, Hluhluwe GR, Kambula, Mfongosi, Mooi River District, Pietermaritzburg, Redlands, Rosslea Estate, Town Bush, Umdoni Park, Umtwalume, Umtwalume River – north bank;
SAM: Mfongosi, Sibudene, Willbrook;
SI: Drakensburg Gardens (3 km west), Hardingdale, Sani Pass (Makhake Store), Sibhudeni;
TM: Ashburton, Botha's Hill, Cathedral Peak, Chard Farm, Craigadam (Itala GR), Darvill Sewerage Works, Dukuduku Forest, Giant's Castle GR, Happy Hills Farm, Hilton Road, Hilton, Hluhluwe GR, Injasuti NR, Itala GR, Kamberg NR, Karkloof Forest, Kilgobbin Farm, Lake Sibayi, Mgeni Vlei, Midmar Dam, Mposa, Ngome Forest Reserve, Nkhandla, Oribi Gorge NR, Pietermaritzburg, Qudeni, Queen Elizabeth Park, Ronan Egg Farm, Royal Natal NP, Sani Pass, Stillerust Cottage (Mooi River), Thornville, Town Bush Valley, Umfolozi Flats, Vernon Crookes NR.

ADDITIONAL RECORDS:
O. Wirminghaus (private collection): Karkloof, Nottingham Road;
Avery (1991): Border Cave, Collingham, Mgede, Umhlatuzana.

Myosorex cafer (Sundevall 1846)
Dark-footed forest shrew
Donkerpoot-bosskeerbek

Taxonomic status
Meester et al. (1986) recognised two subspecies, *M. cafer cafer* and *M. c. sclateri*. However, recent biochemical (Maddalena and Bronner 1992) and morphological (Kearney 1993) evidence indicates that *M. sclateri* is distinct from both *M. cafer* and *M. varius*. Thus, no subspecies of *M. cafer* are currently recognised.

Distribution
In KwaZulu-Natal, the species is largely confined to southern KwaZulu-Natal (Figure 5), although it has been recorded from Afromontane forest at Ngome in northern KwaZulu-Natal. It seems to be replaced by *M. sclateri* in northern-eastern Zululand (north of about 29°S), and by *M. varius* in the Drakensberg. Lee (1976) reported on the presence of *M. cafer* from forest habitats at Royal Natal National Park; however, no museum specimens from this locality were available to validate the record (Kearney 1993).

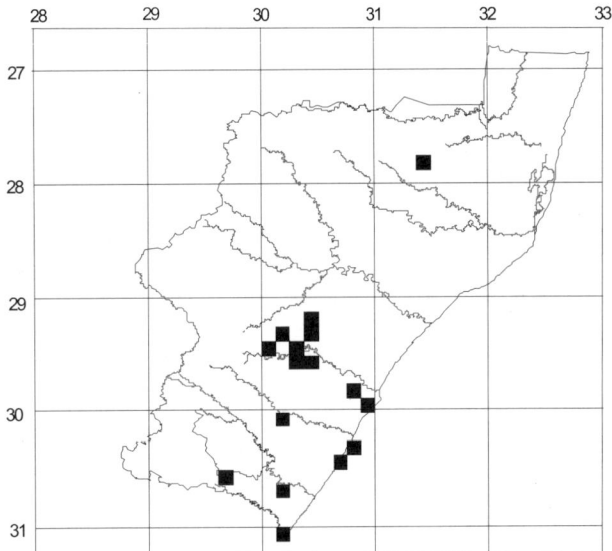

Figure 5

Protected areas
Karkloof, Oribi Gorge, Umtamvuna.

Conservation status
Reasonably common and widespread in forest habitats in KwaZulu-Natal.

Habitat
In KwaZulu-Natal, the species was collected almost exclusively from Afromontane and coastal forest habitats, being very common in the former and less so in the latter. At Umtamvuna Nature Reserve and the adjacent Clearwater Farm, *M. cafer* was collected from forest along the

Umtamvuna Gorge as well as from a recently burnt fire-break and a waterlogged wetland comprising long grasses and shrubs, on the plateau above the gorge.

Habits

M. cafer appears to be active throughout the 24-hour period. Like *M. varius*, they build round nests made of soft grass, usually well concealed and well protected by vegetation or overhanging boulders. In a study in Zimbabwe, their diet was found to comprise 19 different items including beetles, caterpillars, termites, spiders, worms, and a small amount of seeds and green plant material (Skinner and Smithers 1990).

Breeding

Five pregnant females and two lactating females in the Durban Natural Science Museum's collection were collected in KwaZulu-Natal during November. One pregnant female in the Transvaal Museum was collected during April. Because of its restricted distribution, very little is known about breeding in this species. In Zimbabwe, pregnant females were taken in October and November. In greater Transvaal, Rautenbach (1982) recorded a lactating female in March. Of the six pregnant females collected in KwaZulu-Natal, the mean number of foetuses per pregnant female was 3.5 (range 2–5), compared with 3.0 (*n*=3; range 2–4) in Zimbabwe.

Linear measurements (mm) and mass (g)

	Males					Females				
	x̄	s.d.	n	Min	Max	x̄	s.d.	n	Min	Max
TL	125.9	6.7	21	115	138	120.1	11.1	31	101	143
HB	83.7	6.2	21	73	94	77.7	8.3	31	58	94
T	42.2	4.3	21	30	48	42.4	5.1	31	32	51
Hf	14.4	0.9	5	13	15	14.5	0.8	14	13	16
E	9.0	3.0	6	4	12	9.3	1.8	15	7	13
Mass	13.4	3.4	5	10	18	13.4	3.4	11	9	20

Records of occurrence

SPECIMEN RECORDS:
DM: Blinkwater NR, Clearwater Farm, Hillcrest, Hilton (12 Forest Lane), Karkloof NR, Kwamagwaza Mission, Ngome Forest Reserve, Renishaw, Umtamvuna NR, Weza State Forest;
NM: Belfort, Knoops Farm, Oribi Gorge, Redlands, Umtwalume River;
SAM: Zululand (no co-ordinates or grid references given);
SI: Buxton, Hardingdale, Kilgobbin;
TM: Between Pietermaritzburg and Hilton, Carter's Nursery, Happy Hills Farm, Karkloof Forest, Kilgobbin Farm, Lower Stinkwood Forest (=Weza State Forest), Ngome Forest Reserve, Oribi Gorge NR, Town Bush Cave, University of Natal (Pietermaritzburg) campus.

ADDITIONAL RECORDS:
O. Wirminghaus (private collection): Karkloof, Weza.

Myosorex sclateri (Thomas and Schwann 1905)
Sclater's forest shrew
Sclater-bosskeerbek

Taxonomic status

See under *M. varius.* No subspecies are currently recognised.

Distribution

Endemic to Zululand, KwaZulu-Natal, where it occupies Coast lowland and Coast hinterland bioregions (Figure 6). An isolated possible record of *M. sclateri* from Oribi Gorge (**TM** 34135), recognised by Kearney (1993) on morphometric grounds, appears anomolous and may prove to be a misidentification, given the established distribution of *M. sclateri* and the morphometric overlap between *M. varius*, *M. cafer* and *M. sclateri*.

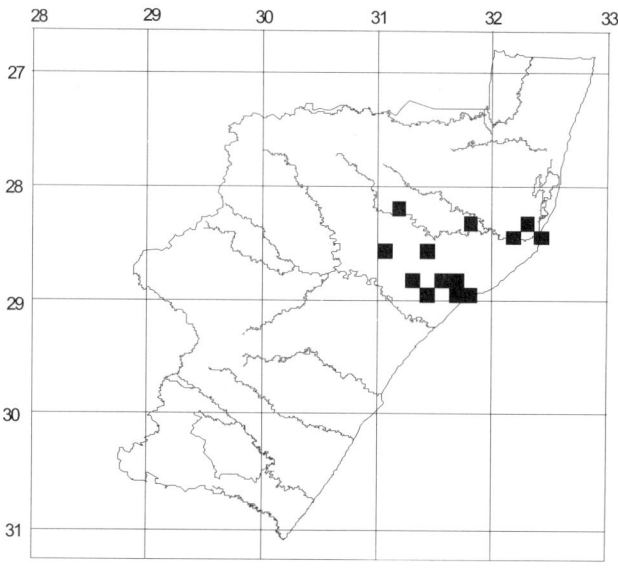

Figure 6

Protected areas

Dlinza, Entumeni, Mapelane, Mhlatuze, Ngoye Forest, Umfolozi, Umlalazi.

Conservation status

Fairly common in, but endemic to, KwaZulu-Natal.

Habitat

Since this species was previously considered to be a subspecies of *M. cafer*, one would consider it to have a similar habitat to *M. cafer*, i.e. restricted mainly to forests in the vicinity of water. However, series of *M. sclateri* from Rutledge Park in Eshowe (*n*=7) and from Umlalazi Nature Reserve (*n*=5), were collected in grassland, wetland and reedbed habitats. Other specimens were collected in either forest or grassland close to grassland/forest ecotones, and one specimen was trapped in brick and concrete rubble next to a stream located in grassland (Eckard, unpublished data). These observations suggest that this species may be more similar to *M. varius* than to *M. cafer* in its habitat association.

Habits

Similar to *M. cafer* and *M. varius*.

Linear measurements (mm) and mass (g)

Data obtained by Kearney (1993) for 58 *M. sclateri* from five localities in KwaZulu-Natal are summarised below (in each case the grand mean for the five localities is given, and sexes are combined):

	$\bar{x}$	n	Min	Max
HB	93.8	57	53	118
T	48.0	58	40	74
H*fcu*	16.6	56	14	18
E	9.8	56	8	13

Records of occurrence

SPECIMEN RECORDS:

DM: Dlinza NR, Entumeni NR, Eshowe, Lake Eteza, Mapelane, Umlalazi NR;

SI: Eshowe;

TM: Dlinza NR, Dukuduku Forest, Entumeni NR, Mapelane NR, Ngoye Hills, Nkandhla Forest, Oribi Gorge, Twin Streams Farm, Umfolozi, Umlalazi NR.

Genus	***Crocidura***	**Wagler 1832**

KEY TO SPECIES (Meester et al. 1986):

1. Larger, condylo-incisive skull length (CI) normally greater than 25 mm . . . .
   C. flavescens, p. 11
 — Smaller, CI normally less than 25 mm 2

2. Larger, CI *c.* 22–25 mm . . . . *C. hirta*, p. 13
 — Smaller, CI less than 22 mm 3

3. Dorsally blackish-brown or black, ventrally dark brown . . . . *C. mariquensis*, p. 13
 — Dorsally greyish-fawn, grey-brown, buffy-brown or brown, ventrally silvery-grey or grey 4

4. Larger, CI *c.* 19–22 mm 5
 — Smaller, CI <19 mm 6

5. Dorsally buffy-brown, ventrally silvery-grey, M_3 normally lacking entoconid . . . .
   C. cyanea, p. 15
 — Dorsally brown, ventrally light grey, M_3 normally with traces of entoconid . . . .
   C. silacea, p. 16

6. Slightly smaller, CI *c.* 15.5–18.0 mm; dorsally grey-brown or greyish-fawn, ventrally silvery-grey; M_3 lacking entoconid . . . . *C. fuscomurina*, p. 17
 — Slightly larger, CI *c.* 17.5–19.0 mm; dorsally grey-brown, ventrally grey; M_3 with well-developed entoconid . . . .
   C. maquassiensis, p. 17

Crocidura flavescens (I. Geoffroy 1827)   PLATE 2
Greater musk shrew
Groter skeerbek

Taxonomic status

While many earlier authors regarded *C. flavescens* as conspecific with *C. occidentalis*, Meester et al. (1986) regarded it to be a separate species, endemic to southern African, a view supported by later karyotypic work by Maddalena et al. (1987). No subspecies are recognised.

Distribution

Crocidura flavescens is widespread throughout KwaZulu-Natal, occupying all bioregions (although absent from much of northern Zululand) (Figure 7). The species was recorded from late Quaternary archaeological remains at Border Cave (in the Lebombo Mountains), Umhlatuzana (near Durban) and at a number of sites in the Tugela Basin (see under **Records of occurrence**; Avery 1991).

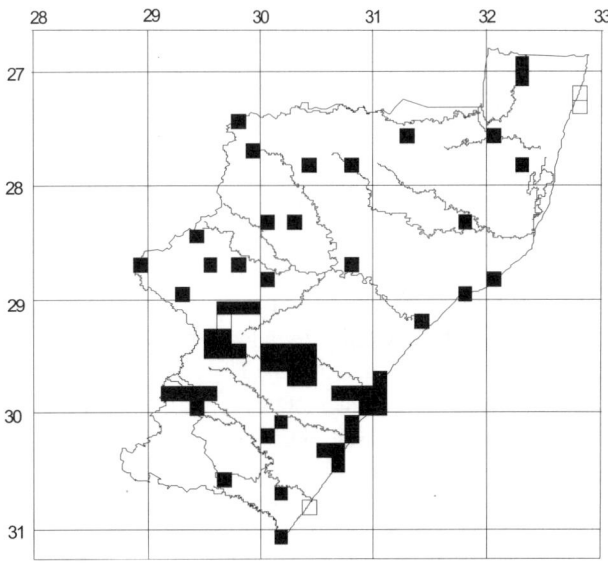

Figure 7

Protected areas

Albert Falls, Beachwood Mangroves, Bluff, Cathedral Peak, Coleford, Giant's Castle, Harold Johnson, Highmoor, Itala, Kamberg, Kenneth Stainbank, Krantzkloof, Loteni, Mgeni Vlei, Midmar, Oribi Gorge, Royal Natal, Spoienkop, Umlalazi, Umtamvuna, Vernon Crookes, Vryheid, Weenen.

Conservation status

Widespread and commonly collected in KwaZulu-Natal.

Habitat

This species has a wide habitat range, with a preference for close proximity to water, adequate ground cover (grass, shrubs), rocky terrain, and ecotones on the edge of forests or riverine bushveld thickets. Although preferring lush habitats, *C. flavescens* was occasionally collected from sparse, short, recently burnt grassland habitats, for example at Loteni Nature Reserve and at Clearwater Farm. Greater musk shrews are often found in suburban gardens and near buildings in built-up areas. Around Durban they are particularly common in *Lantana* scrub bordering on forest or grassland habitats.

Habits

While Skinner and Smithers (1990) reported that *Crocidura flavescens* is sporadically active throughout the 24-hour period, trapping results from KwaZulu-Natal show that the species is mainly nocturnal or crepuscular (of 22 *C. flavescens* collected by myself, all were collected in early morning trap checks, and none in afternoon checks).

Open, cup-shaped, untidy nests are constructed in dense vegetation (Skinner and Smithers 1990). This shrew has been found to feed mainly on arthropods and also eats earthworms (Lynch 1989; Rowe-Rowe 1986).

Breeding

Pregnant females were collected during October, November, December, January, February and May (specimens in **DM** and **TM** collections). One individual collected during December was both lactating and pregnant indicating that they have at least two litters per year. Mean litter size was six (*n*=4; range 4–8).

Very little is known of their reproduction (Skinner and Smithers 1990). Skinner and Smithers (1990) reported the occurrence of pregnant females during summer and winter months, and a maximum litter size of five. However, in the Eastern Cape (Lynch 1989), Lesotho (Lynch 1994), and the Drakensberg (Rowe-Rowe and Meester 1982a), pregnant females were collected between October and March, with litter sizes ranging from four to eight. Lynch (1989) recorded a lactating female in April. These data agree well with the results obtained by the present study.

Like most if not all musk shrews, the young are born hairless, open their eyes at about two weeks of age, and remain in their nest until they are sexually mature at the age of about two to three months. The young display the curious habit of 'caravanning' whereby the first one holds on to the hind end of its mother, and the rest follow, each holding on the hind end of the one in front (Nowak and Paradiso 1983).

Linear measurements (mm) and mass (g)

	Males					Females				
	$\bar{x}$	s.d.	n	Min	Max	$\bar{x}$	s.d.	n	Min	Max
TL	153.8	13.3	41	125	180	152.8	11.8	49	128	177
HB	100.9	10.0	41	79	120	101.8	10.6	49	76	121
T	52.9	5.6	41	40	61	51.0	5.6	49	40	65
Hf	14.5	1.3	10	12	17	15.4	1.6	13	12	19
E	9.6	1.4	18	7	12	10.0	2.1	31	7	12
Mass	30.0	6.9	10	20	40	25.2	7.7	17	20	38

Records of occurrence

SPECIMEN RECORDS:

CM: Pennington;

DM: Albert Falls NR, Arden Estate (Merrivale), Beachwood Mangroves NR, Bluff (76 Sharwell Road), Bosch Hoek Farm, Burman Bush, Clearwater Farm, Durban, Durban North, Empisini NR, Harold Johnson NR, Highmoor, Hilton (25 Hillside Road, Golden Pond), Hlatikulu Vlei, Howick (12 Miller Street), Kenneth Stainbank NR, Linwood Forest Estate, Loteni NR, Merrivale (Arden Estate), Mfongosi, Mgeni Vlei NR, Mount Moreland (12 Williams Street), Newcastle (10 Loerie Street), Pigeon Valley NR, Puntans Hill (12 Silver Maple Road), Royal Natal NP, Shongweni Dam, Spoienkop NR, Umhlanga Rocks, Umkomaas, Umtamvuna NR, University of Natal (Durban), Vernon Crookes NR, Vryheid NR, Weenen NR, Westville (Atholl Heights, 55a Jan Hofmeyr Road), Willbrook;

KM: Bellair, Underberg;

NM: Botanic Gardens (Pietermaritzburg), Burman Bush, Castle View, Drumclog, Geluk Farm, Giant's Castle NR, Leliefontein, Mfongosi, Pietermaritzburg, Redlands, Rosslea Estate, Shemula (=Shemula's Pont), Town Bush, Umdoni Park, Umgeni River, Umgeni Valley Game Ranch, Umtwalume River;

SAM: Durban, Estcourt, Mfongosi, KwaZulu-Natal (locality unspecified);

SI: Drakensberg Gardens, Hardingdale, Nkomkoni, Petchaye;

TM: Albert Falls NR, Ashburton, Bisley, Bluff NR, Cathedral Peak NR, Cornhill Farm, Dargle State Forest, Dundee (8 km from), Durban, Durban North (Gainsborough Drive), Estcourt, Highmoor, Hillview, Hilton, Hilton Road, Howick, Klipkuil, Klipspruit, Itala GR (Craigadam), Malvern, Manor Gardens (71 Elland Road), Midmar Dam NR, Mkondeni, Mount Edgecombe, Mtunzini, Ndumu GR, Oribi Gorge NR, Pietermaritzburg, Pietermaritzburg Girls High School, Pigeon Valley NR, Pinetown, Potters Hill Farm, Richards Bay, Scottsville, Shepstone Reserve, Spioenkop NR, Sunnyside, The Grange, Theunis Bester GR, Umbilo (16 Brisker Road), Umtamvuna NR, University of Natal (Durban), University of Natal (Pietermaritzburg), Vernon Crookes NR, Weenen NR, Wyford Farm.

ADDITIONAL RECORDS:

Meester (1963): Port Shepstone, Tabamhlope;

Bruton (1978): Lake Sibayi, Manzengwenya, Mbazwane;

Bronner (1985): Bluff NR;
Avery (1991): Border Cave, Collingham, Mbabane, Mgede, Mhlwazini, Nkupe, Umhlatuzana.

Crocidura hirta (Peters 1852)
C. h. hirta (Peters 1852)
Lesser red musk shrew
Klein rooiskeerbek

Taxonomic status
Four subspecies have been recognised throughout Africa, of which two are found in southern Africa: *C. h. hirta* occurring in wetter areas in the above 500 mm annual rainfall zones, and *C. h. deserti* from drier areas in the above 500 mm annual rainfall zone (Meester et al. 1986).

Distribution
The lesser red musk shrew is found throughout most of northern Zululand and Maputaland, extending as far south as Pennington on the South Coast (Figure 8). Based on specimens in the **SI** and **TM** collections, an isolated population of *C. hirta* occurs in the upper reaches of the Tugela River catchment, in the Van Reenen, Ladysmith and Newcastle Districts. Avery (1991) recorded this species from late Quaternary archaeological remains at Border Cave (in the Lebombo Mountains).

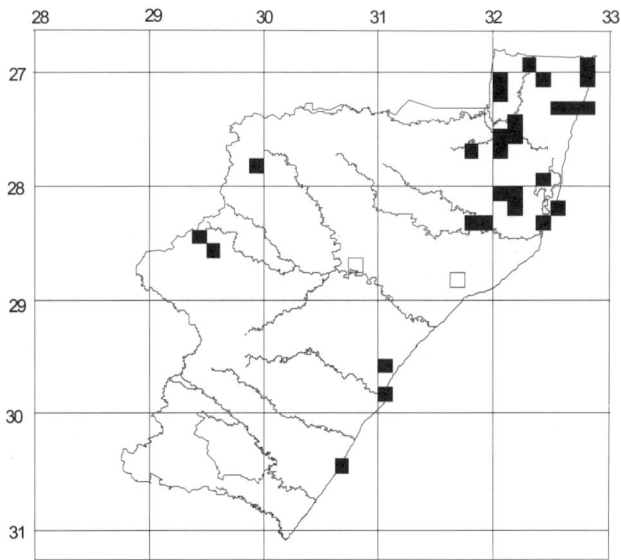

Figure 8

Protected areas
Coastal Forest Reserve (=Kosi Bay), Eastern Shores, False Bay, Hluhluwe, Mkuzi, Pongola, Umfolozi.

Conservation status
Fairly common in north-eastern KwaZulu-Natal.

Habitat
Habitats in which KwaZulu-Natal specimens were collected include grass clumps among reeds, a fig-dominated forest, reedbeds in swampy areas, savanna woodland and tall grassland in the vicinity of rivers and pans. They have also been collected from fallow agricultural lands. *C. h. hirta* is generally associated with a wide variety of habitats, usually offering good ground cover (undergrowth, rocks, fallen logs, disused rodent burrows) and proximity to water. *C. h. deserti*, on the other hand, occurs in much drier habitats including the central Kalahari where water is only seasonally available (Skinner and Smithers 1990).

Habits
They seem to be sporadically active during the night and day and feed on earthworms, insects, and, in captivity, on the carcasses of rodents, bats and even other shrews.

Linear measurements (mm) and mass (g)

	x̄	s.d.	n	Min	Max	x̄	s.d.	n	Min	Max
TL	126.5	10.4	6	114	145	123.5	–	2	117	130
HB	79.3	10.9	6	70	99	78.5	–	2	78	79
T	47.2	5.8	6	40	54	45.0	–	2	39	51
Hf	13.2	1.7	4	11	15	–	–	1	14	14
E	9.4	0.6	5	9	10	8.5	–	2	8	9
Mass	–	–	1	14	14	–	–	–	–	–

Records of occurrence
SPECIMEN RECORDS:
CM: Pennington;
DM: Eastern Shores NR (Lake Bangazi), False Bay Park, Hluhluwe GR, Inanda Game Park (=Inanda Dam), Mkuzi GR;
KM: Confluence of Black and White Umfolozi River (13,5 km north, 21 km west), Mission Rocks, Umbilo (Durban);
NM: Entendweni Bush, Mkuzi GR;
SI: Makatini Flats, Mkyola State Lands, Newcastle (16 km south), Smallhoek Farm;
TM: Coastal Forest Reserve (=Kosi Bay), Eastern Shores NR, Hazelmere Dam NR, Hluhluwe GR (Gontshi Outpost), Ingwavuma, Ingwavuma River, Lake Sibayi (Banana Pan, Banda-Banda Bay, Research Station), Maputa (=Manguzi), Ndumu GR, Pongola NR, Sihangwane, Tete Pan, Umfolozi.

ADDITIONAL RECORDS:
Meester (1963): Mfongosi, Ngoye Hills;
Avery (1991): Border Cave;
T. Bodbijl (private collection): Dukuduku Forest.

Crocidura mariquensis (A. Smith 1844)
C. m. mariquensis (A. Smith 1844)
Swamp musk shrew
Vleiskeerbek

Taxonomic status
Three subspecies are currently recognised (Dippenaar 1979; Meester et al. 1986), of which only one (*C. m. mariquensis*) is found in KwaZulu-Natal. Morphometric evidence (Dippenaar 1979) shows *C. mariquensis* from KwaZulu-Natal to be quite distinct from all other populations, rais-

ing the possibility of a distinct KwaZulu-Natal subspecies separate from *C. m. mariquensis*. Data presented below lend further support to such an argument, but more detailed analysis is necessary before such a subspecies can be formally described.

Distribution

Crocidura mariquensis occurs coastally from Durban in the south to the Mozambique border in the north, as well as throughout much of central KwaZulu-Natal from the Ixopo District in the south to the Newcastle District in the north (Figure 9). It is usually associated with dams, lakes, wetlands and river systems, and occurs in all bioregions with the exception of Coast hinterland, Drier upland and Montane. Avery (1991) recorded this species from archaeological remains at Border Cave, from the upper Pleistocene (>15 000 years BP) and very late Holocene (<1 000 years BP) periods.

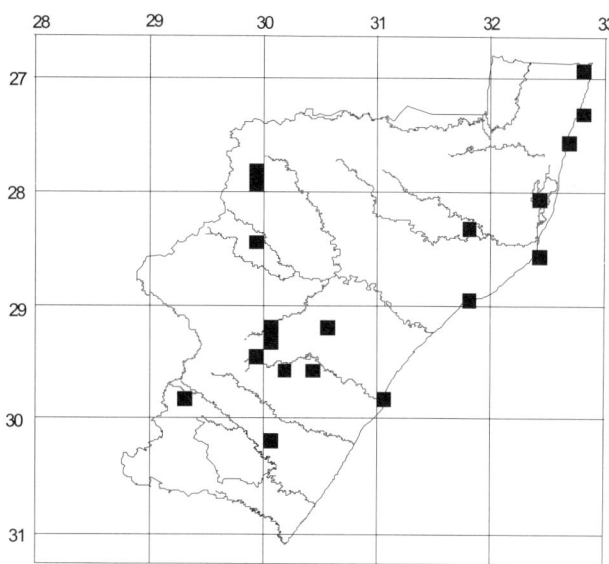

Figure 9

Protected areas

Chelmsford, Coastal Forest Reserve (Kosi Bay), Eastern Shores, Fort Nottingham, False Bay, Midmar, Sodwana Bay, Umfolozi, Umlalazi, Umvoti Vlei.

Conservation status

Fairly common in wetland areas of KwaZulu-Natal.

Habitat

As its English name suggests, this species is typically associated with marshy conditions. In KwaZulu-Natal, the species was usually collected on waterlogged or marshy ground in dense grassy or reedbed wetlands associated with dams and rivers. However, it was occasionally collected from drier grasslands on dam margins, for example at False Bay on the western shores of Lake St Lucia, and on the shores of Chelmsford Dam.

Habits

Active throughout the day and night with a peak during the night (Rautenbach 1982). They use runways of vlei rats (*Otomys*) and water rats (*Dasymys*), but are also trapped in the midst of thick vegetation, suggesting that they do not create their own runways. In captivity they make nests but do not burrow (Meester 1963).

Breeding

Pregnant females were collected in January, September and November (**TM** and **DM** collections). The number of embryos ranged from two to four, with a mean of 3.2 (*n*=4). From published records (Skinner and Smithers 1990), the reproductive period seems to be somewhat extended in *C. m. shortridgei* from Botswana, with pregnant females occurring from August to December and from February to April (mean number of foetuses was 3.3; *n*=22; range 2–5).

Linear measurements (mm) and mass (g)

	Males					Females				
	$\bar{x}$	s.d.	n	Min	Max	$\bar{x}$	s.d.	n	Min	Max
TL	130.2	8.1	5	121	139	130.3	10.5	6	110	140
HB	74.8	5.4	5	68	79	79.8	9.3	6	64	90
T	55.4	3.4	5	53	60	50.5	3.8	6	46	57
Hf	14.8	2.6	4	11	17	14.0	0.8	4	13	15
E	7.0	0.7	5	6	8	6.8	1.3	5	5	8
Mass	13.0	0.8	4	12	14	9.5	–	3	9	10
CIL	22.0	0.4	5	21.5	22.6	20.8	0.4	5	20.5	21.6

Dippenaar (1979) found significant sexual dimorphism in this species. Females were smaller than males in external and cranial measurements. The above data show females to be larger-bodied but shorter-tailed, lighter in mass, and smaller in condylo-incisor skull length (CIL) than are males. The data for CIL agree very closely with the values for a small sample (three males, two females) of KwaZulu-Natal *C. mariquensis* given in Dippenaar (1979). Dippenaar (1979) found that, particularly in the case of males, KwaZulu-Natal populations were conspicuously larger in cranial measurements than was the case in all other *C. mariquensis* populations, and separated from them in a multivariate cluster analysis at an average taxonomic distance of 2.5. While small sample size prevented Dippenaar (1979) from recognising this difference as being indicative of subspecific status, the data presented here provide more substantial support for recognising *C. mariquensis* from KwaZulu-Natal as a separate subspecies.

Records of occurrence

SPECIMEN RECORDS:
DM: Bosch Hoek Farm, Chelmsford NR, Elandslaagte, False Bay Park, Fort Nottingham NR, Lake St Lucia;
NM: Sodwana Bay Park;
SI: Drakensberg Gardens (3 km north-east), Hardingdale;

TM: Chelmsford NR, Darvill Sewage Works, Eersteling Farm, Coastal Forest Reserve (=Kosi Bay), Lake Sibayi Research Station, Midmar Dam NR, Mooi River (11 km west), Newcastle (16 km south), Umfolozi GR, Umlalazi NR, Umvoti Vlei NR.

ADDITIONAL RECORDS:
Avery (1991): Border Cave.

Crocidura cyanea (Duvernoy 1838)   PLATE 3
C. c. infumata (Wagner 1841)
Reddish-grey musk shrew
Rooigrysskeerbek

Taxonomic status
Nine subspecies occur throughout Africa, of which two occur in southern Africa and one (*C. c. infumata*) occurs in KwaZulu-Natal (Meester et al. 1986).

Distribution
The species is widespread in KwaZulu-Natal, having been collected in all bioregions except the Drier upland region of northern KwaZulu-Natal in the vicinity of Ladysmith, Dundee and Newcastle (i.e. upper Tugela Basin) (Figure 10). Avery (1991) recorded this species from late Quaternary archaeological remains at Border Cave (in the Lebombo Mountains) as well as at Mgede and possibly some other Tugela Basin sites (see under **Records of occurrence**).

Protected areas
Coleford, Eastern Shores, False Bay, Garden Castle, Giant's Castle, Hluhluwe, Itala, Kamberg, Karkloof, Loteni, Mhlatuze, Mkuzi, Oribi Gorge, Spioenkop, St Lucia Park, Umfolozi, Umlalazi, Vernon Crookes, Vryheid.

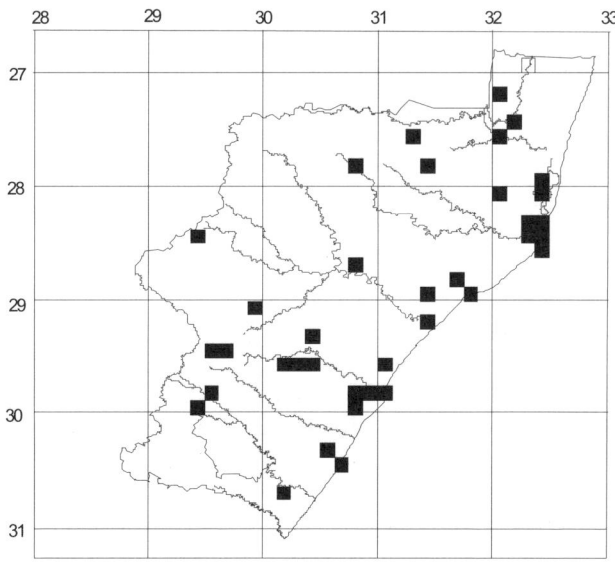

Figure 10

Conservation status
Widespread but not very abundant in KwaZulu-Natal.

Habitat
This species was collected in habitats ranging from moist, dense grassy habitats, bordering reedbeds fringing the western shores of Lake St Lucia to drier bushveld habitats in the Vryheid area. They are found in both grassland and forest habitats. Specimens were also taken from gardens (often drowned in swimming pools) and houses (in boxes or old timber in storage areas). This reflects the wide habitat tolerance by this species. Outside KwaZulu-Natal, it has been found in a variety of habitats, including montane forest as well as rocky areas within drier, sparsely vegetated areas such as the Karoo.

Habits
Very little is known of the habits of this species (Skinner and Smithers 1990), although they can be assumed to be similar to other species of musk shrews.

Breeding
Pregnant females were collected in March, July, October and December (**DM** and **TM**). The number of foetuses per pregnant female ranged from two to four, with a mean of 3.3 (*n*=3). Based on a small sample, Skinner and Smithers (1990) suggested that the young are born during the warm wet summer months, an argument that is not supported by the present data. Based on published accounts (Lynch 1983, 1989, 1994; Skinner and Smithers 1990), litter size varies from two to six with a mean of 3.4 (*n*=10).

Linear measurements (mm) and mass (g)
Males are distinctly larger than females in all of the above measurements except Hf*su*. These measurements agree closely with measurements of *C. cyanea* from greater Transvaal (Rautenbach 1982), but exceed measurements from the Free State (Lynch 1983; males vary in TL from 110 to 129), Eastern Cape (Lynch 1989; females vary in TL from 106 to 120) and Lesotho (Lynch 1994; single male and female measured 124 mm and 126 mm in TL respectively).

	Males					Females				
	x̄	s.d.	n	Min	Max	x̄	s.d.	n	Min	Max
TL	131.5	7.3	17	120	145	126.1	7.0	15	112	138
HB	75.9	4.8	17	69	86	72.4	6.4	15	61	82
T	55.5	3.6	17	50	62	53.7	4.6	15	46	60
Hf	13.0	1.0	3	12	14	13.0	1.0	5	12	14
E	9.0	2.1	7	5	12	7.8	2.6	5	5	11
Mass	11.0	1.0	3	10	12	7.7	2.0	8	4	11

Records of occurrence:
SPECIMEN RECORDS:
DM: False Bay Park, Harold Johnson NR, Krantzkloof NR, Lake St Lucia, Mapelane, Merrivale (Arden Estate),

Mfongosi, Ngome, Northdene, Shallcross, Umbilo (Durban), Umlalazi NR, Vryheid NR, Winterskloof (Pietermaritzburg);
KM: Mission Rocks, St Lucia, Underberg area (no specific locality);
NM: Drumclog (Pietermaritzburg), Hluhluwe GR (Egodeni), Hluhluwe GR, Leliefontein (Pietermaritzburg), Mkuzi GR, Oribi Gorge, Town Bush, Umtwalume River;
SI: Makatini Flats, Nkonkoni, Petchaye;
TM: Coleford NR, Dukuduku Forest Station, Eshowe Railway Station, Everton (Durban), Hazelmere Dam NR, Hluhluwe GR (Egodeni), Ingwavuma, Itala GR (Craigadam), Kamberg NR, Karkloof Forest, Loteni NR, Loteni River, Ngome State Forest, Ngoye Hills, St Lucia village, Umlalazi NR, Vernon Crookes NR, Wyford Farm.

ADDITIONAL RECORDS:
Avery (1991): Border Cave;
T. Bodbijl (private collection): Dukuduku Forest;
Bourquin et al. (1971): Hluhluwe GR (Hilltop);
Dixon (1966): Ndumu GR;
O. Wirminghaus (private collection): Ashburton.

Crocidura silacea (Thomas 1895)
C. s. silacea (Thomas 1895)
Lesser grey-brown musk shrew
Kleiner grysbruinskeerbek

Taxonomic status
Of the four subspecies listed by Heim de Balsac and Meester (1977), only one, *C. s. silacea*, occurs in southern Africa.

Distribution
The lesser grey-brown musk shrew has a wide but scattered distribution in KwaZulu-Natal, concentrated in northern Zululand (Figure 11). Skinner and Smithers (1990) gave Umlalazi NR in Zululand as the southernmost limit of the species' distribution in KwaZulu-Natal: however, the present map shows the species to occur as far south as Vernon Crookes NR on the South Coast and inland as far west as the southern Drakensberg in the south and the Newcastle District in the north. The species occupies a wide range of bioregions from Drier upland and Lowveld to the higher-rainfall Montane and Mistbelt regions (Figure 11).

Protected areas
Eastern Shores (Cape Vidal), Giant's Castle, Itala, Mkuzi, Spioenkop, Umfolozi, Vernon Crookes.

Conservation status
Rather uncommon in KwaZulu-Natal, judging by the scarcity of available specimens.

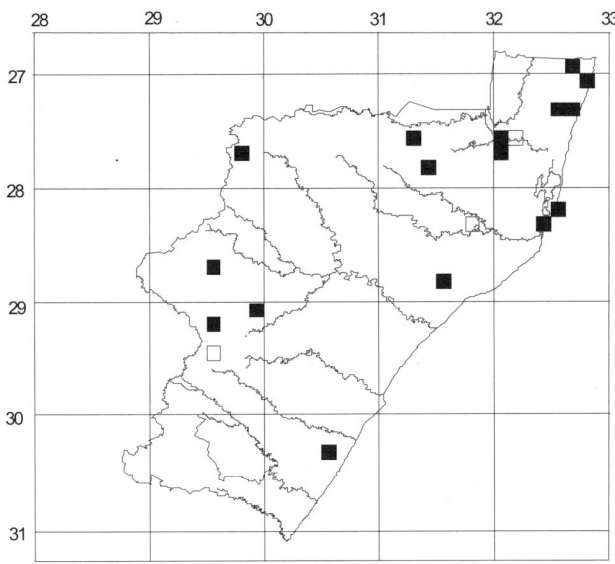

Figure 11

Habitat
A specimen from Maputa was collected from 'open veld'. Specimens from KwaZulu-Natal in the Transvaal Museum were collected from 'forest', 'forest edge' and 'palm bush'. I have never collected this species, and, because they are seldom collected, very little is known of their habitat requirements. However, they presumably have a broad habitat tolerance as they have been recorded elsewhere in coastal forest, savanna woodland, montane forest, and grassland (Skinner and Smithers 1990).

Habits and breeding
No information is available.

Linear measurements (mm) and mass (g)

Cat. No.	Males					
	TL	HB	T	Hf	E	Mass
DM 1020	141	78	63	–	–	9
NM 50	116	69	47	13	9	–

Records of occurrence
SPECIMEN RECORDS:
DM: Ngome, Spioenkop;
KM: Mission Rocks;
NM: Donkerhoek Farm, Maputa (=Manguzi);
SI: Buxton, Nkonkoni, Petchaye;
TM: Cape Vidal, Giant's Castle GR (Hillside), Itala GR (Craigadam), Lake Sibayi (Banana Pan), Maputa (=Manguzi), Vernon Crookes NR.

ADDITIONAL RECORDS:
Meester (1963): Loteni River;
Dixon (1964): Mkuzi GR;
Bourquin et al. (1971): Umfolozi GR.

Crocidura fuscomurina (Heuglin 1865)
C. f. bicolor (Bocage 1889)
Tiny musk shrew
Dwergskeerbek

Taxonomic status
Of the three subspecies recognised, two occur in southern Africa and one (*C. f. bicolor*) occurs in KwaZulu-Natal (Meester et al. 1986). The species' name was only recently changed to *C. fuscomurina*, having been known for 50 years previously as *Crocidura bicolor* (Meester et al. 1986).

Distribution
This species is uncommon in KwaZulu-Natal, with only three scattered records from the eastern region of the province (Figure 12). The distribution of *C. fuscomurina* in southern Africa is generally well correlated with the Southern Savanna Biome of Meester (1965), an extension of which comprises the Lowveld, Valley bushveld, Coast lowlands and Coast hinterland bioregions of the coastal portion of KwaZulu-Natal. Skinner and Smithers (1990) did not record the species as occurring in KwaZulu-Natal. However, Bourquin (1988) listed two records, one from Umfolozi and one from Hilton. A third record (from Durban) is added to the current distribution map (Figure 12). The specimen, although collected in 1974, was only catalogued in the Transvaal Museum subsequent to Bourquin's study (**TM** 42291). Avery (1991) recorded this species from archaeological remains at Border Cave in the Lebombo Mountains, from the upper Pleistocene (>15 000 years BP) and very late Holocene (<1 000 years BP) periods.

Protected areas
Umfolozi.

Conservation status
Uncommon in KwaZulu-Natal judging from the scarcity of records.

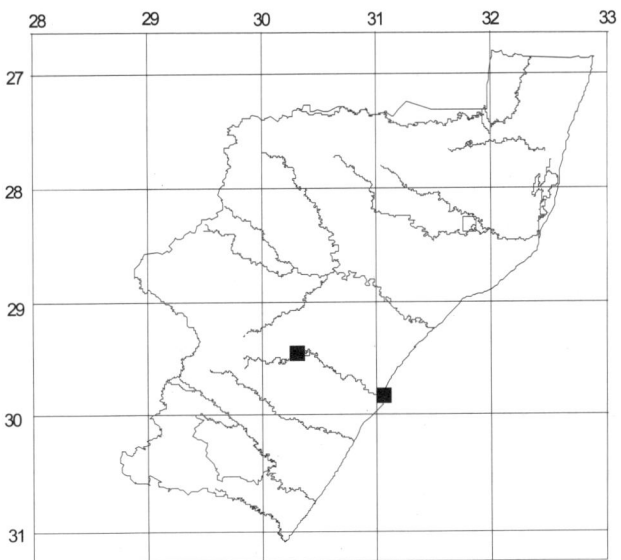

Figure 12

Habitat, habits, breeding and measurements
Virtually nothing is known of these aspects of the life history of this species (Skinner and Smithers 1990).

Records of occurrence
SPECIMEN RECORDS:
TM: Durban, Hilton.

ADDITIONAL RECORDS:
Bourquin et al. (1971): Umfolozi GR (Junction of Madhlozi and Umfolozi Rivers);
Avery (1991): Border Cave, Collingham, Mbabane, Mgede, Nkupe.

Crocidura maquassiensis (Roberts 1946)
Makwassie musk shrew
Makwassie-skeerbek

Taxonomic status
This species is very difficult to separate from both *C. cyanea* and *C. silacea*, and there is still some doubt as to its validity, as it may be a variant of *C. cyanea* (Meester et al. 1986). Specimens from the Transvaal Museum collection from Injasuti and Coastal Forest Reserve (=Kosi Bay) were provisionally identified as *C.* cf. *maquassiensis* by N. J. Dippenaar. However, the identification of these specimens, as well as of recent KwaZulu-Natal specimens in the Durban Natural Science Museum, were verified by means of multivariate morphometric analysis including known samples of *C. maquassiensis*, *C. cyanea* and *C. silacea* (Taylor et al. 1994).

Distribution
Subsequent to Bourquin's (1988) report where the species was not listed as occurring in KwaZulu-Natal, the Makwassie musk shrew has been recorded from five localities in KwaZulu-Natal: Kosi Lake, Lake Sibayi, Injasuti (Giant's Castle NR), Royal Natal National Park, and Chase Valley Heights (Taylor et al. 1994; Figure 13), thus indicating a wide but scattered distribution in KwaZulu-Natal.

Protected areas
Coastal Forest Reserve (Kosi Bay), Giant's Castle, Royal Natal.

Conservation status
Listed in the South African Red Data Book as Indeterminate. Although only known previously from six specimens, and listed as 'very rare' in an earlier South African Red Data Book (Meester 1976), the addition of several new records which increase the known range of the species provide evidence that the species is not as rare as previously thought. It is nevertheless uncommon and should be retained in the South African Red Data Book under the Indeterminate category.

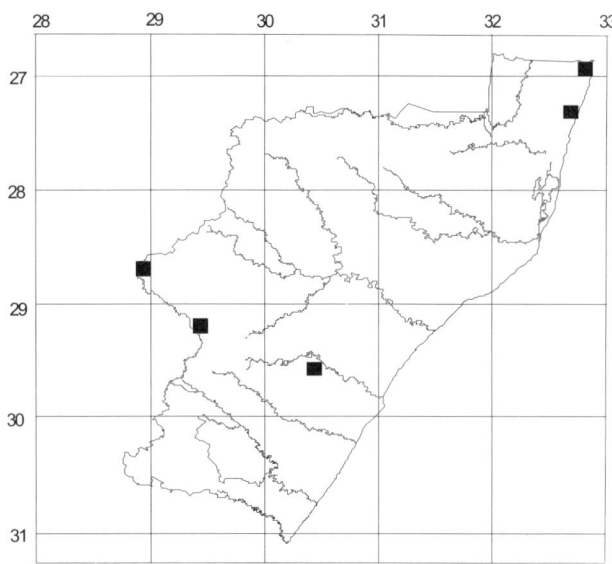

Figure 13

Habitat
The Chase Valley specimen was caught by a cat in a garden on a south-facing hillslope. The Royal Natal National Park specimen was trapped in mixed bracken and grassland along the Tugela River, at an altitude of 1 500 m. Most of the available specimens of *C. maquassiensis* were taken from either montane or rocky habitats, mostly at higher altitudes (>1 500 m) (Skinner and Smithers 1990). However, the two specimens from Coastal Forest Reserve (=Kosi Bay) indicate a broader habitat preference in KwaZulu-Natal, encompassing Montane, Coast hinterland and Coast lowland bioregions.

Habits
Nothing is known of the habits of this species (Skinner and Smithers 1990).

Breeding
One pregnant female collected in October had three foetuses. No comparative information on breeding is available for this species (Skinner and Smithers 1990).

Linear measurements (mm) and mass (g)

Cat. No.	Sex	TL	HB	T	Hf*su*	E	CI	Mass
DM1110	?	95	55	40	12	6	18.6	–
DM2392	F	110	66	44	11	7	18.9	6

Records of occurrence
SPECIMEN RECORDS:
DM: Chase Valley, Royal Natal NP;
TM: Coastal Forest Reserve (=Kosi Bay), Giant's Castle NR (Injasuti Outpost), Lake Sibayi.

Genus	*Suncus*	Ehrenberg 1833

KEY TO SPECIES (Meester et al. 1986):

1. Larger, condylo-incisive skull length (CI) normally 19–21 mm, upper toothrow (UTR) 7.4–8.8 mm, mandible and incisor (M&I) 11.0–12.6 mm; colour greyish above, paler below, dorsal and ventral colour inter-grading gradually . . . . *S. lixus*, p. 18
—   Smaller, CI 13.9–17.5 mm, UTR 5.3–7.1 mm, M&I 8.1–104 mm 2

2. Larger, CI 15.1–17.5 mm, UTR 6.0–7.1 mm, M&I 8.7–10.4 mm; colour greyish-chestnut above, silvery-fawn below, dorsal and ventral colour sharply demarcated . . . . *S. varilla*, p. 19
—   Smaller, CI 13.9–15.2 mm, UTR 5.3–6.3 mm, M&I 8.1–9.2 mm; colour greyish-brown above, greyish below, dorsal and ventral colour gradually intergrading . . . . *S. infinitesimus*, p. 20

Suncus lixus (Thomas 1898)
S. l. gratulus (Thomas and Schwann 1907)
Greater dwarf shrew
Groter dwergskeerbek

Taxonomic status
Meester et al. (1986) listed two subspecies of the greater dwarf shrew, of which one (*S. l. gratulus*) occurs in KwaZulu-Natal.

Distribution
Skinner and Smithers (1990) did not record the species from KwaZulu-Natal. Bourquin (1988) noted the occurrence of *S. lixus* at False Bay Park. Numerous subsequent records document the occurrence of the species at a further eight localities (Wirminghaus and Nanni 1989; Baker and Meester 1991; additional Durban Natural Science Museum and Transvaal Museum specimens), concentrated in two foci, one west of Lake St Lucia, and one from Westville to Pietermaritzburg (Figure 14). Avery (1991) recorded this species from late Quaternary archaeological remains at Border Cave in the Lebombo Mountains.

Protected areas
Entumeni, False Bay Park, Hluhluwe, Mkuzi.

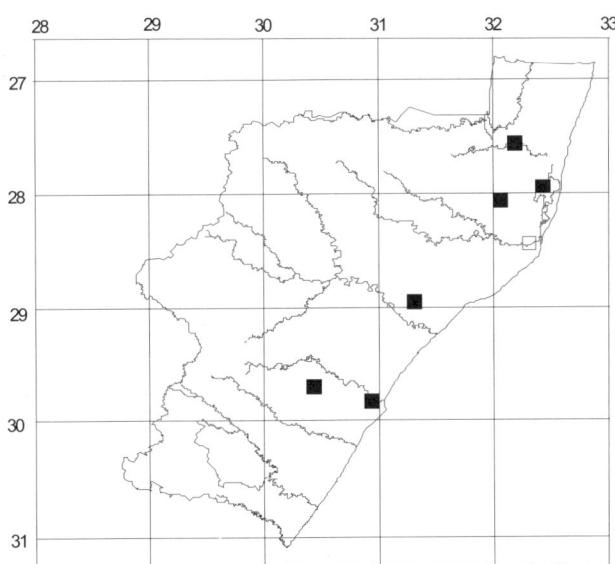

Figure 14

Conservation status

Listed as Indeterminate in the South African Red Data Book. The addition of several recent records from throughout KwaZulu-Natal indicate that this species is more common in the province than previously thought.

Habitat

Suncus lixus has been recorded from a wide range of habitats in KwaZulu-Natal, including coastal lowland forest, open grassland, *Acacia* woodland, and suburban gardens (Wirminghaus and Nanni 1989; Baker and Meester 1991). Elsewhere, they have been taken from dense riverine forest (greater Transvaal), savanna woodland (Zimbabwe) and dry open scrub (Makgadikgadi Pan) (Skinner and Smithers 1990).

Habits

No information is available.

Breeding

Wirminghaus and Nanni (1989) reported a lactating female collected in August. Apart from this observation, no other information is available.

Linear measurements (mm) and mass (g)

Cat. No.	Sex	TL	HB	T	Hf	E	Mass
DM1322	M	120	70	50	12	8	4.9
DM2239	M	111	63	48	12	7	–
DM2447	F	107	63	44	12	5	–

Records of occurrence

SPECIMEN RECORDS:
DM: Entumeni NR, Mkuzi GR, Roosfontein NR, Westville (Jan Hofmeyr Road), Westville (39 Springdale Ave);
KM: False Bay Park (Lister Point);
TM: False Bay Park, Hluhluwe GR (Egodeni), Poly Shorts road bridge.

ADDITIONAL RECORDS:
Wirminghaus and Nanni (1989): Dukuduku, Kingthorpe Farm;
Avery (1991): Border Cave;
T. Bodbijl (private collection): Dukuduku.

Suncus varilla (Thomas 1895)
S. v. orangiae (Roberts 1924)
Lesser dwarf shrew
Kleiner dwergskeerbek

Taxonomic status

Four subspecies were listed by Meester et al. (1986), of which one (*S. v. orangiae*) occurs in KwaZulu-Natal.

Distribution

The lesser dwarf shrew is widely distributed in KwaZulu-Natal throughout all bioregions (Figure 15). Avery (1991) recorded *S. varilla* from late Quaternary archaeological remains at Border Cave (in the Lebombo Mountains) and Umhlatuzana (near Durban) (see under **Records of occurrence**).

Protected areas

Hluhluwe, Itala, Spioenkop, Vernon Crookes, Weenen.

Conservation status

Scattered distribution in KwaZulu-Natal.

Habitat

Roberts (1951) noted the association of *S. varilla* and *S. infinitesimus* with dead mounds of the snouted harvester termite *Trinervitermes trinervitermes*. Lynch (1986) found that lesser dwarf shrews occupied (shrew present) or had occcupied (nest present) 56 per cent of dead termitaria in a study site near Bloemfontein, giving an average density of 1/16 865 m². Termitaria provide a suitable form of cover

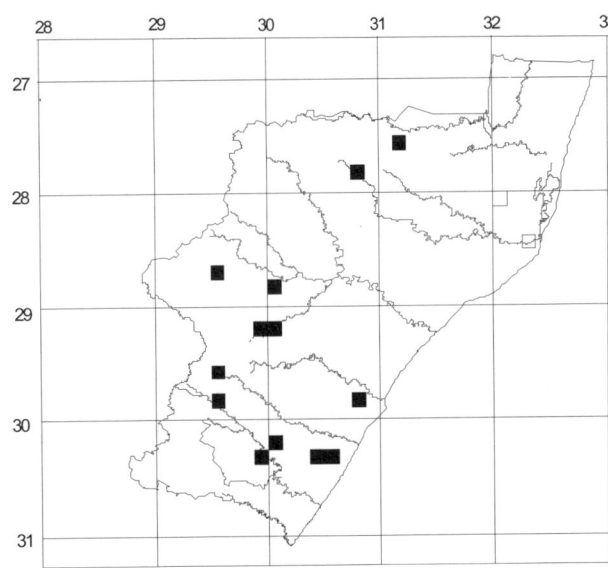

Figure 15

for these shrews in open grasslands where alternative cover is often lacking. In greater Transvaal (Rautenbach 1982), Free State (Lynch 1983), and Eastern Cape (Lynch 1989), the species is limited to grassland habitats where termitaria are present. In KwaZulu-Natal, the species has been collected in all bioregions, and does not seem to be restricted to termitaria in grassland habitats, as it was found in a suburban garden at Hillcrest, under a pile of firewood at Weston Agricultural College, in pitfall traps in coastal forest at Dukuduku, in open savanna at Spioenkop NR, and on the secondary margins of coast forest at Hluhluwe. Although *Trinervitermes trinervitermes* termites are apparently more common in drier habitats, they do occur throughout KwaZulu-Natal (Richardson 1987). However, it is possible that, in better wooded habitats, the presence of termitaria for cover is not an absolute habitat requirement.

Habits

The species builds 100 mm-diameter, ball-shaped nests of grass within termitaria. Usually more than one nest (and up to three) is present in a given termite mound, and nests are situated at different depths and at different orientations (i.e. north versus south-facing) within the mound, suggesting that different nests may be selected during the dry, cold Highveld winter and the warm, wet summer (Lynch 1986). These shrews may occupy termitaria throughout most of the day. Termitaria offer a stable microclimate which in turn minimises the substantial physiological demands of thermoregulation in such a small mammal. Captive lesser musk shrews are active during the day and night, and have been observed to feed on grasshoppers, crickets, termites and ox-liver.

Linear measurements (mm)

		Sexes combined			
	x̄	s.d.	n	Min	Max
TL	83.1	5.4	8	75	91
HB	54.5	4.2	8	49	60
T	28.6	2.7	8	25	34
Hf	8.0	–	1	–	–
E	5.3	2.3	3	4	8

Records of occurrence

SPECIMEN RECORDS:
DM: Hillcrest, Spioenkop NR, Weenen NR, Weston Agricultural College;
KM: Umzimkulu;
NM: Castle View, The Hoek;
TM: Doornpan (Itala GR), Ixopo (40 km south-east), Spioenkop NR, Umzimkulu, Vernon Crookes NR, Vryheid East.

ADDITIONAL RECORDS:
Bourquin et al. (1971): Hilltop (Hluhluwe GR);
Avery (1991): Border Cave, Umhlatuzana;
T. Bodbijl (private collection): Dukuduku.

Suncus infinitesimus (Heller 1912)   PLATE 4
S. i. chriseos (Kershaw 1921)
Least dwarf shrew
Kleinste dwergskeerbek

Taxonomic status

Three subspecies were recognised by Heim de Balsac and Meester (1977), of which only one (*S. i. chriseos*) is found in southern Africa. The name *gracilis*, which was previously used for this species, was shown by Meester and Lambrechts (1971) to be indeterminable as no type specimen existed, and could equally apply to *S. varilla*. For this reason, its use should be discontinued.

Distribution

This species occurs throughout much of the lower-lying central and eastern parts of KwaZulu-Natal, being absent from the Drier upland bioregion of northern KwaZulu-Natal, as well as from the Montane and Highland bioregions associated with the Drakensberg (Figure 16). The recent data show this species to be more widely distributed in KwaZulu-Natal than suggested by Bourquin (1988). There is some doubt over the identity of the Ndumu record (Dixon 1966) as no specimen exists and the identification given by Dixon is *S. etruscus*, a separate species which had earlier (Ellison, Morrison-Scott and Hayman 1953) been recognised as including both *S. varilla* and *S. infinitesimus*. While Bourquin (1988) referred the record to *S. varilla*, Avery (1991) referred it to *S. infinitesimus*. As *S. etruscus* is even smaller than *S. infinitesimus*, it is perhaps safer to assume that the specimen belonged to the smallest species, *S. infinitesimus*, a view that is tentatively followed here. *S. infinitesimus* has been recorded from the late Quaternary archaeological remains at Border Cave (in the Lebombo Mountains) as well as tentatively identified from Nkupe in the north-western area of the Tugela Basin.

Protected areas

Midmar, Ndumu, Vernon Crookes, Weenen.

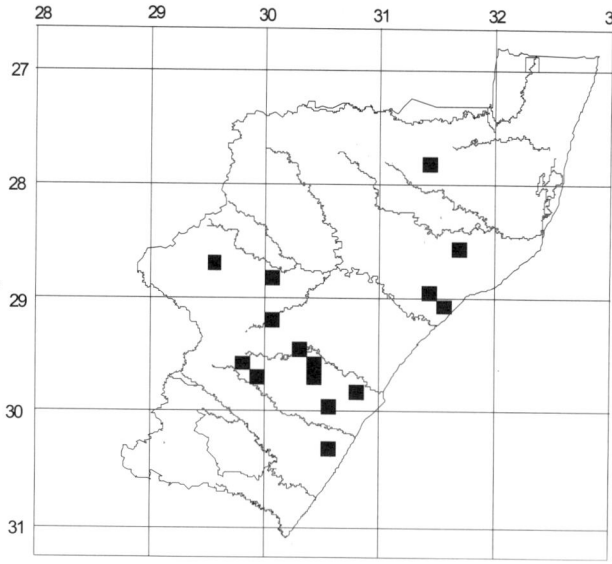

Figure 16

Conservation status
Listed as Indeterminate in the South African Red Data Book. Fairly widespread in KwaZulu-Natal but infrequently collected because of its small size.

Habitat
Most specimens in the Durban Natural Science Museum collection were caught by cats or found dead in suburban or farm gardens. A specimen in the Transvaal Museum was taken from a hole under a large boulder on the border of a sugar cane field. Other specimens were collected from grassland and from a horse paddock. As is the case with *S. varilla*, the species has elsewhere been collected from termitaria in open grassland or woodland habitats (Rautenbach 1982; Lynch 1983).

Habits
They construct round ball-shaped grass nests of some 100 mm diameter within disused termitaria (Lynch 1983).

Breeding
On two occasions pairs of juveniles or subadults (basioccipital-basisphenoid suture unfused or teeth unworn) were collected simultaneously or a few days apart from the same locality, and on one occasion three juveniles were collected together, suggesting a minimum litter size of two or three. Many shrews remain in their parental nest until they have reached adult size. One pregnant female collected during the present study had three embryos, and another pregnant female in the Transvaal Museum collection was collected during December. Hardly any other information is available on the reproduction of this species, although Lynch (1983) reported a lactating female, having three pairs of inguinal mammae, in January.

Linear measurements (mm) and mass (g)

		Sexes combined			
	$\bar{x}$	s.d.	n	Min	Max
TL	70.5	6.9	8	61	77
HB	46.2	4.9	8	40	53
T	24.2	2.4	8	21	27
Hf	7.9	0.6	8	7	9
E	4.8	1.8	8	3	7
Mass	3.0	–	2	2.8	3.1

Records of occurrence
SPECIMEN RECORDS:
DM: Boston (13 km north-west), Eshowe (84 Melmoth Road), Eston, Ferncliff, Hilton shopping centre, Hilton (12 Forest Lane), Impendle, Kloof (16 Springdale Road), Pietermaritzburg (17 Frankish Road), Spioenkop NR, Umkomaas, Virginia Farm;
TM: Ashburton, Clarendon (Pietermaritzburg), Kingthorpe Farm, Mtunzini, Ngome Forest Reserve, Ntambanana, Vernon Crookes NR, Weenen NR, Weston Agricultural College.

ADDITIONAL RECORDS:
Dixon (1966): Ndumu GR;
Avery (1991): Border Cave, Nkupe (?).

Family CHRYSOCHLORIDAE — Golden moles

KEY TO GENERA (Bronner 1995):

1. Malleus enlarged with a spherical or club-like shape; epitympanic recess housing the head of the malleus visible externally as a bony vesicle, or as a swelling on the lateral face of the squamosal; if no vesicle is present, an M^3 is present and well-developed talonids are present on the lower molars (Subfamily CHRYSOCHLORINAE) 2
— Malleus not expanded, with a typical mammalian shape; M^3 absent but molar talonids present and well developed or M^3 present but molar talonids are absent or feeble (Subfamily AMBLYSOMINAE) 3

2. Size larger, greatest skull length 32 mm or more, zygomatic breadth greater than 20 mm; head of malleus housed in a bony vesicle visible externally as a lateral swelling of the squamosal; zygomatic arch produced upwards posteriorly to meet lambdoidal crest at back; fur long and coarse without a pronounced sheen . . . . . . *Chrysospalax*, p. 22
— Size smaller, greatest skull length less than 30 mm, zygomatic breadth less than 19 mm . . . . *Chlorotalpa*, p. 23

3. P1 molariform; rostrum wider, anterior rostral breadth more than 21 per cent of greatest skull length; claws gracile, basal width of the third foreclaw less than 14 per cent of greatest skull length; skull broader, width/length index greater than 70 per cent . . . . *Calcochloris*, p. 23
— P1 sectorial; rostrum narrow, anterior rostral breadth less than 20 per cent of greatest skull length; claws more robust, basal width of the third foreclaw greater than 16 per cent of greatest skull length; skull narrower, width/length index less than 70 per cent . . . . *Amblysomus*, p. 24

Subfamily	**CHRYSOCHLORINAE**	
Genus	*Chrysospalax*	**Gill 1883**

Chrysospalax villosus (A. Smith 1833)
C. v. villosus (A. Smith 1833)
C. v. leschae (Broom 1918)
C. v. dosbsoni (Broom 1918)
Rough-haired golden mole
Growwehaargouemol

Taxonomic status
Of six subspecies listed by Meester et al. (1986), three occur in KwaZulu-Natal: *C. v. villosus* from the Durban area, *C. v. leschae* from Griqualand East and *C. v. dobsoni* from the KwaZulu-Natal midlands.

Distribution
Southern and central KwaZulu-Natal (Figure 17). Outside KwaZulu-Natal, they are known only from restricted areas in the Eastern Cape and the Wakkerstroom area of Mpumalanga. The two most recent specimens available from KwaZulu-Natal were collected in 1973 and 1974 in Pietermaritzburg. The specimen from Ladysmith in the **SAM** collection was collected in 1889. No recent specimens have been obtained despite intensive trapping attempts by G. Bronner, G. Hickman and A. C. Duckworth. The species has been recorded from variously dated Holocene sites (from >10 000 years BP to <1 000 years BP) at Umhlatuzana (near Durban), Collingham, Gehle, Mgede and Nkupe (the last four localities are in the upper Tugela Basin) (Avery 1991).

Protected areas
Giant's Castle (but may no longer occur there).

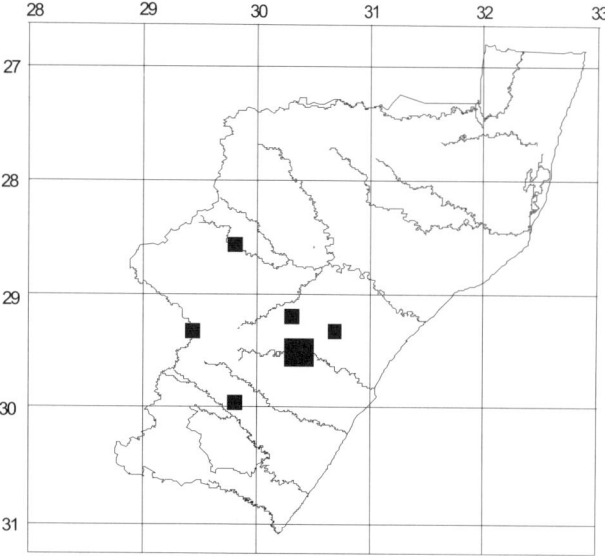

Figure 17

Conservation status
It is listed as Vulnerable in the South African Red Data Book. The species is very uncommon and may be endangered in KwaZulu-Natal. The rough-haired golden mole was documented from a number of widespread archaeological sites in the late Quaternary (Avery 1991; see under **Distribution**). Although respectable numbers of specimens were collected incidentally up until the 1950s and 1960s, none have been collected more recently in spite of intensive efforts. These facts suggest that the species has been on the decline since prehistoric times but that this has been accelerated in recent years by anthropogenic factors, such as hunting by domestic dogs in suburban Pietermaritzburg where most of existing museum specimens originated.

Habitat
Skinner and Smithers (1990) give the habitat as grassland with a preference for drier soils bordering on vleis. One specimen from Pietermaritzburg was killed by a dog in a suburban garden.

Habits
Unlike many other golden moles they do not make subsurface runways but excavate burrows with open entrances marked by piles of loose sand. According to Roberts (1951), they leave cleared tracks between their burrows and preferred feeding areas owing to the rooting action of the horny nose pad. They forage above ground after rains, feeding mainly on insects and earthworms.

Breeding
Roberts (1951) recorded a pregnant female with two foetuses, but did not indicate the month.

Linear measurements (mm)
Head and body length of three males ranged from 150 to 160 mm with a mean of 156 mm (s.d.=5.1), while four females varied from 120 to 152 mm with a mean of 135 mm (s.d.=16.9). Two unsexed individuals had head and body lengths of 127 mm and 180 mm. These data suggest the possibility of sexual dimorphism.

Records of occurrence
SPECIMEN RECORDS:
DM: Dalton;
NM: Allerton Laboratories (Pietermaritzburg), Bisi River, Botanic Gardens (Pietermaritzburg), Crammond, Giant's Castle GR (Suicide Corner), Meadow Farm, Pietermaritzburg (14 Pepworth Road), Pietermaritzburg, Town Bush;
SAM: Botanic Gardens (Pietermaritzburg), Ladysmith;
TM: Pietermaritzburg (Jesmond Road), Pietermaritzburg, Scottsville Racecourse (Pietermaritzburg), Shafton House (Karkloof).

ADDITIONAL RECORDS:
Avery (1991): Collingham, Gehle, Mgede, Nkupe, Umhlatuzana.

Genus	*Chlorotalpa*	Roberts 1924

Chlorotalpa sclateri (Broom 1907)
C. s. sclateri (Broom 1907)
Sclater's golden mole
Sclater-gouemol

Taxonomic status
Four subspecies were listed by Meester et al. (1986), of which only one, *C. s. sclateri*, occurs in KwaZulu-Natal.

Distribution
Sclater's golden mole was not listed as occurring in KwaZulu-Natal by Bourquin (1988). Two records exist for the KwaZulu-Natal Drakensberg (Taylor et al. 1994; Figure 18) and, as the species is widely distributed in the Lesotho Highlands (Lynch 1994), it is likely to be more widely distributed in the KwaZulu-Natal Drakensberg than the current paucity of specimens suggests.

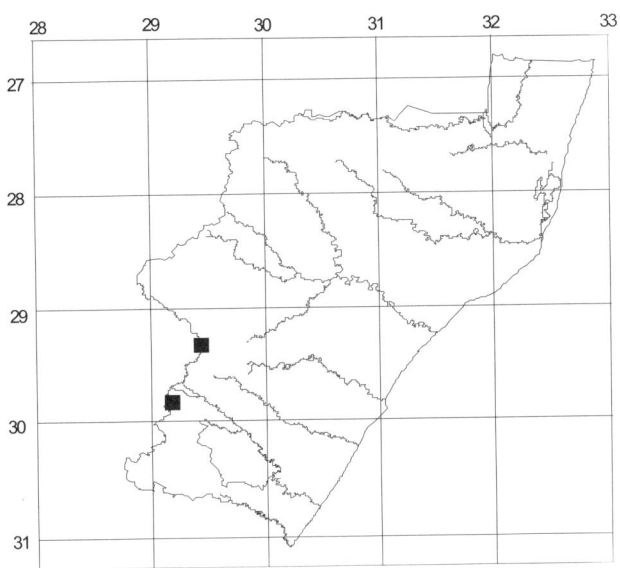

Figure 18

Protected areas
Garden Castle and Giant's Castle.

Conservation status
Chlorotalpa sclateri is listed as Indeterminate in the South African Red Data Book. As mentioned above it is likely that the species is fairly common in the KwaZulu-Natal Drakensberg. However, this need not be true and needs to be verified by further trapping studies.

Habitat
The specimen from Garden Castle was collected from a lawn at an altitude of 1 800 m (Taylor et al. 1994). In Lesotho they were collected at altitudes ranging from 1 750 m to over 3 000 m, in black turf soils amongst alpine shrubs and sedges, or in short grassland on mountain slopes, often near streams (Lynch 1994).

Habits
Lynch (1994) found that Sclater's golden mole is sometimes found in association with *Cryptomys hottentotus*. The stomach contents of the Garden Castle specimen consisted exclusively of earthworms.

Breeding
Bernard et al. (1994) gave the mean litter size to be two (*n*=3).

Linear measurements (mm) and mass (g)
The specimen from Garden Castle NR had a total length of 105 mm and a mass of 39 g. The specimen from Langalibalela Pass (Giant's Castle GR) had a total length of 97 mm. These values fall well within the range reported for Lesotho (Lynch 1994).

Records of occurrence
SPECIMEN RECORDS:
DM: Garden Castle;
NM: Giant's Castle GR (Langalibalela Pass).

Subfamily	**AMBLYSOMINAE**	
Genus	*Calcochloris*	Mivart 1867

Calcochloris obtusirostris (Peters 1851)
C. o. chrysillus (Thomas and Schwann 1905)
Yellow golden mole
Geelgouemol

Taxonomic status
Of three subspecies listed by Meester et al. (1986), one, *C. o. chrysillus*, is found in KwaZulu-Natal. This subspecies is smaller is size than other subspecies and was considered a distinct species by Roberts (1951).

Distribution
Northern Zululand, from the Mozambique border to the Mkuzi River (27°30´S) in the south and the Pongola Floodplain (32°15´E) in the west (Figure 19). While Meester et al. (1986) stated that the subspecies ranges to the Umfolozi River in the south, there are no available specimen records to substantiate this claim.

Protected areas
Coastal Forest Reserve (Kosi Lake, Manzengwenya Forest Station), Ndumu.

Conservation status
Because of its marginal distribution in South Africa (Pafuri area of the Kruger NP and northern Zululand), the species is recorded as Rare in the South African Red Data Book (Smithers 1986).

Habitat
They are confined to light sandy soils, sandy alluvium, and coastal sand dunes (Skinner and Smithers 1990).

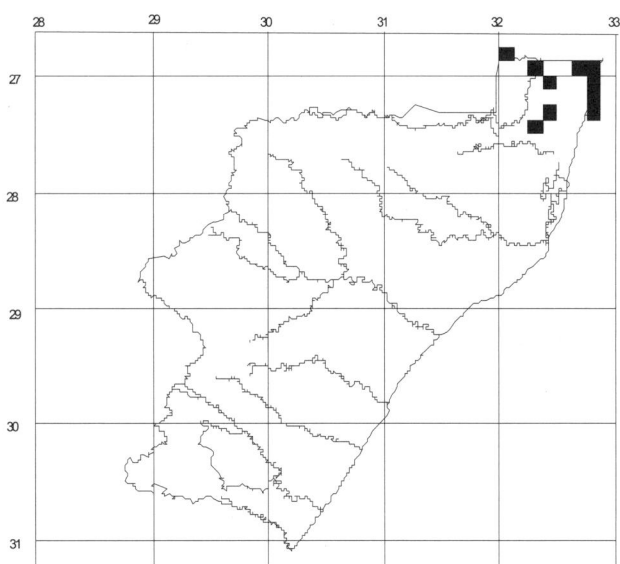

Figure 19

KEY TO SPECIES (Bronner 1995):

1. Size smaller, greatest skull length
    24.8 mm or less; mid-interorbital width
    less than 7.2 mm, and less than 108 per
    cent of palate width across P^2; claws
    gracile, basal claw width less than
    18 per cent of skull length . . . . *A. marleyi*, p. 26
–   Size larger, greatest skull length more
    than 25 mm; mid inter-orbital width
    greater than 7.5 mm, and more than
    110 per cent of palate width across P^2;
    claws robust, basal claw width more
    than 20 per cent of skull length . . . .
    . . . . *A. hottentotus*, p. 24

Habits

They occupy subsurface runs, and, in captivity, will burrow to a depth of 200 mm. Mounds have not been noted. Yellow golden moles have been observed to construct chambers around the base of trees from which their runs radiate for large distances. In captivity, they take earthworms, cockroaches, grasshoppers, moths, mealworms, flies, and small beetles.

Breeding

Two lactating females were collected in January (**DM** 406 and 408), while pregnant females were collected during October (series of five from the **TM**) and September (**NM** 126). In four pregnant females in which the number of embryos was recorded, two contained two embryos, and two contained one.

Linear measurements (mm)

Three females collected in Zululand had head and body lengths of 102, 102 and 98, and hind foot (*su*) lengths of 10, 9, and 11.

Records of occurrence

SPECIMEN RECORDS:

DM: Ndumu;

KM: Ndumu;

NM: Coastal Forest Reserve (=Kosi Bay), Lake Sibayi, Maputa (=Manguzi);

SI: Coastal Forest Reserve (=Kosi Bay);

TM: Between Enseleni and Ubombo, Kosi Lake, Lalanek, Maputa (=Manguzi), Manaba, Manzengwenya, Sihangwane.

Amblysomus hottentotus (A. Smith 1829)   PLATE 5

A. h. iris (Thomas and Schwann 1905)

A. h. longiceps (Broom 1907)

A. h. pondoliae (Thomas and Schwann 1905)

Hottentot golden mole

Hottentot-gouemol

Taxonomic status

This species was recently revised by Bronner (1995), resulting in the recognition of five subspecies, of which the three listed above are found in KwaZulu-Natal. The subspecies, *A. h. iris*, was previously considered to be a full species (Meester et al. 1986).

Distribution

The Hottentot golden mole is distributed widely in KwaZulu-Natal throughout all bioregions with the exception of Bushveld, where it is replaced by *Calcochloris obtusirostris* (Figure 20). Approximate distributions of the three subspecies in KwaZulu-Natal, based on Bronner's (1995) revision, are as follows. *Amblysomus hottentotus iris* occurs along the northern KwaZulu-Natal coastal belt from the Tugela River in the south northwards to St Lucia, and then inland to Eshowe, Hluhluwe, Umfolozi, Hlabisa and possibly Ngome (Figure 20: open squares). *Amblysomus hottentotus longiceps* occurs throughout the Highland, Mistbelt, Moist upland and Drier upland bioregions at altitudes exceeding 600 m, from Underberg in the south to the Newcastle and Vryheid Districts in the north, and Pietermaritzburg in the east (Figure 20: closed squares). *Amblysomus hottentotus pondoliae* occurs along the southern coastal belt of KwaZulu-Natal from the Tugela

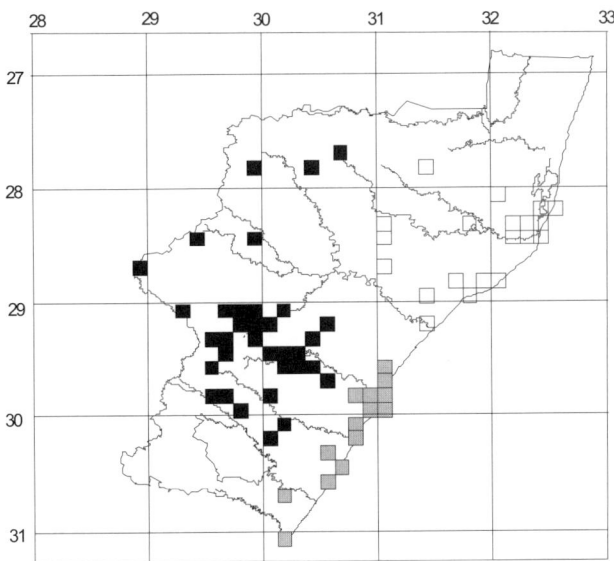

Figure 20

River in the north to the KwaZulu-Natal border in the south (Figure 20: hatched squares). Colour is highly variable within each subspecies, and intergradation occurs between the subspecies. Avery (1991) recorded *A. hottentotus* from several Holocene sites in the midlands and Tugela basin (see under **Records of occurrence**).

Protected areas

Bluff, Cathedral Peak, Chelmsford, Eastern Shores, Garden Castle, Giant's Castle, Hazelmere, Hluhluwe, Kamberg, Krantzkloof, Oribi Gorge, Royal KwaZulu-Natal, St Lucia Park, Umlalazi, Umtamvuna, Vernon Crookes.

Conservation status

Very common and widespread in KwaZulu-Natal. *Amblysomus iris* was listed as Indeterminate in the South African Red Data Book (Smithers 1986). However, this taxon is treated here as a subspecies of *A. hottentotus*.

Habitat

Occurs throughout coastal and montane forest types and various grassland veld types, as well as commonly in suburban gardens. Bronner (1995) listed the following Acocks (1975) veld types for each subspecies: *A. h. iris*: Coastal Forest and Thornveld (1), Zululand Thornveld (6) and Ngonigoni Veld (5); *A. h. longiceps*: Highland Sourveld and Dohne Sourveld (44), Southern Tall Grassland (65) and North-eastern Sandy Highveld (57); *A. h. pondoliae*: Coastal Forest and Thornveld (1), Pondoland Coastal Plateau Sourveld (3).

Habits

The Hottentot golden mole uses both shallow subsurface runs as well as burrows of up to 500 mm in depth. The characteristic dome-like ridges produced by the subsurface runs distinguishes it from the common molerat (*Cryptomys hottentotus*) which may at times share the same burrow systems. They frequently come above ground to forage, particularly in loose, moist soils such as are found in gar-

den beds. At least 60 moles were collected by hand by a mole-catcher as they moved above ground during early mornings at the Durban Botanic Gardens over a period of two years. Hottentot golden moles are frequently found in pellets of barn owls (*Tyto alba*). At St Lucia, samples of owl pellets were found to contain nothing but *A. hottentotus* (R. Taylor, personal communication). Like all golden moles, *A. hottentotus* feeds on insects and earthworms, but, unlike other golden moles, some of them will also eat snails and vegetable matter. In captivity, they are very fond of mealworms and earthworms but, with patience, can be made to feed on raw liver or minced beef.

Breeding

Based on a large sample of 88 specimens originating from Durban (voucher specimens lodged in the Durban Natural Science Museum and Transvaal Museum), Bernard et al. (1994) determined that breeding is aseasonal and continues throughout the year, possibly owing to the constant, buffered microclimate within their burrow systems. Mean litter size was 1.9 (*n*=15), with 14 litters of two (one foetus in each uterine horn), and a single litter of one.

Linear measurements (mm) and mass (g)

	Males					Females				
	$\bar{x}$	s.d.	n	Min	Max	$\bar{x}$	s.d.	n	Min	Max
HB	123.0	10.6	76	101	150	121.1	8.2	78	105	139
Hf	14.3	1.6	43	10	18	14.0	1.8	41	9	18
Mass	55.5	13.6	11	31	78	49.7	17.3	9	34	90

Records of occurrence

SPECIMEN RECORDS:

BM: Durban (probably Malvern), Estcourt, Illovo;

CM: Pennington;

DM: Bellair (Durban: Sarnia Road), Berea (Durban), Bosch Hoek Farm, Botanic Gardens (Durban), Brighton Beach (Durban), Cape Vidal, Cape Vidal (5 km south), Chelmsford NR, Cowies Hill (10 Penpard Grove), Doreen Clark NR, Durban North (23 Clarendon Ave, 108 Edgeley Road, 35 Ellis Park Drive), Durban (Ridge Road), Eastern Shores (Iphiva), Escombe School (Durban), Eshowe, Giant's Castle NR, Harold Johnson NR, Highmoor, Hillary (325 Stella Road), Hilton (12 Forest Lane), Inchanga Farm, Kharwastan (Durban: 83 Penguin Street), Kloof, Kloof (16 Springdale Road), Krantzkloof NR, Mariannhill, Mount Edgecombe, Mount Moreland (12 William Street), New Germany (7 Ridge Road), Pietermaritzburg, Pinetown, Redcliff Farm, Royal Natal NP, St Lucia Forest Station, St Lucia (Natal Parks Board House No. 2), St Winifreds (Kingsburgh), Thomas More College (Kloof), Ubombo village, Umbilo (15 Boxley Place), Umhlanga Rocks, Umhlanga (60 Hilken Drive), Umtamvuna NR, University of Natal (Durban), Victoria (Lower Umgeni), Warner Beach, Westville (Jan Hofmeyr/ Blair Athol Roads, 7 River Drive, 39 Springvale Road), Winterskloof (13 Crompton Road);

KM: St Lucia, St Lucia Forest Station;

NM: Babanango, Balgowan, Carter's Nursery (Pietermaritzburg), Castle View, Firle Farm, Hilton, Hluhluwe GR, Kamberg NR, Kambula, KwaZulu-Natal (unspecified), Loteni GR, Meadow Farm, Mooi River District, Ngoye Forest, Oribi Gorge (Shalimar), Pietermaritzburg, Prestbury (Pietermaritzburg), Rosetta, Spring Grove, Sweetwaters (Pietermaritzburg), Royal Natal NP (Tendele Camp), The Hoek (Mooi River), Town Bush (Pietermaritzburg), Tumble Inn, Tweedie, Umdoni Park, Umgeni Valley Game Ranch, Umtwalume River (north bank), Underberg area, Wembley (Pietermaritzburg), Wylde Holme Farm;

SAM: Howick, KwaZulu-Natal (unspecified locality), Umkomaas River;

SI: Hardingdale, Kilgobbin Farm, Petchaye;

TM: Blackridge (Pietermartizburg), Bluff NR, Bosch Hoek Farm (=Michaelhouse Golf Course), Chard Farm, Charter's Creek, Clairwood, Coastal Forest Reserve (Kosi Bay, Madhlanzula), Cornhill Farm, Curry's Post, Dargle, Dukuduku Forest Station, Durban Beach, Durban Country Club, Durban Botanic Gardens, Durban (51 Hartley Road), Empangeni, Eshowe, Estcourt, Futululu Forest Station, Giant's Castle GR (Hillside), Gillitts, Glenwood (Durban), Goudhoek Farm, Hazelmere NR, Hibberdene, Hilton (12 Forest Lane), Hluhluwe GR (Staff Camp, Egondeni), Hluhluwe Flats, Howick, Isipingo Flats, Kamberg NR, Karkloof, Ketelfontein (Pietermarizburg), Kilgobbin Farm, Klipspruit, Kloof, Kula, Mapelane, Maputa (=Manguzi), Monzi, Moore Park NR, Mtubatuba, Mtunzini, Ngome Forest Reserve, Oribi Gorge NR, Pietermaritzburg, Pinetown, Richards Bay, Shepstone Reserve, St Lucia Estuary, St Lucia Forest Station, St Lucia village, Sweetwaters (Pietermartizburg), Sydenham (Durban), Tabamhlope, Two Streams Farm, Umbilo (Durban), Umdoni Park, Umfolozi, Umgoye Forest, Umhlanga Rocks (Hilken Drive), Umhloti River near Verulum, Umlalazi River, Umtamvuna NR, University of Natal (Durban), Vernon Crooks NR, Weston Agricultural College, Wyford Farm.

ADDITIONAL RECORDS:
Avery (1991): Collingham, eSinhlonhweni, Mgede, Nkupe, Umhlatuzana.

Amblysomus marleyi (Roberts 1931)
Marley's golden mole
Marleyse-gouemol

Taxonomic status
While Meester et al. (1986) recognised *marleyi* as a subspecies of *A. hottentotus*, Bronner (1995) recognised it as a full species (following Roberts 1951), as it is more distinct morphometrically from *A. hottentotus* than *A. julianae* is from *A. hottentotus*. *Amblysomus marleyi* is distinctly

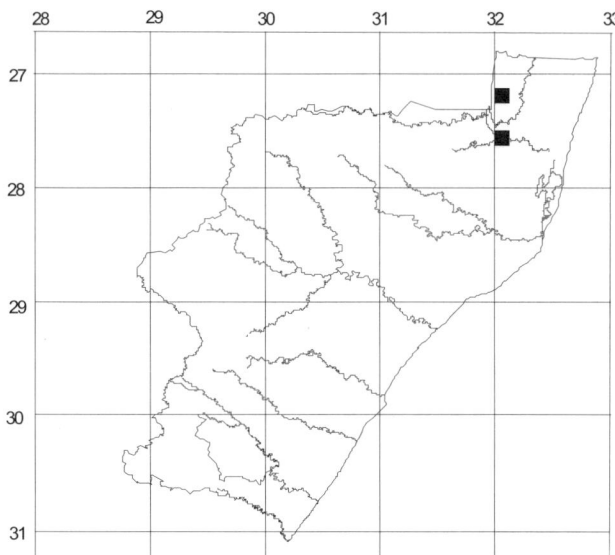

Figure 21

smaller than *A. hottentotus*, with little or no overlap in certain skull characters (Bronner 1995). The type locality is Ubombo.

Distribution
This species is known from only two localities in northern Maputaland, Ubombo and Ingwavuma (Figure 21; Bronner 1995). Avery (1991) considered that *Amblysomus* from Pleistocene samples of Border Cave would be referable to *marleyi*.

Conservation status
As this species is endemic to Maputaland, where its habitat is being degraded, and is not known from any reserves, it should be afforded Vulnerable status (Bronner 1995). As it was not previously considered to be a full species, it was not listed in the current South African Red Data Book (Smithers 1986).

Habitat
The two localities from which *A. marleyi* have been collected fall in Acock's veld type 66 (Zululand thornveld), are characterised by a relatively arid climate, and are separated from St Lucia–Hluhluwe populations of *A. hottentotus iris* by both bushveld (veld type 10), and lowveld and Zululand palm veld (veld type 1b) vegetation types (Bronner 1995).

Records of occurrence:
SPECIMEN RECORDS:
TM: Ingwavuma, Ubombo.

ADDITIONAL RECORDS:
Avery (1991): Border Cave.

Order CHIROPTERA

KEY TO FAMILIES and GENERA (Meester et al.1986):

1. Tail reduced or absent; face dog-like; second digit ending in claw; cheekteeth simple without well developed cusp pattern (Suborder MEGACHIROPTERA) 2
—   Tail well developed; face not dog-like; second digit not ending in claw; cheekteeth forming well developed cusp pattern 4

2. Ears with white basal tufts; no tail .... *Epomophorus*, p. 28
—   Ears without white basal tufts; reduced tail 3

3. Forearm generally 110–130 mm; colour generally tawny, with dorsal fur restricted to a narrow median band, sharply demarcated from naked wing membranes .... *Eidolon*, p. 30
—   Forearm 65–102 mm, ruff of stiff hair present or absent on lower neck of adult males .... *Rousettus*, p. 31

4. Terminal portion of tail projecting past edge of inter-femoral membrane 5
—   Tail almost entirely enclosed within interfemoral membrane 9

5. Free terminal portion of tail emerges above middle of upper surface of membrane (Family EMBALLONURIDAE) .... *Taphozous*, p. 32
—   Free terminal portion of tail projects considerably beyond hind margin of membrane (Family MOLOSSIDAE) 6

6. Size very large, forearm 62–73 mm; ears very large, 38–40 mm, conjoined on extended snout; conspicuous bicoloured dorsal coat with pale neck collar .... *Otomops*, p. 58
—   Size smaller, forearm less than 60 mm; ears smaller; dorsal coat not bicoloured 7

7. Size small, forearm 36–39 mm .... *Chaerephon*, p. 62
—   Size larger; forearm greater than 40 mm 8

8. Size smaller, forearm 45–50 mm; colour dark above and paler below, paler band in front of shoulders .... *Mops*, p. 64
—   Size larger, forearm 46–54 mm; colour dark sooty-brown almost black above, same colour or a shade paler below, no band in front of shoulders ... *Tadarida*, p. 60

9. Muzzle with nose-leaves or deep central slit 10
—   Muzzle without nose-leaves or deep central slit (Family VESPERTILIONIDAE) 15

10. Muzzle with deep, central longitudinal slit (Family NYCTERIDAE) .... *Nycteris*, p. 33
—   Face without slit, but with well-developed nose-leaves covering muzzle 11

11. Posterior leaflet of noseleaf with upper edge either elliptical in outline or tridentate (three-pronged) (Family HIPPOSIDERIDAE) 12
—   Posterior leaflet of noseleaf triangular with erect, pointed tip (Family RHINOLOPHIDAE) .... *Rhinolophus*, p. 35

12. Posterior leaflet of noseleaf with upper edge elliptical in outline; size larger, forearm greater than 40 mm .... *Hipposideros*, p. 40
—   Posterior leaflet of noseleaf with tridentate shape; size smaller, forearm 31–35 mm .... *Cloeotis*, p. 41

13. Second phalanx of third digit about three times as long as first (Subfamily MINIOPTERINAE) .... *Miniopterus*, p. 41
—   Second phalanx of third digit not especially elongated 14

14. Ears not funnel-shaped, without deep emargination below tip; tragus not sharply pointed; braincase not high and rounded (Subfamily VESPERTILIONINAE) 15
—   Ears funnel-shaped, with deep emargination below tip; tragus long, narrow, sharply pointed; braincase high and rounded (Subfamily KERIVOULINAE) .... *Kerivoula*, p. 57

15. Six upper, six lower cheekteeth on each side .... *Myotis*, p. 44
—   Fewer than six upper and six lower cheekteeth 16

16. Five upper, five lower cheekteeth; two upper incisors on each side .... *Pipistrellus*, p. 46
—   Four upper, five lower cheekteeth; one or two upper incisors on each side 17

17. Two upper incisors on each side 18
–   One upper incisor on each side 20

18. Rostrum very short and broad; braincase
     very high, more than two thirds of condy-
     lobasal length; conspicuous reticulation
     of wing membranes . . . . *Chalinolobus*, p. 49
–   Rostrum not shortened; braincase not
     elevated; no wing pattern 19

19. Ears 18 mm or more, about half length of
     forearm . . . . *Laephotis*, p. 49
–   Ears less than 18 mm, much less than
     half length of forearm . . . . *Eptesicus*, p. 50

20. Size larger, forearm greater than 46 mm;
     first and second upper molars with
     W-pattern obsolescent; tragus long and
     tapering . . . . *Scotophilus*, p. 53
–   Size smaller, forearm less than 46 mm;
     first and second upper molars with
     normal W-pattern; tragus either sickle-
     shaped or short and broad 21

21. Wings dark-coloured; skull with rostrum
     not broadened; palatal emargination
     narrow; anterior lower premolar about
     half the crown area of the next premolar;
     upper canines with rounded anterior
     surface; tragus halfmoon-shaped; penis
     not enlarged; forearm 29–33 mm . . . .
                                          Nycticeius, p. 56
–   Wings pale-coloured; skull with rostrum
     notably broadened, particularly across
     lachrymals; palatal emargination
     broader; lower premolars equal in
     crown area; upper canines with broad,
     flat anterior surface; tragus short and
     broad; penis greatly lengthened;
     forearm 28–38 mm . . . . *Scotoecus*, p. 56

Suborder	**MEGACHIROPTERA**	
		Fruit bats
Family	**PTEROPODIDAE**	
Genus	*Epomophorus*	Bennett 1836

Epomophorus wahlbergi (Sundevall 1846)   PLATE 6
E. w. wahlbergi (Sundevall 1846)
Wahlberg's epauletted fruit bat
Wahlberg-witkolvrugtevlermuis

Taxonomic status
Two subspecies have been recognised of which one (*E. w. wahlbergi*) occurs in southern Africa (Meester et al. 1986).

Distribution
This species is widespread throughout the lower-lying eastern parts of KwaZulu-Natal. Its distribution seems to pene-

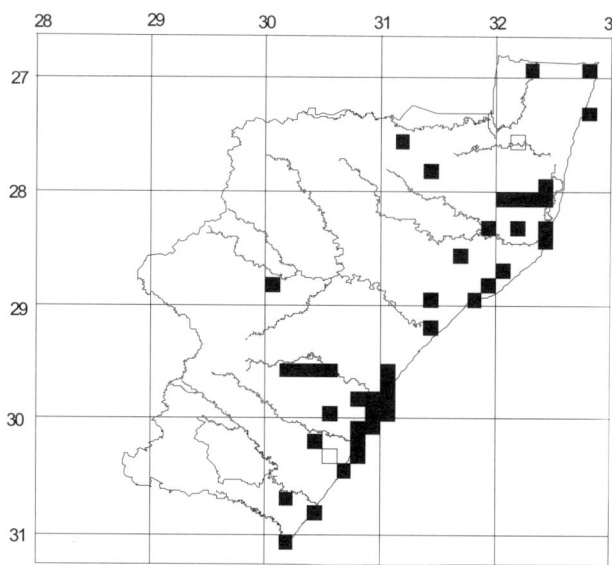

Figure 22

trate inland from the coast along river valleys such as the Mkuze, Umfolozi, Tugela and Umgeni, reaching as far west as Weenen Nature Reserve at 30°E (Figure 22).

Protected areas
False Bay, Harold Johnson, Hluhluwe, Itala, Kenneth Stainbank, Mkuzi, Oribi Gorge, St Lucia, Umfolozi, Umlalazi, Vernon Crookes.

Conservation status
Common in the eastern half of KwaZulu-Natal, including suburban areas where they are often regarded as pests.

Habitat
Coastal forest, riverine forest, Afromontane forest and well wooded urban and suburban areas, associated with fruit-bearing trees. They can occupy extensively built-up areas as long as fruiting trees and suitable daytime roosting trees are available. Preferred roosting trees include certain tall, exotic tree species such as *Casuarina equisetifolia* and *Roystonia elata*. They occasionally roost in man-made structures; in the Durban central business district, colonies roost year-round in concrete recesses under a freeway, as well as under the eaves of a double storey house.

Habits
Colonies roost in well-vegetated trees during the daytime, in groups varying in size from single individuals up to about 100. Males are territorial and emit a monotonous, honking call to attract females. This vocalisation is associated with an elaborate visual display to attract females whereby the long, white epaulette hairs are erected and the wings flutter rapidly, possibly to waft scents from their epaulette glands to the females. Their habit of defecating in flight means that, in well-wooded suburbs, the walls of residents' homes can become severely plastered with droppings. In Yellowwood Park, for example, this problem is particularly prevalent because of the abundance of yellowwood trees (*Podocarpus sp.*), whose fruit are par-

ticularly attractive to bats. Limited research has shown that the use of bright lights to illuminate the walls can act as a deterrent in such cases.

They prefer figs and other wild fruits as well as over-ripe, soft-fleshed cultivated fruits such as mango, pawpaw, avocado and guava. In a study of fruiting trees represented in 'spit-outs' from fruit bats at Mtunzini, in KwaZulu-Natal, C. Sapsford (personal communication) recorded the following identified species:

Ficus sur (=capensis) Cape fig
F. trichopoda Swamp fig
F. natalensis Natal fig
Voaconga thouarsii Wild frangipani
Tabernaemontana ventricosa Forest toad tree
Syzigium cordatum Umdoni
Bridelia mucrantha Coastal goldenleaf
Euclea natalensis Natal ebony
Eugenia capensis Dune myrtle
Ekebergia capensis Cape ash
Annona senegalensis Wild custard-apple
Podocarpus (latifolius and *falcatus)* Yellowwood
Sideroxylon inerme White milkwood
Rauvolfia caffra Quinine Tree (from observation of bats)
Halleria lucida Tree fuschia
Sclerocarya caffra Marula
Trichelia emetica Natal mahogany
Harpephyllum caffrum Wild plum
Mimusops caffra Red milkwood. Highly favoured
Tabermontana elegans Not positively identified

Breeding

Based on a study at Mtunzini (C. Sapsford, personal communication), in KwaZulu-Natal, breeding occurs throughout the year with peaks in July and in the summer. Breeding males display particularly long epaulette hairs as well as a brown discoloration of the skin in the vicinity of the testes (C. Sapsford, personal communication). Postlactating females have been recorded in February, and pregnant females in February and March (E. Seamark, unpublished data; **TM** collection).

Linear measurements (mm) and mass (g)

	Males					Females				
	x̄	s.d.	n	Min	Max	x̄	s.d.	n	Min	Max
HB	144.0	19.8	19	103	180	136.0	15.2	17	110	168
Hf	19.8	1.5	7	18	22	19.0	2.0	7	15	21
E	24.3	2.5	10	20	28	23.9	4.3	8	15	28
FA	83.8	5.5	7	75	90	81.3	4.3	7	75	87
Mass	98.0	22.3	4	70	124	87.0	21.5	5	50	104

Records of occurrence

SPECIMEN RECORDS:
CM: Pennington;
DM: Amanzimtoti, Beaumont Sugar Estate, Bonamanzi PGR, Clairwood Park, Durban City Hall, Durban (170 Brand Road, Walnut/Commercial Roads), Empisini NR, Enseleni NR, False Bay Park, Fynnlands (261 Lighthouse Road), Glenashley (6 Adrienne Avenue), Glenwood (35 St John's Avenue), Greenwood Park, Hillary, Hluhluwe GR (Mansiya Valley), Kloof (18 Stormont Avenue), Mtunzini, Northdene, Pietermaritzburg, Pinetown (6 John White Road), Sarnia, Seaview (5 Armadale Road), St Lucia village, Westville;
NM: Durban Road, Empangeni, Nagle Dam, Ntombeni, Pietermaritzburg, Scottsville, Town Bush, Umdoni Park, Umfolozi River, Umgeni Valley Game Ranch;
SAM: Pinetown;
SI: Pietermaritzburg;
TM: Clearwater Farm, Dukuduku Forest Station, Eshowe, Howick, Hluhluwe GR (Mansiya Valley), Itala GR, Kosi Lake, Lake Sibayi, Malvern, Ndumu GR, Ngome Forest Reserve, Ntambanana, Oribi Gorge NR, Pigeon Valley Reserve (Durban), Scottburgh Golf Course, University of Natal (Durban: Shepstone Reserve), St Lucia Estuary, St Lucia village, Umdoni Caravan Park, Weenen NR.

ADDITIONAL RECORDS:
Dixon (1964): Mkuzi GR;
Bourquin et al. (1971): Hluhluwe GR (Nsizwa);
Bourquin and Sowler (1980): Vernon Crookes NR;
E. Seamark (unpublished sight data): Harold Johnson NR, Tropicale Restaurant (Durban);
Sapsford (unpublished data: specimen records and roost sightings):
Specimens: 9 Arbingdon Road (Umbilo), 69 Abrey Road (Kloof), 14 Ashley Road (Hillcrest), 549 Bluff Road (Bluff), 12 Broadwood Drive (Umhlanga Rocks), Brochlee Farm (Eshowe), 23 Carlton Avenue (Westville), Cedar Road (Congella), 6 Chatman Place (Hillary), 9 Chearsley Road (Westville), 39 Cherry Avenue (Durban), 12 Chirol Drive (Westville), 72 Dawncliff Road (Westville), 4 De Villiers Road (Northdene), Dirkie Uys School (Bluff), Doonside Station (Amanzimtoti), 1 Douglas Road (Pinetown), 3 Drayton Place (Westville), 25 Fourth Avenue (Reunion), 7 Fife Avenue (Westville North), 12 Glenugie Road (Pinetown), Hullett's Hospital (Mount Edgecombe), 7 Jackson Road (Farningham Ridge), 103 Kennelworth Road (Overport), 200 Kensington Drive (Durban North), 297 Kenton Howden Road (Yellowwood Park), 49 Loudoun Road (Berea), 8 Madiera Road (Berea), 143 Manning Road (Glenwood), 76 Manor Drive (Manor Gardens), 28 Margaret Crescent (Forest Hills), 18 Melrose Place (Durban North), 91 Mimosa Road (Greenwood Park), Mitchell Village (Tongaat), 37 Mobile Road (Berea), 22 New Germany Gardens (Pinetown), 15 Nipper Road (New Germany), 9 Oak Avenue (St Winifreds), 82 Oakley Drive (Durban), Pot Luck Caravan Park (Doonside), 90 Riley Road (Overport), 15 Sharp Place (Montclair), 12 Starling Place (Yellowwood Park), 8 St Johns Avenue (Glenwood), 18 Sunbird Crescent (Yellowwood Park), 17 Trotter Road (Pinetown), Umhloti, 311 Vause Road (Berea), 20 Venice Road (Morningside), 71 Welfrere Road (Bluff);
Roost sightings: Old Caravan Park (Amanzimtoti), Beach Road/Inyoni Rocks Road (Amanzimtoti), Grant Road (Amanzimtoti), Ingram Mansions (Amanzimtoti), King George V Avenue (Berea), 9 Overdale Road (Berea), Galloway House – Musgrave Road (Berea), Doonside Station

(Amanzimtoti), Mitchell Park (Morningside), Old Fort Place (Durban), Botha Park (Durban), Botanic Gardens (Durban), 84 Wall Road (Escombe), 48 Parkside Road (Hillary), 76 Currie Crescent (Escombe), Queen Mary Avenue/Bartle Road (Glenwood), 6 Ijuba Place (Kloof), 58 Coronation Road (Malvern), Orchard Gardens – Dawnlea Road (Malvern), Becton Avenue Park (Malvern), Mtunzini Hotel, Main Road (Mtunzini), John Carr Farm (Mtunzini), 87 Seaview Road (Mtunzini), Seaview Road Circle (Mtunzini), Salt Rock Caravan Park, Salt Rock turnoff (Umhlali), Italian Club (Umkomaas), Barrow Street (Umkomaas), Westville Civic Centre;
Sight record (J. H. Grobler, 14/2/85): Umlalazi NR.

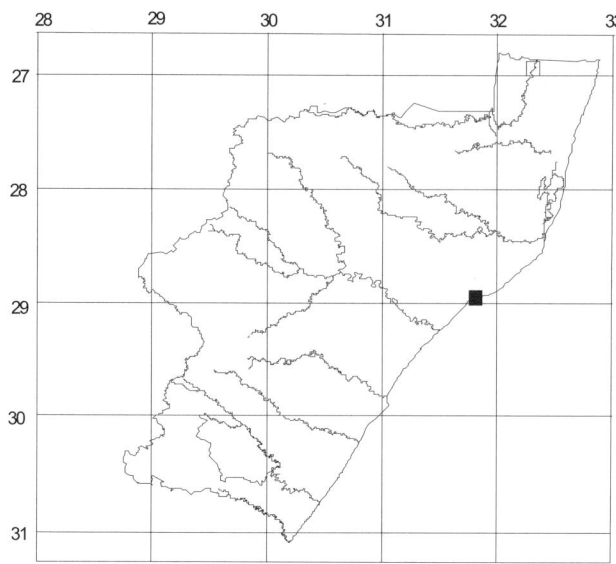

Figure 23

? *Epomophorus gambianus* (Ogilby 1835)
E. g. crypturus (Peters 1852)
Peters's epauletted fruit bat
Peters-witkolvrugtevlermuis

This species was recorded by Bourquin (1988) as occurring at two widely separated localities in KwaZulu-Natal: Pietermaritzburg and Ngome. Two specimens from Ngome in the Transvaal Museum (**TM** 39131 and 39132) have since been re-identified by I. L. Rautenbach as *E. wahlbergi* and not *E. g. crypturus*. According to personal notes of O. Bourquin, the specimen identified as *E. g. crypturus* from Pietermaritzburg was sent to the USA by J. Meester during 1977 or 1978. No specimen is therefore available to corroborate this record. Accurate separation of *E. g. crypturus* from *E. wahlbergi* is difficult, being based on the presence of two ridges on the palate behind the last molars in *E. g. crypturus* as opposed to only one in *E. wahlbergi*. For the above reasons, the presence of *E. g. crypturus* in KwaZulu-Natal cannot be accepted at present, even though their general distribution as far south as the Mpumalanga Province of South Africa indicates that they may possibly occur in KwaZulu-Natal. Apparent records of *E. g. crypturus* from the Eastern Cape (Skinner and Smithers 1990) may also turn out to be misidentifications.

Genus	*Eidolon*	Rafinesque 1815

Eidolon helvum (Kerr 1792)   PLATE 7
E. h. helvum (Kerr 1792)
Straw-coloured fruit bat
Geelvrugtevlermuis

Taxonomic status
Of the three recognised subspecies, only the nominate race is found in southern Africa (Meester et al. 1986).

Distribution
The species occurs as a sporadic but widespread migrant in southern Africa. Its breeding range is limited to the tropi-cal forests of West, Central and East Africa. In KwaZulu-Natal, the straw-coloured fruit bat had been recorded from only two localities in Zululand: Ndumu Game Reserve and Mtunzini (Figure 23).

Conservation status
As this species is a non-breeding migrant in KwaZulu-Natal, it is difficult to gauge its conservation status, or adequacy of protection in reserves. Nevertheless, preservation of suitable coastal and riverine forests in KwaZulu-Natal are necessary to ensure its continued survival as a migrant in KwaZulu-Natal.

Habitat
Evergreen forest habitats in the form of coastal and riverine forest.

Habits and breeding
Straw-coloured bats are gregarious, occupying large colonies within their breeding range. For example, a colony of 200 000 individuals is known to roost in a grove of saligna gums *Eucalyptus saligna* at Kampala in Uganda. In southern Africa, smaller groups of one up to 12 have been observed. Colonies migrate large distances in search of food. Apart from soft-fleshed cultivated fruits they have been known to eat a variety of wild fruits as well as the flowers and leaves of certain trees. The species does not breed in southern Africa, but observations from West Africa have shown that the young (usually one, but sometimes two) are born around November (Skinner and Smithers 1990).

Records of occurrence
Dixon (1966): Ndumu GR;
Sight record (C. Sapsford): Mtunzini.

Genus	*Rousettus*	Gray 1821

Rousettus aegyptiacus (E. Geoffroy 1810)   PLATE 8
R. a. leachii (A. Smith 1829)
Egyptian fruit bat
Egiptiese vrugtevlermuis

Taxonomic status
Of the two described subspecies, only one is found in southern Africa (Meester et al. 1986).

Distribution
Resident colonies of these cave-dwelling bats are known from caves at Mission Rocks on the Eastern Shores of Lake St Lucia and a deep crevasse at Ismont Farm near Eston. There are undoubtedly other, as yet undiscovered, roosts of this species throughout KwaZulu-Natal. These bats are capable of large scale migrations, and individuals marked by N. Jacobsen in the Northern Province have been subsequently recovered by C. Sapsford at Mtunzini in Zululand (Figure 24).

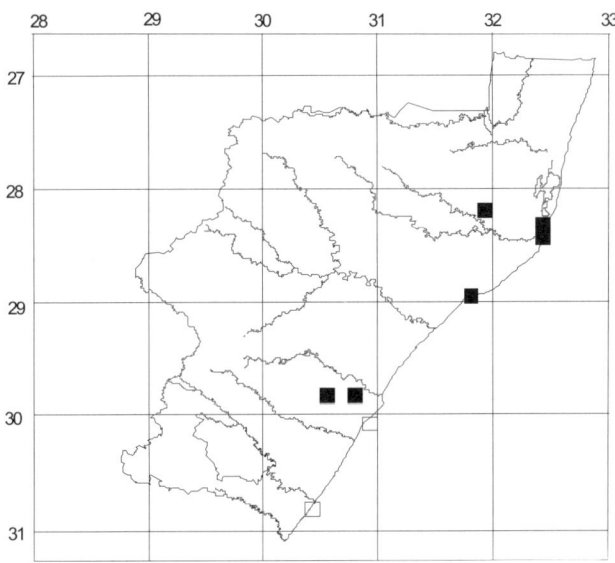

Figure 24

Protected areas
Eastern Shores, Umlalazi.

Conservation status
As this species is: (1) limited to the coastal belt of southern Africa; (2) dependent on adequate protection of caves for day roosting sites; and (3) recorded from only two protected areas in KwaZulu-Natal, an attempt should be made to ensure the co-operation of landowners to protect known roosts of this species on privately-owned land.

Habitat
Habitat requirements include suitable caves as well as proximity to fruiting trees. In KwaZulu-Natal they are associated with forest and savanna habitats within the Coast lowland, Valley bushveld and Lowveld bioregions. They are often found in suburban areas, although not reported in KwaZulu-Natal, and have become a nuisance in the southern suburbs of Cape Town, after the city council planted numerous yellowwood trees in the area.

Habits
Egyptian fruit bats are gregarious and cave-dwelling. Colonies of up to thousands occur. They roost in the darkest parts of the cave, closely packed together, usually hanging by one of the hind feet. Although their eyesight is excellent they can also use echolocation to navigate in the pitch darkness of caves. They are the only fruit bat known to echolocate. Low frequency clicks produced by the tongue are used in echolocation. They are known to feed on the fruit of a variety of indigenous trees including Cape fig (*Ficus capensis*), Cape ash tree (*Ekebergia capensis*), saffronwood (*Cassine crocea*), water berry (*Syzigium cordatum*) and wild plum (*Harpephyllum caffrum*), as well as on fleshy garden and orchard fruits (Skinner and Smithers 1990).

Breeding
A single young (or occasionally twins) is born during the summer months, after a gestation period of around 106 days. The young are carried on the ventral surface of the mother for the first six weeks, and they learn to fly at about nine to ten weeks of age. During the mating season the males tend to form groups, ignoring the females and juveniles, which form separate nursery colonies (Skinner and Smithers 1990).

Linear measurements (mm) and mass (g)

	Males					Females				
	x̄	s.d.	n	Min	Max	x̄	s.d.	n	Min	Max
HB	112.0	–	2	105	119	118.0	24.8	3	90	137
T	12.0	–	2	11	13	18.0	0.0	3	18	18
Hf	21.5	–	2	21	22	21.3	0.6	3	21	22
E	21.0	–	2	20	22	22.0	1.0	3	21	23
FA	92.5	–	2	91	94	93.7	7.8	3	85	100

Records of occurrence
SPECIMEN RECORDS:
CM: Mission Rocks;
DM: Mission Rocks, Umlalazi NR;
KM: Hlabisa, Mission Rocks;
NM: Eston;
TM: Mission Rocks.

ADDITIONAL RECORDS:
Bergmans (1994): Amanzimtoti, Durban, Ismont Crevasse, 'Natal', Port Shepstone, Mission Rocks, St Lucia;
Laycock (1976): Ismont Crevasse;
Sapsford (unpublished specimen data): 7 Hilldene Road (Hillcrest).

Suborder	MICROCHIROPTERA
	Insect-eating bats
Family	**EMBALLONURIDAE**
	Sheath-tailed bats
Genus	***Taphozous*** **E. Geoffroy 1818**

Taphozous mauritianus (E. Geoffroy 1818)   PLATE 9
Mauritian tomb bat
Witlyfvlermuis

Taxonomic status
No subspecies are recognised (Meester et al. 1986).

Distribution
Scattered but wide distribution in KwaZulu-Natal (Figure 25). Apparent absence from large areas is almost certainly a result of lack of collecting effort, as numerous records are available from the larger population centres of Durban and Pietermaritzburg where people are more likely to encounter specimens and report them to the local museum.

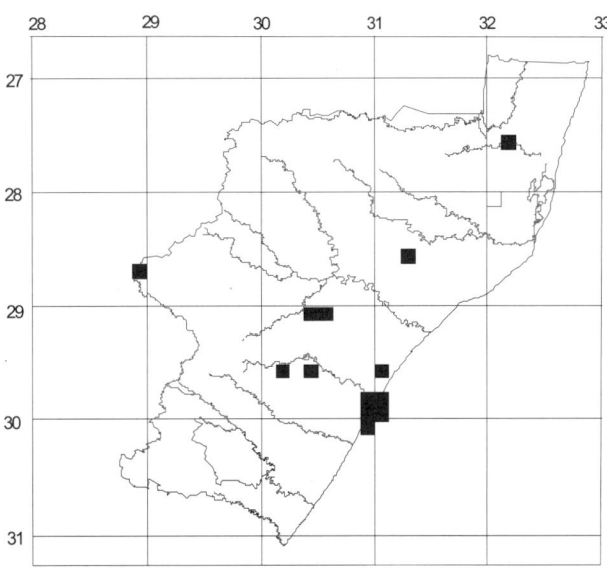

Figure 25

Protected areas
Hluhluwe, Mkuzi, Royal Natal NP.

Conservation status
Fairly widespread but known from few specimens, probably owing to lack of sampling. The species appears to be fairly common in the suburban areas of Durban and is probably reasonably common throughout its range.

Habitat
This species roosts on the bark of tall trees, or on the outside walls of buildings, usually with some form of cover, such as overhanging foliage or the eaves of a roof. A specimen from Royal Natal National Park was recovered from an accumulation of owl pellets found in a gum plantation surrounded by Montane grassland and shrubs. Of 21 roosts observed in the Durban area, six were from trees, three from houses with painted walls, and 12 from houses with face brick walls (F. Mackenzie, personal communication). In KwaZulu-Natal, they tolerate a wide range of bioregions from moist Montane and Mistbelt to drier Coast lowland, Valley bushveld and Lowveld.

Habits
Groups of between one and five have been observed in 21 colonies in the Durban area (F. Mackenzie, personal communication). Males and females either occur in separate colonies or are separated from each other by at least 100 mm. They cling face down to the vertical surface of a tree or wall by their hind feet claws, with their belly in contact with the surface. When disturbed they crawl rapidly sideways in a crab-like motion, or they will fly away and alight on a nearby tree.

Breeding
Limited data for southern Africa suggest that a single young is born during the warm, wet summer months (Skinner and Smithers 1990). Recent observations of tomb bat roosts in the Durban area have shown that females give birth to a single young, and that a single female can give birth to up to two young per year (during February or March and then again during October to December (F. Mackenzie, personal communication)).

Linear measurements (mm) and mass (g)

	Males					Females				
	x̄	s.d.	n	Min	Max	x̄	s.d.	n	Min	Max
TL	102.5	3.7	4	99	107	–	–	0	–	–
HB	82.0	4.8	4	77	87	–	–	0	–	–
T	20.6	2.1	5	17	22	–	–	0	–	–
Hf	11.4	1.5	5	10	13	–	–	0	–	–
E	16.9	1.8	4	15	19	–	–	0	–	–
FA	61.6	1.0	6	60	63	–	–	1	61	61
Mass	28.0	–	2	26	30	–	–	1	27	27

Some forearms and masses provided by F. Mackenzie

Records of occurrence
SPECIMEN RECORDS:
DM: Durban (Umbilo Road), Durban North, Howick (12 Miller St), Mhlopeni Ranch, Royal Natal NP;
TM: Durban, Greytown, Howick (Phillpot Street), Mkuzi GR (Mantuma), Scottsville (26 Hutchinson Road).

ADDITIONAL RECORDS:
Bourquin et al. (1971): Hluhluwe GR (Maphumula);
Sapsford (unpublished specimen data): 40 Arnold Road (Springfield), 58 Bank Terrace (Howardene), 16 Sackville Place (Durban North), Salisbury Island (Durban Naval Command), 53 Second Avenue (Pinetown), Tongaat General Hospital;
Sight records (Durban Bat Interest Group, 1994–1997): Athlone Park (3 Chelsea Road), Bluff (14 Pitman Road), Clare Estate, Durban Botanic Gardens, Durban North (14

Buckleigh Place), Glenmore (15 Melesina Avenue), Glenwood (Bath Road, 16 Chelmsford Avenue, Haig Avenue), Isipingo (12 Bonito Avenue), Morningside (99 Gordon Road, 225 Lambert Road, 77 Marriot Road), Newlands (117 Parlock Drive), Newlands West (17 Crowndale Mews), Old Fort Place (Durban), Ridge Park College (Durban), Umbilo (41 Bottomley Road, 83 Frere Road);

D. Leach (personal communication): Saxony Farm (1 km from Melmoth).

Family	**NYCTERIDAE**	**Slit-faced bats**
Genus	*Nycteris*	G. Cuvier and E. Geoffroy 1795

KEY TO SPECIES (After Meester et al. 1986):

1. Upper incisors trifid; tragus falciform or semilunate; ears smaller, 25 mm or less . . . . . .
    N. hispida, p. 33

— Upper incisors bifid; tragus pyriform; ears elongated, 28–37 mm . . . .
    N. thebaica, p. 33

Nycteris hispida (Schreber 1774)
N. h. villosa (Peters 1852)
Hairy slit-faced bat
Harige spleetnuisvlermuis

Taxonomic status
Two subspecies have been recognised of which only one occurs in southern Africa (Meester et al. 1986).

Distribution
This species has been recorded from the vicinity of Lake St Lucia and Lake Sibayi in the coastal region of northern Zululand (Figure 26).

Protected areas
Eastern Shores

Conservation status
Localised and uncommon; known from five specimens in KwaZulu-Natal.

Habitat
They occupy a wide range of habitats. Roosting sites include low dense bushes, huts, holes in termitaria, hollow trees, and caves.

Habits
They are predominantly solitary but have been observed in groups of up to 20. Hairy slit-faced bats are attracted to lights to catch insects and frequently enter buildings when foraging.

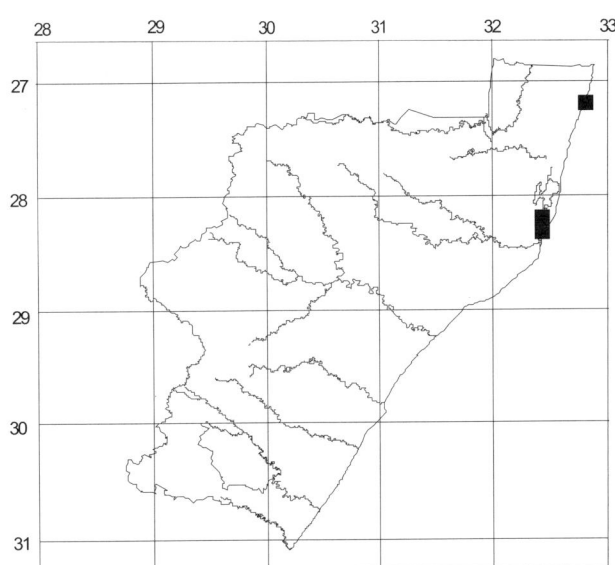

Figure 26

Breeding
No information is available for KwaZulu-Natal or for southern Africa (Skinner and Smithers 1990).

Linear measurements (mm)

Cat. No.	Sex	TL	HB	T	Hf	E	FA
DM2431	M	81	47	34	7	20	38
DM2432	M	87	47	40	8	20	40
DM2458	M	93	46	47	7	20	43

Records of occurrence
SPECIMEN RECORDS:
DM: Eastern Shores NR (Mission Rocks), St Lucia Forest Station;
KM: Charter's Creek, Manzengwenya Forest Station.

Nycteris thebaica (E. Geoffroy 1813)   PLATE 10
N. t. capensis (A. Smith 1829)
Egyptian slit-faced bat
Egiptiese spleetnuisvlermuis

Taxonomic status
The above-named subspecies is one of three that were recognised by Meester et al. (1986).

Distribution
Widespread in KwaZulu-Natal, occurring in all but the colder, higher-altitude Highland, Montane, and Mistbelt bioregions (Figure 27). Avery (1991) recorded this species from the upper Pleistocene deposits at Border Cave.

Protected areas
Eastern Shores, False Bay, Hazelmere Dam, Hluhluwe, Itala, Mkuzi, Ndumu, Spioenkop, Stainbank, Umfolozi, Umlalazi, Umtamvuna, Vernon Crookes, Weenen.

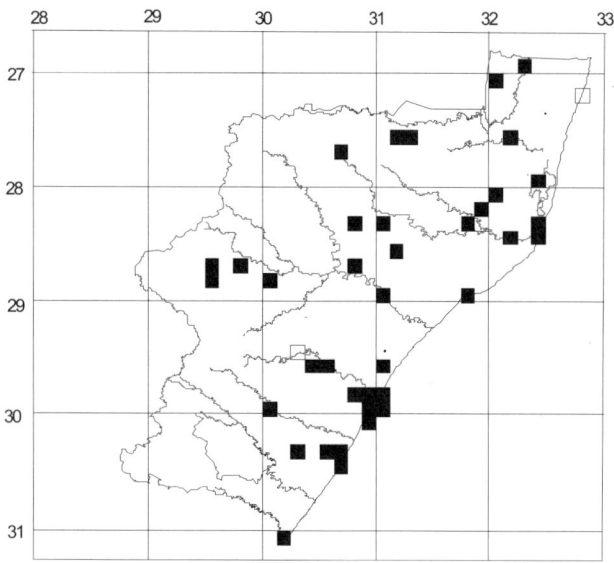

Figure 27

Conservation status

Widespread and reasonably common.

Habitat

They are essentially a cave-roosting species, but also roost during the day in mine adits, aardvark holes, rock crevices, culverts under roads, roofs, and hollow trees, typically in open savanna woodland (Skinner and Smithers 1990). In KwaZulu-Natal they have been recorded from a range of habitats from fairly open bushveld habitats to dense coastal forest. Many of the inland records originate from the Valley bushveld bioregion, associated with the Mfolozi, Tugela and Mgeni Rivers.

Habits

They are gregarious, occurring in small to medium-sized colonies numbering up to hundreds. They are often seen foraging inside buildings at night where they establish night-time feeding roosts. Thatched viewing hides within Mkuzi Game Reserve are used as night-roosts by large numbers of *N. thebaica* in close association with groups of night-foraging *Hipposideros caffer*. When a colony of slit-faced bats is disturbed during the day, individuals tend to evacuate the roost and take refuge in cover outside, unlike most cave-dwelling bats which remain within the cave. Bowie, Jacobs and Taylor (in press) found the diet of this species during summer at Mkuzi Game Reserve to consist primarily of non-volant prey such as orthopterans (44 per cent of volume, based on faecal samples) and arachnids (36 per cent). Crickets (Gryllidae) formed an important part of the diet, suggesting that *N. thebaica* was exploiting male cricket mating calls for target detection. The species is known to feed on stationary prey, relying partly on auditory stimuli from the prey. La Val and La Val (1980) demonstrated seasonal variation in the diet of this species at Umdoni Park near Scottburgh, with moths (Lepidoptera) and termites (Isoptera) increasing in importance in late winter and spring.

Breeding

In a study of several cave sites in the KwaZulu-Natal midlands, Bernard (1982a) found that Egyptian slit-faced bats show no evidence of hibernation. However, *N. thebaica* was found to occupy coastal caves throughout the year, while inland sites were occupied only during winter. Migration did not occur between coastal and inland sites. According to Bernard's (1982a) study, copulation and fertilisation occurs in early June. After a five month gestation, the young (a single young per female) are born in early November, and the mothers suckle the young for two months, until mid-January, after which the young are weaned. In a colony observed at Nagle Dam on 28 December, females were observed to be carrying their (single) young, although the young were capable of flight. During March, juveniles and post-lactating females were observed together in an abandoned goldmine at Mooiplaas Plantation (SAPPI) in the Babanango District (E. Seamark and T. Kearney, personal communication).

Linear measurements (mm) and mass (g)

	Males					Females				
	x̄	s.d.	n	Min	Max	x̄	s.d.	n	Min	Max
TL	101.8	8.3	22	83	113	103.8	8.1	14	92	126
HB	53.0	5.8	22	42	65	52.1	4.0	14	45	60
T	48.7	5.3	22	38	56	51.8	6.3	14	44	66
Hf	10.0	0.5	13	9	10	10.0	0.9	10	8	11
E	29.7	2.6	14	25	34	30.0	2.2	11	27	34
FA	45.0	3.2	14	37	52	46.8	1.8	7	45	49
Mass	10.6	1.1	7	9	12	8.7	0.6	3	8	9

Bowie, Jacobs and Taylor (in press) recorded a mean mass of 12 g and a mean forearm length of 46 mm for nine individuals from Mkuzi Game Reserve (sexes combined).

Records of occurrence

SPECIMEN RECORDS:

DM: Bluff (Durban), Eastern Shores NR (Mission Rocks), Game Valley Ranch, Hluhluwe GR, Mkuzi GR (Bube Pan, Mantuma Camp, Msinga Pan), Mtunzini, Nagle Dam, Sezela (3 Msinzi Avenue), Spioenkop NR (Ntabamnyama), Stainbank NR, Umlalazi NR, Umtamvuna NR, Weenen NR;

KM: Doornhoek Mine Tunnel (Pietermaritzburg), Winterton;

NM: Geluk Farm, Hluhluwe GR (Maganda), Kambula, Mfongosi, Ntombeni, Pietermaritzburg, Scottsville, Umdoni Park;

SAM: Durban, Insusi Valley, Mfongosi;

TM: Bishopstowe Cave (Pietermaritzburg), Dumisa Mine, Durban, False Bay Park (Lister Point), Gillitts, Goudhoek Farm, Hazelmere Dam NR, Hlabisa, Ingwavuma River, Itala GR (Craigadam, Doornkraal), Malvern, Mkuzi GR (Bube Pan, Malibali Pan), Mtubatuba, Ndumu GR, St Lucia village (Crocodile Centre), Umfolozi GR (Masinda Lodge), Vernon Crookes NR.

ADDITIONAL RECORDS:
Avery (1991): Border Cave;
Bernard (1980): Reunion (=Bluff);
Bourquin et al. (1971): Umfolozi GR (Black Umfolozi River, Hilltop), Hluhluwe GR (Mahlungula);
Bruton (1978): Manzengwenya Forest Station;
Laycock (1976): Beaumont Adit (Camperdown), Saxony Cave;
Rautenbach et al. (1980): Ndumu GR;
Sapsford (unpublished specimen data): 24 Haslam Road (Escombe), 622 Lighthouse Road (Bluff), 70 Seventh Avenue (Pinetown), 12 Syringa Place (Amanzimtoti).

Family	**RHINOLOPHIDAE**	Horseshoe bats
Genus	*Rhinolophus*	Lacépède 1799

KEY TO SPECIES (After Meester et al. 1986):

1. Anterior upper premolar, when present, external to toothrow; canine and fourth upper premolar in contact; connecting process bluntly pointed 2
   - Anterior upper premolar in toothrow; canine and fourth upper premolar not in contact; connecting process blunt or sharply pointed 3

2. Forearm 50–57 mm . . . . *R. clivosus*, p. 35
   - Forearm 45–50 mm . . . . *R. darlingi*, p. 36

3. Connecting process rises to an erect point; inter-pterygoid groove usually shallow, not clearly defined by bordering ridges; anterior upper premolar usually not crowded between canine and fourth upper premolar, longer than wide 4
   - Connecting process with low, bluntly pointed tip; inter-pterygoid groove deep and clearly defined by bordering ridges; anterior upper premolar usually crowded between canine and fourth upper premolar, at least as wide as it is long 5

4. First phalanx of fourth finger notably shortened in relation to metacarpal length; connecting process pointed, molar width less than half width of palate between molars . . . . *R. landeri*, p. 37
   - First phalanx of fourth finger not notably shortened in relation to meta-carpal length; connecting process rises to a high narrow horn, molar width more than half width of palate between molars . . . . *R. blasii*, p. 37

5. Sella broader; ears longer, 20–22 mm; condylocanine length more than 15.5 mm . . . . .
   R. simulator, p. 38
   - Sella narrower; ears shorter, 18–20 mm; condylocanine length less than 15.5 mm . . . . *R. swinnyi*, p. 39

? *Rhinolophus hildebrandtii* (Peters 1878)
Hildebrandt's horseshoe bat
Hildebrandt-saalneusvlermuis

Hildebrandt's horseshoe bat has been recorded from late Pleistocene deposits at Border Cave in northern Maputaland (Avery 1991). However, no modern records have as yet been recorded from KwaZulu-Natal. Its distribution in the northern parts of the Northern Province of South Africa may mark the southern limit of its modern distribution (Rautenbach 1982).

Rhinolophus clivosus (Cretzschmar 1828)   PLATE 11
R. c. zuluensis (Andersen 1904)
Geoffroy's horseshoe bat
Geoffroy-saalneusvlermuis

Taxonomic status
Two subspecies were recognised in southern Africa by Meester et al. (1986). Only the above-mentioned subspecies occurs in KwaZulu-Natal.

Distribution
Geoffroy's horseshoe bat is widely distributed in KwaZulu-Natal, occurring throughout all of the bioregions from Coast lowland to Montane (Figure 28). Avery (1991) recorded this species from the upper Pleistocene and late Holocene deposits at Border Cave.

Protected areas
Garden Castle, Hluhluwe, Itala, Kamberg, Mkuzi, Spioenkop.

Conservation status
Widespread and fairly common.

Habitat
While Skinner and Smithers (1990) indicate that they are predominantly associated with savanna woodlands, they clearly have a wide habitat tolerance in KwaZulu-Natal, occurring in wooded habitats as well as more open grassland habitats in the Drakensberg. In Lesotho, Lynch (1994) remarked on their occurrence in open montane grasslands, but noted that this might explain why individuals rather than larger groups were encountered. The availability of cover in the form of abundant sandstone caves and crevices in the Drakensberg may explain their presence. In KwaZulu-Natal, they were found to roost during the day in old mine adits and sandstone caves.

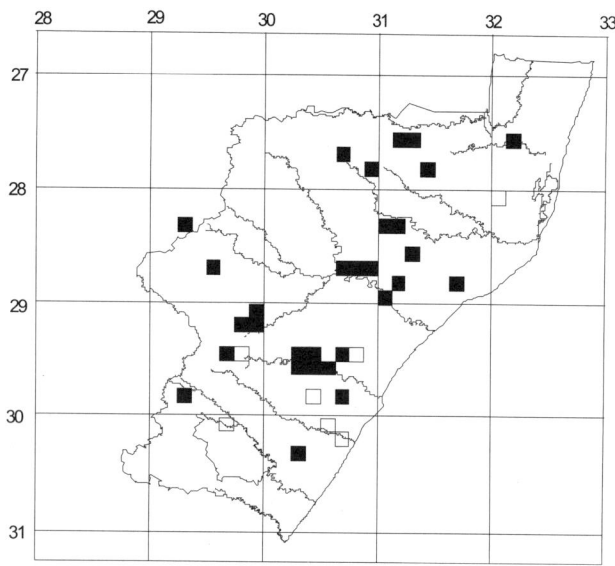

Figure 28

Habits

They are gregarious, with colonies of up to 10 000 individuals recorded from the Western Cape and Eastern Cape (Herselman and Norton 1985). During the day they roost in caves and mine adits, hanging from their hind claws from the roof and walls in clusters, with the individuals slightly separated from one another. They leave their day roosts shortly after sunset to forage, and return to them just before dawn. During the night they establish feeding stations in trees, or even under the eaves or verandas of buildings. Their feeding stations can often be located from the piles of discarded insect remains underneath. They are known to eat moths and beetles.

Breeding

In KwaZulu-Natal, mating occurs in early winter (May), and females are known to store sperm in their reproductive tract (oviducts and uterus) during the period of winter hibernation until the following spring (August), when ovulation and fertilization occur (Bernard 1983). A single young is born in December. Females are capable of reproducing during their first year of life.

Linear measurements (mm) and mass (g)

	Males					Females				
	x̄	s.d.	n	Min	Max	x̄	s.d.	n	Min	Max
TL	91.2	4.7	18	83	100	93.9	8.8	14	78	110
HB	61.4	4.4	18	55	70	63.1	5.6	14	56	74
T	29.7	1.9	18	27	35	30.8	4.7	14	22	40
Hf	12.1	0.6	4	11	13	–	–	–	–	–
E	17.8	1.9	7	16	21	20.2	2.1	4	18	23
FA	52.7	2.2	6	49	55	–	–	2	52	54
Mass	16.4	1.5	5	14	18	–	–	2	21	21

Records of occurrence

SPECIMEN RECORDS:

BM: Estcourt, Ngoye Hills, Willbrook;
DM: Estcourt, Fort Yolland Farm, Garden Castle NR, Itala

NR, Melmoth area (Mondi Forests), Mkuzi GR, Spioenkop NR, Willbrook;
KM: Denny Dalton Mine, Doornhoek Mine Tunnel, Goedehoop Farm, Saxony Cave, Speedwell Mines (Kranskop), Table Mountain, Town Bush Cave;
NM: Babanango, Domleos Cave (=Laager Cave), Kambula, Langewacht, Mfongosi, Mooi River, Pietermaritzburg, Town Bush;
TM: Bishopstowe Cave, Doornhoek, Dumisa Gold Mine, Goudhoek Farm, Insuzi Valley, Itala GR (Doornpan, Wonderfontein), Kamberg NR, Ngome Forest Reserve, Ngome Police Station, Ngubevu Mine, Otto's Bluff (Kok's Farm), Strathfieldsay Farm (near Eston), Town Bush Cave.

ADDITIONAL RECORDS:

Avery (1991): Border Cave;
Bourquin et al. (1971): Hluhluwe GR (Hilltop);
Laycock (1976): Beaumont Adit (Camperdown), Coopers Falls Cave, Hilton, Ismont Crevasse (near Eston), Mfume Copper Mine, Pietermaritzburg storm drains, Shongweni Dam, Spioenkop Cave, Tala Cave (Camperdown).

Rhinolophus darlingi (K. Andersen 1905)   PLATE 12
R. d. darlingi (K. Andersen 1905)
Darling's horseshoe bat
Darling-saalneusvlermuis

Taxonomic status

Two subspecies were recognised by Meester et al. (1986), of which one occurs in KwaZulu-Natal.

Distribution

Darling's horseshoe bat is confined to the Lowveld bioregion of northern Zululand (Figure 29). Avery (1991) recorded this species from the upper Pleistocene and late Holocene deposits at Border Cave. The KwaZulu-Natal records represent the southernmost extent of the species' range.

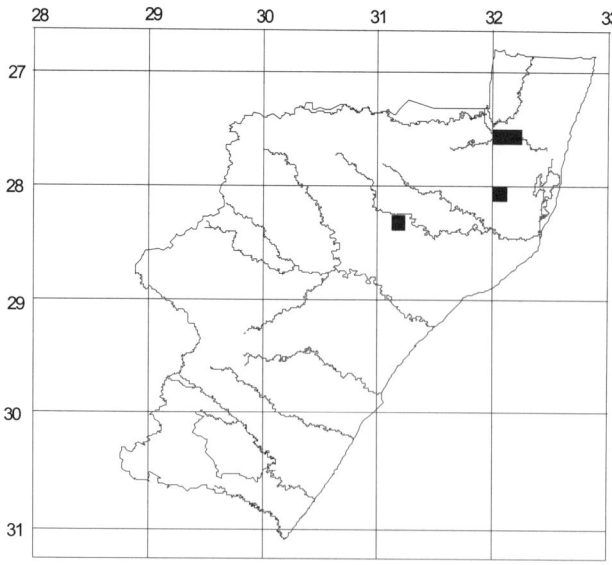

Figure 29

Protected areas
Hluhluwe, Mkuzi.

Conservation status
Uncommon in KwaZulu-Natal; known from only eight specimens.

Habitat
Savanna woodland where suitable daytime roosts are available in the form of caves, mine adits and disused buildings. A colony of some 20 individuals was found roosting in a series of interconnecting tunnels behind a shallow rock overhang on Ghost Mountain near the town of Mkuze.

Habits
They are gregarious, occurring in colonies of up to dozens.

Breeding
No data available for KwaZulu-Natal.

Linear measurements (mm) and mass (g)

Cat. No.	Sex	TL	HB	T	Hf	E	FA	Mass
DM4573	F	80	55	25	7	18	48.3	10
DM4574	F	80	53	27	8	20	48.3	9

Forearm lengths of a male and female from Hluhluwe Game Reserve were 50 mm and 47 mm respectively. This species is somewhat smaller than *R. clivosus*, averaging a total length of about 85 mm and a mass of about 9 g (Skinner and Smithers 1990).

Records of occurrence
SPECIMEN RECORDS:
DM: Ghost Mountain, Hluhluwe GR;
KM: Denny Dalton Mine;
TM: Mkuzi GR.

ADDITIONAL RECORDS:
Avery (1991): Border Cave.

Rhinolophus landeri (Martin 1838)
R. l. lobatus (Peters 1852)
Lander's horseshoe bat
Lander-saalneusvlermuis

Taxonomic status
Two subspecies were recognised by Meester et al. (1986), of which only one occurs in southern Africa.

Distribution
Known from only Denny Dalton Mine in the Babanango District (Figure 30). The species has not previously been recorded from KwaZulu-Natal (Bourquin 1988; Skinner and Smithers 1990).

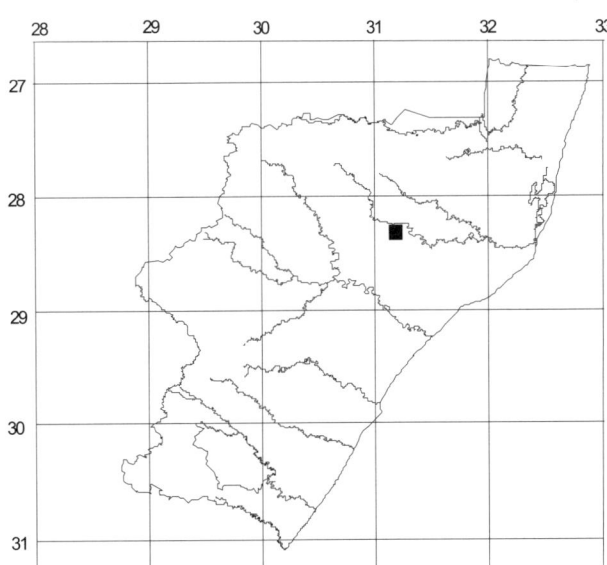

Figure 30

Conservation status
Rare in KwaZulu-Natal, due to its marginal occurrence here.

Habitat
The locality from which this species has been collected is situated within the Drier upland bioregion, in close vicinity to Lowveld to the east and Highland to the west.

Habits and breeding
No data available for KwaZulu-Natal. In Zimbabwe, Fenton (1985) found their prey to consist predominantly (92 per cent) of Lepidoptera.

Records of occurrence
KM: Denny Dalton Mine.

Rhinolophus blasii (Peters 1867)
R.b. empusa (K. Andersen 1904)
Peak-saddle horseshoe bat
Spitssaalneusvlermuis

Taxonomic status
The nominate form of this widespread species was described from Italy. Only one subspecies is found in South Africa (Meester et al. 1986).

Distribution
Rautenbach et al. (1981) reported this species from Itala GR in northern KwaZulu-Natal and Ngubevu Gold Mine in the Tugela Valley. A further record (identified as cf. *blasii* by L. Wingate, Kaffrarian Museum) originates from the Umvoti Vlei Nature Reserve (Figure 31). These records constitute the southernmost extent of the species' range.

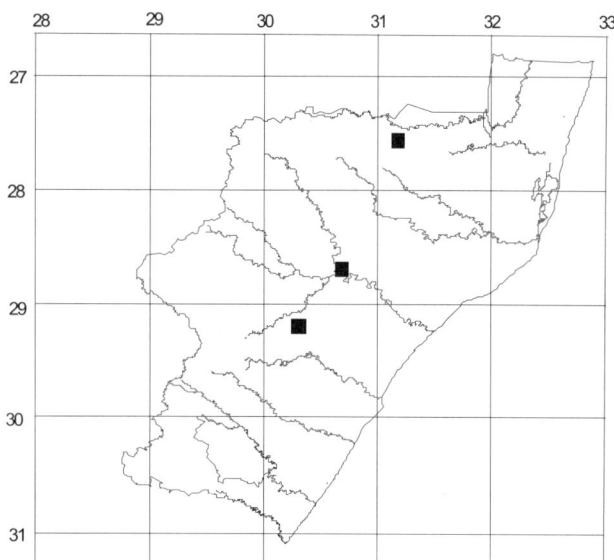

Figure 31

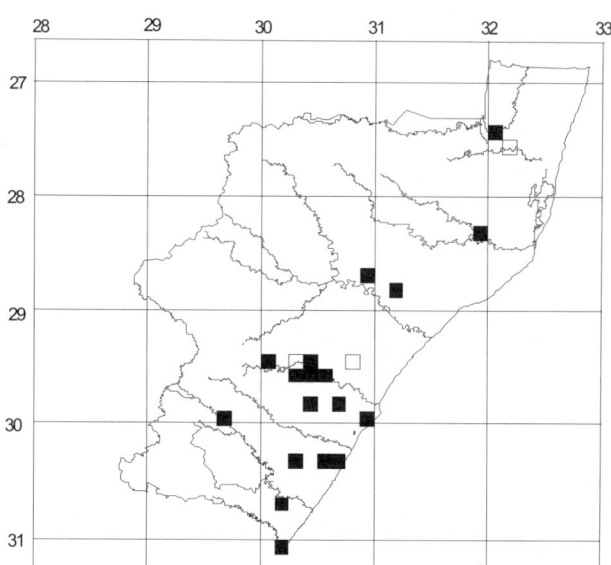

Figure 32

Protected areas
Itala GR, Umvoti Vlei NR.

Conservation status
Uncommon in KwaZulu-Natal, having a patchy distribution.

Habitat
Rautenbach et al. (1981) recorded them from deep within sandstone caves in open woodland at Itala. Most of the localities in KwaZulu-Natal are associated with major river valleys running into the Indian Ocean, encompassing Lowveld, Valley bushveld and Mistbelt bioregions.

Habits
Rautenbach (1982) noted from isolated observations that this species behaves in much the same way as other rhinolophids, i.e. they roost in caves on the roof and walls, hanging in clumps with individuals slightly separated. In greater Transvaal, they are known to hibernate in caves.

Linear measurements (mm) and mass (g)
Based on measurements from greater Transvaal, they have a total length of about 80 mm, with tails of 30 mm and a mass of about 4.0 g. No data available for KwaZulu-Natal.

Records of occurrence
SPECIMEN RECORDS:
SAM: Umvoti Vlei NR(?);
TM: Itala GR (Doornpan), Ngubevu Gold Mine.

Rhinolophus simulator (K. Andersen 1904)
Bushveld horseshoe bat
Bosveld-saalneusvlermuis

Taxonomic status
No subspecies were recognised by Meester et al. (1986).

Distribution
The bushveld horseshoe bat occurs fairly widely over the eastern half of KwaZulu-Natal, often associated with Valley bushveld habitats along major, east-flowing river valleys such as the Mfolozi, Tugela and Mgeni (Figure 32). The Umtamvuna Nature Reserve record represents the known southernmost extent of the species' distribution in Africa. Avery (1991) recorded this species (with some query as to its identification) from the late Holocene deposits at Nkupe in the upper Tugela Basin.

Protected areas
Mkuzi, Oribi Gorge, Umfolozi, Umtamvuna, Vernon Crookes.

Conservation status
Widespread and fairly common where suitable day-roosting sites are available in the form of mines, disused tunnels and caves.

Habitat
In KwaZulu-Natal, bushveld horseshoe bats are associated with Valley bushveld along the lower reaches of the major, east-flowing river valleys. Elsewhere they are associated with savanna woodland, particularly *Brachystegia* woodland in Zimbabwe (Skinner and Smithers 1990). Availability of day-roosts in the form of caves and mine adits is a necessary component of their habitat.

Habits
They occupy caves and mine adits in colonies typically numbering up to dozens. Rautenbach (1982) noted colonies of approximately 300 individuals in greater Transvaal. Wingate (1983) recorded a colony of 150 bushveld horseshoe bats at Doornhoek Tunnel near Pietermaritzburg. In the Doornhoek colony, Wingate (1983) found that females migrated to maternity roosts elsewhere in spring, returning to Doornhoek at the end of summer. Males remained throughout the year.

Breeding

In greater Transvaal, pregnant females have been taken in September and lactating females in January (Rautenbach 1982). At Doornhoek Tunnel near Pietermaritzburg, Wingate (1983) noted sexual activity in males (increased size of scrotal rudiments) from April to July indicating that males probably copulated during winter.

Linear measurements (mm) and mass (g)

	Males					Females				
	$\bar{x}$	s.d.	n	Min	Max	$\bar{x}$	s.d.	n	Min	Max
TL	66.7	2.1	7	64	70	72.5	–	2	71	74
HB	45.7	2.5	7	42	50	48.5	–	2	47	50
T	21.0	2.4	7	18	24	24.0	–	2	24	24
Hf	7.1	1.6	6	4	8	6.0	–	2	5	7
E	19.0	2.0	7	16	22	20.0	–	2	20	20
FA	44.0	1.0	7	42	45	45.5	–	2	43	48
Mass	–	–	1	7	7	–	–	1	9	9

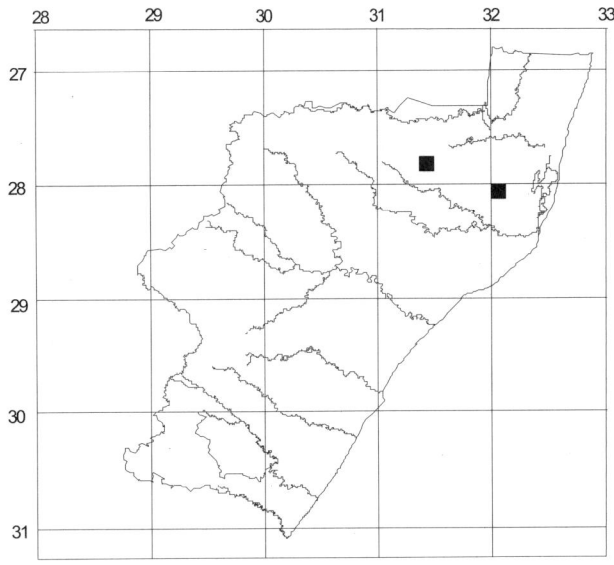

Figure 33

Seamark and Kearney (personal communication) recorded mean forearm lengths of 44 mm (s.d.=1.3; *n*=16; males) and 44.5 mm (s.d.=0.9; *n*=11; females) and mean masses of 10.0 (s.d.=2.9; *n*=16; males) and 10.5 (s.d.=3.2; *n*=11; females) for two colonies of bushveld horseshoe bats on the South Coast of KwaZulu-Natal.

Records of occurrence

SPECIMEN RECORDS:

DM: Esperanza, Hlabeni Forest, Jozini Dam, Shongweni Dam, Umtamvuna Gorge;

KM: Doornhoek Mine Tunnel, Dumisa Mine, Speedwell Mines, Table Mountain Pothole, Umfolozi GR (Confluence of Black and White Mfolozi River: 17 km north, 22 km west);

TM: Bishopstowe Caves, Doornhoek Mine, Dumisa Mine, Kilgobbin Farm, Oribi Gorge NR, Otto's Bluff, Strathfieldsay Farm, Town Bush Cave, Vernon Crookes NR.

ADDITIONAL RECORDS:

Avery (1991): Nkupe;

Dixon (1964): Mkuzi GR;

Laycock (1976): Doornhoek, Laager Cave (= Domleo's Cave), Saxony Cave, Town Bush Cave.

Rhinolophus swinnyi (Gough 1908)
Swinny's horseshoe bat
Swinny-saalneusvlermuis

Taxonomic status

This species may be conspecific with *R. denti* (Meester et al. 1986, Bronner 1990). No subspecies were recognised by Meester et al. (1986).

Distribution

Bronner (1990) recorded Swinny's horseshoe bat from Ngome Forest Reserve in northern KwaZulu-Natal, this being the first published record for KwaZulu-Natal. This locality provided a distributional link between isolated records from the Eastern Cape and Northern Province. Previously misidentified specimens from Hluhluwe Game Reserve (**DM** and **TM** collections) have been recently identified as *R. swinnyi* (Figure 33).

Protected areas

Hluhluwe.

Conservation status

Very uncommon in KwaZulu-Natal, with localised distribution.

Habitat

Bushveld and possibly forest.

Habits

In Zimbabwe they are known to roost in caves in small colonies of no more than five individuals.

Breeding

No information available.

Linear measurements (mm) and mass (g)

Swinny's horseshoe bats are very small, with a total length of about 70 mm, a mass of about 7.6 g and a forearm length of 43–44 mm. No data available for KwaZulu-Natal.

Records of occurrence

SPECIMEN RECORDS:

DM: Hluhluwe GR (Egodeni);

TM: Ngome Forest Reserve, Hluhluwe GR (Rest Camp).

Family	**HIPPOSIDERIDAE**	
		Trident and leaf-nosed bats
Genus	***Hipposideros***	**Gray 1831**

Hipposideros caffer (Sundevall 1846)   PLATE 13
H. c. caffer (Sundevall 1846)
Sundevall's leaf-nosed bat
Sundevall-bladneusvlermuis

Taxonomic status
Two subspecies occur in southern Africa (Meester et al. 1986), of which only the above is found in KwaZulu-Natal.

Distribution
Widely distributed in the eastern half of KwaZulu-Natal, from sea level up to altitudes of about 535 m, mostly associated with major river valleys (Figure 34).

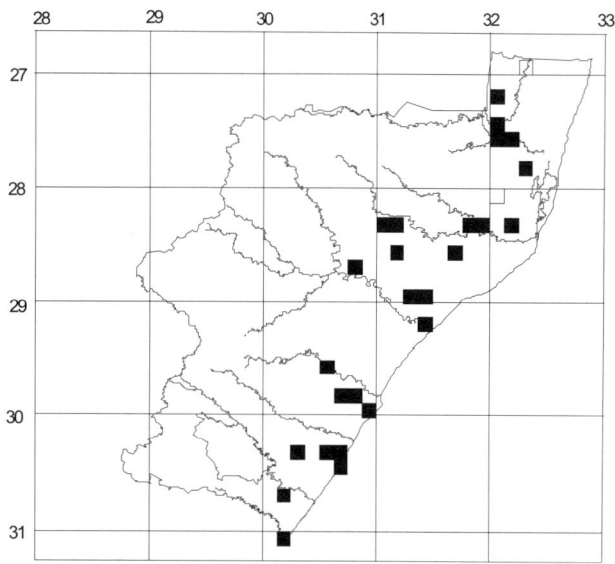

Figure 34

Protected areas
Dlinza Forest, Entumeni, Hluhluwe, Krantzkloof, Mkuzi, Oribi Gorge, Pongolapoort, Umfolozi, Vernon Crookes.

Conservation status
Fairly common and widely distributed over eastern KwaZulu-Natal.

Habitat
Bushveld and coastal forest. They are apparently dependent on water (Smither 1971), which explains the association of collecting localities in KwaZulu-Natal with major river valleys (Figure 34).

Habits
Under optimum conditions, they congregate in colonies of several hundreds. Smaller colonies may be found in the roofs of old houses, culverts under roads, hollows in trees and rock crevices. In KwaZulu-Natal, specimens were collected mainly while day-roosting in caves and mine tunnels. An individual was collected during the day hanging from the branch of a tree next to a river in Kloof Gorge. Individuals were taken from a hollow tree at Ndumu GR (Dixon 1966) and from outbuildings at Mkuzi GR (Dixon 1964). Night-roosting bats have been taken from under the eaves of houses and thatched game viewing hides and from under the watertower at the Dukuduku Forest Station (Rautenbach and Bronner 1988). Sundevall's leaf-nosed bat has well developed echolocation capabilities and is capable of slow but manoevrable flight, allowing it to hover and to glean insects from branches. They are also capable of aerial pursuit of their prey and their echolocation call is highly sensitive to the detection of fluttering insect wings. At Mkuzi Game Reserve, Bowie, Jacobs and Taylor (in press) found their diet to consist mainly (80 per cent) of moths (Lepidoptera), followed in importance by Trichoptera (10 per cent) and Coleoptera (8 per cent).

Breeding
In KwaZulu-Natal, copulation and ovulation occurs in April, followed, after a gestation of 220 days, by parturition (birth) of a single young early in December, and lactation/anoestrus during December and January (Bernard and Meester 1982). In a colony observed in the dam wall at Nagle Dam on December 28, females were still carrying their single young, even though the young were capable of flying by themselves. The gestation is prolonged by a further 100 days, compared with tropical populations of this species, by means of a period of retarded embryonic development which corresponds to the winter period. Individuals do not appear to hibernate during winter, but enter short periods of torpor during cold spells.

Linear measurements (mm) and mass (g)

	Males					Females				
	x̄	s.d.	n	Min	Max	x̄	s.d.	n	Min	Max
TL	78.4	3.7	22	70	87	81.1	4.2	21	73	89
HB	49.5	3.6	22	39	54	51.6	2.8	21	45	55
T	28.9	3.1	22	24	36	29.4	2.2	21	26	34
Hf	6.2	1.5	12	4	9	7.0	2.1	15	5	8
E	12.9	1.9	18	10	16	13.9	1.3	19	11	16
FA	47.6	1.0	12	46	49	47.9	1.3	15	45	50
Mass	8.0	0.6	6	7	9	8.6	1.2	8	7	10

Records of occurrence
SPECIMEN RECORDS:
DM: Beneva Tunnel, Dlinza Forest NR, Entumeni NR (Mandatana Estates), Jozini Dam, Krantzkloof NR, Mkuzi GR (Bube, Msinga Pans), Mooiplaas Plantation (SAPPI),

Nagle Dam, Ngxwala Hill, Sezela (3 Msinsi Avenue), Shongweni Dam;
KM: Denny Dalton Mine, Doornhoek Mine, Ntambanana, T. Bester Game Farm (Mtubatuba), Umdoni Park, Umfolozi GR (Confluence of Black and White Umfolozi Rivers);
NM: Mfongosi, Umdoni Park;
SAM: Mfongosi;
TM: Dukuduku Forest Station, Dumisa Gold Mine, Goudhoek Mine, Ingwavuma, Malvern (Durban), Mkuzi GR (Bube Pan), Oribi-Lind Valley Game Ranch, Umdoni Park, Umfolozi GR, Vernon Crookes NR.

ADDITIONAL RECORDS:
Dixon (1966): Ndumu GR;
Bourquin et al. (1971): Hluhluwe GR (Hilltop).

Genus	*Cloeotis*	Thomas 1901

Cloeotis percivali (Thomas 1901)　　　PLATE 14
C. p. australis (Roberts 1917)
Short-eared trident bat
Drietandbladneusvlermuis

Taxonomic status
Of the two recognised subspecies (Meester et al. 1986), only *C. p. australis* is found in KwaZulu-Natal.

Distribution
The short-eared trident bat is known in KwaZulu-Natal from two specimens collected at Mkuzi Game Reserve, as well as from a colony of some 100–200 individuals roosting in inspection tunnels in the wall of the Jozini Dam in the Pongolapoort Biosphere Reserve (Figure 35). Its occurrence in KwaZulu-Natal is not suprising as it has been recorded from Swaziland. The KwaZulu-Natal records represents the southernmost extent of the species' known distribution, which extends as far north as Kenya. The species has a scattered distribution, with only 15 locality records in southern Africa (six in Zimbabwe, two in Botswana, four in greater Transvaal, one in Swaziland and the two KwaZulu-Natal localities).

Protected areas
Mkuzi.

Conservation status
They are listed in the South African Red Data Book as Indeterminate. Because of their restricted distribution in KwaZulu-Natal, they have been listed under Schedule 6 of the Provincial Conservation Ordinance of 1974 as 'Endangered'.

Habitat
Not much is known about their habitat requirements, but presumably they occur in savanna areas where sufficient cover is present in the form of caves and mine tunnels for day-roosting.

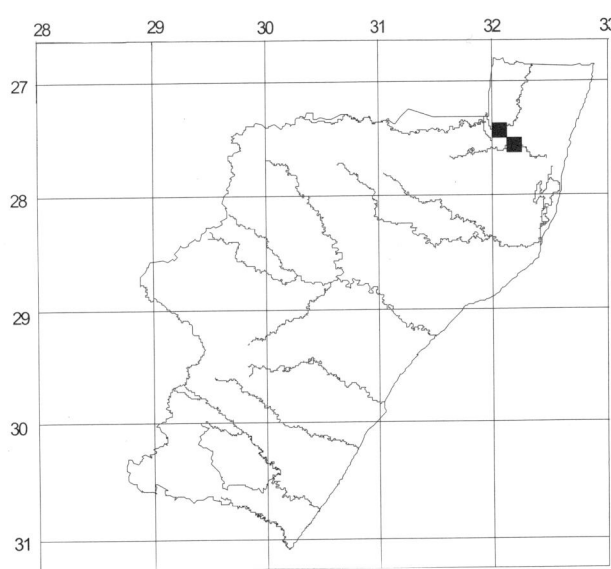

Figure 35

Habits
They are gregarious occurring in small colonies of around 10, as well as in larger colonies numbering hundreds (Skinner and Smithers 1990). The Mkuzi specimen was caught inside a house at night, presumably while foraging.

Breeding
In Zimbabwe, pregnant females were collected in October. Females with young were recorded in early December at Jozini Dam in KwaZulu-Natal.

Linear measurements (mm) and mass (g)

	Males					Females				
	x̄	s.d.	n	Min	Max	x̄	s.d.	n	Min	Max
TL	65.4	1.9	11	62	69	65.5	1.0	4	65	67
HB	40.7	2.9	11	37	47	38.0	3.6	4	35	42
T	24.7	3.2	11	20	31	27.5	2.9	4	25	30
Hf	5.4	0.9	11	3	7	4.2	0.5	4	4	5
E	8.4	1.0	11	7	10	8.5	0.6	4	8	9
FA	34.1	1.1	11	33	36	34.5	1.0	4	33	35
Mass	4.8	0.7	8	4	6	4.0	0.0	3	4	4

Records of occurrence
DM: Jozini Dam, Mkuzi GR.

Family	**VESPERTILIONIDAE**	
		Vesper bats
Subfamily	**MINIOPTERINAE**	
		Long-fingered bats
Genus	*Miniopterus*	**Bonaparte 1837**

KEY TO SPECIES (After Meester et al. 1986):

1.　Forearm 42–44 mm; skull length
　　c. 14 mm; colour brownish-black
　　and/or russet . . . . *M. fraterculus*, p. 42

—   Forearm 42–47 mm; skull length *c.*
    15 mm; colour slaty-black . . . .
                                                 M. schreibersii, p. 42

Miniopterus fraterculus (Thomas and Schwann 1906)
Lesser long-fingered bat
Klein grotvler

Taxonomic status
No subspecies were recognised by Meester et al. (1986).

Distribution
The lesser long-fingered bat is quite widely distributed in
KwaZulu-Natal (Figure 36). Most collecting localities seem
to be associated with major river valleys such as the Tugela,
Mgeni and Mfolozi.

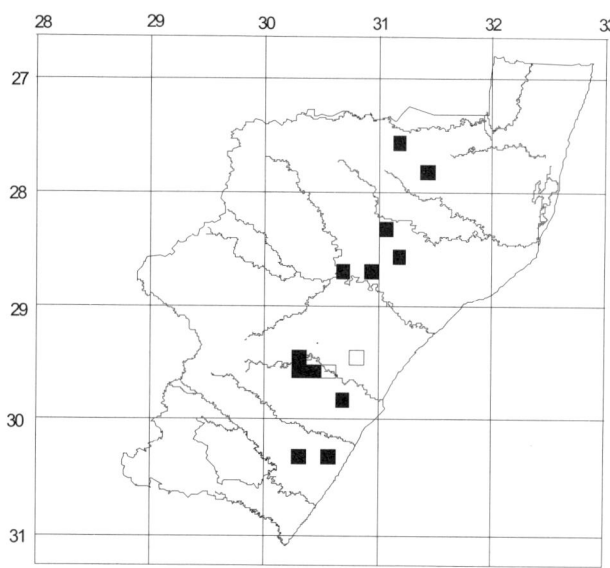

Figure 36

Protected areas
Itala, Vernon Crookes.

Conservation status
Widespread and quite common in suitable caves.

Habitat
Their distribution in KwaZulu-Natal indicates a wide range
of bioregions and habitats, from drier Valley bushveld and
Lowveld to moister Mistbelt (including forest habitats),
where suitable cover is present in the form of caves, overhangs, and disused mine and railway tunnels.

Habits
Like the following species, *M. schreibersii*, this is a cave-
dwelling species. In KwaZulu-Natal, day-roosting specimens have been collected from damp sandstone caves, a
solution cave of poorly consolidated glacio-fluvial boulder
clay, a rocky overhang over a forest stream, a rock fissure,

a disused railway tunnel, as well as from disused mine
adits.

Breeding
Like *M. schreibersii*, delayed implantation of the fertilized
blastocyst occurs during the winter hibernation. Mating in
M. fraterculus occurs in May, some two months later than
in *M. schreibersii*, with the result that the period of delayed implantation is somewhat shorter in *M. fraterculus*
than in *M. schreibersii*. Births occur in November and
December (Bernard 1980).

Linear measurements (mm) and mass (g)

Cat. No.	Sex	TL	HB	T	Hf	E	FA
DM3513	M	97	55	42	8	9	41.8
DM5108	M	92	54	37	7	7	42.5
NM 593	?	105	55	50	–	–	–

Mean monthly mass of *M. fraterculus* from caves in the
KwaZulu-Natal midlands varied from 6 g to 9.1 g for females, and from 6.4 g to 10.8 g for males (Laycock 1976).

Records of occurrence
SPECIMEN RECORDS:
DM: Mooiplaas Plantation, Shongweni Dam;
KM: Speedwell Mines;
NM: Town Bush Cave;
TM: Dumisa Gold Mine, Goudhoek, Hilton Railway Tun-
nel, Itala GR (Doornpan), Karkloof Forest, Ngome Forest
Reserve, Ngubevu Gold Mine, Town Bush Cave, Vernon
Crookes NR.

ADDITIONAL RECORDS:
Laycock (1976): Doornhoek Adit, Laager Cave (= Domleo's
Cave), Saxony Cave.

Miniopterus schreibersii (Kuhl 1819)   PLATE 15
M. s. natalensis (A. Smith 1833)
Schreiber's long-fingered bat
Schreiber-grotvlermuis

Taxonomic status
All specimens from southern African were included in
M. s. natalensis by Meester et al. (1986).

Distribution
Widespread throughout the eastern half of the province,
occurring as far west as Underberg Hatchery in the
Drakensberg foothills (Figure 37). Avery (1991) recorded
M. schreibersii from archaeological sites at Border Cave
(upper Pleistocene and very late Holocene) and Nkupe
(middle to late Holocene).

Protected areas
Eastern Shores, Hluhluwe, Mkuzi, Ngoye, Umfolozi,
Vernon Crookes.

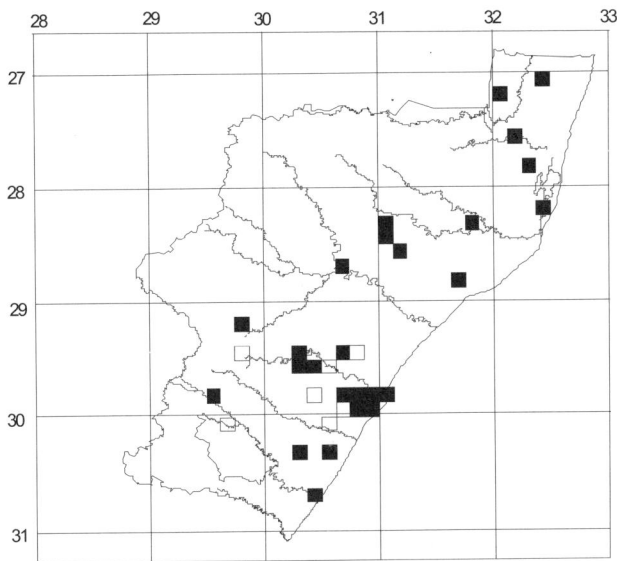

Figure 37

Conservation status
Fairly common and widespread in KwaZulu-Natal.

Habitat
Schreiber's long-fingered bats are cave-dwellers which occupy a range of vegetational associations. In KwaZulu-Natal, although they are mostly confined to the lower-lying bushveld and coastal forest habitats, a few have been collected from grassland habitats associated with the Highland and Coast hinterland bioregions. However, they are generally absent from the Montane, Highland, Drier upland and Moist upland bioregions (Figure 37). Rautenbach (1982) also found them to be largely restricted to bushveld, and absent from highveld grasslands in greater Transvaal, and Lynch (1994) recorded only one specimen from the mountain valleys of Lesotho (<2 500 m).

Habits
While this species occurs mostly in caves, they have been reported from houses and in crevices in rocks and trees (Skinner and Smithers 1990). Rautenbach (1982) found that they regularly used night-time roosts later at night, utilising beams under open verandahs, pumphouses, open garages, etc. The presence of 13 widespread suburban records of *M. schreibersii* from the Durban metropolitan area (Sapsford, unpublished) suggests that they use houses and outbuildings as night roosts (one specimen was collected hanging from one leg from a garage wall; time of day not recorded).

Based on observations in KwaZulu-Natal, colony sizes of this species fluctuate dramatically from year to year, as well as throughout the year (Laycock 1976). For example, at Laager Cave, Laycock (1976) recorded fluctuations between 0 and 3 600 over four years. Dramatic influxes and exoduses often occurred over periods of less than a month (e.g. from 3 600 on 21 May 1972 to 40 on 2 June 1972). A maximum colony size of 8 000 was recorded by Laycock at Shongweni Tunnel in October 1975.

Elsewhere in South Africa, seasonal movements have been recorded in this species, between warmer climate summer maternity roosts (e.g. lowveld of the Northern Province) and colder winter hibernacula (e.g. highveld of Gauteng) (Van der Merwe 1975; Brown and Bernard 1994).

Breeding
In the KwaZulu-Natal midlands, copulation occurs in March and is followed by a prolonged gestation of eight months, the first four months comprising a period of delayed implantation of the fertilised blastocyst in the female uterus (coinciding with the period of winter hibernation), and the latter four months comprising normal fetal growth and development (Bernard 1980). Young are born early in December. Although Laycock (1976) identified certain caves in the KwaZulu-Natal midlands which served as both maternity caves (e.g. Saxony cave) and hibernating caves (hibernacula) for small, transient groups of Schreiber's long-fingered bats, no caves have been identified which serve as major maternity roosts or hibernacula.

Linear measurements (mm) and mass (g)

	Males					Females				
	x̄	s.d.	n	Min	Max	x̄	s.d.	n	Min	Max
TL	106.5	9.5	10	88	120	105.2	5.6	13	95	114
HB	55.8	4.8	10	50	63	55.0	6.6	13	39	63
T	50.7	6.1	10	40	60	50.2	3.8	13	45	58
Hf	8.8	0.5	5	8	9	9.0	1.1	8	7	10
E	9.7	1.5	7	8	11	9.5	0.9	9	8	11
FA	45.4	1.4	5	43	47	47.4	2.9	8	44	52
Mass	10.2	1.5	6	8	12	12.3	2.0	18	10	16

Masses from E. Seamark (personal communication)

Records of occurrence
SPECIMEN RECORDS:
BM: Ngoye Hills, Town Bush Valley, Willbrook;
DM: Eastern Shores NR (Mission Rocks), Shongweni Dam;
KM: Pietermaritzburg, Theunis Bester Game Farm, Town Bush Cave, University of Natal (Pietermaritzburg);
NM: Domleo's Cave (= Laager Cave), Ferncliff, Town Bush Cave;
SAM: KwaZulu-Natal (unspecified locality);
TM: Drakensberg Gardens Hotel (Trout Hatchery), Dumisa Gold Mine, Ferncliff, Goudhoek Farm, Hilton Railway Tunnel, Ingwavuma, Mkuzi GR (Nhlonhlela Pan), Ngubevu Mine, Sihangwane Farm, Town Bush Cave, Umfolozi GR, Vernon Crookes NR.

ADDITIONAL RECORDS:
Avery (1991): Border Cave, Nkupe;
Laycock (1976): Coopers Falls Cave, Doornhoek Adit, Hilton Tunnel, Laager Cave, Pietermaritzburg Storm Drains, Saxony Cave, Shongweni Dam, Spioenkop Cave, Tala Cave, Town Bush Cave;
Sapsford (unpublished): 16 Alida Place (Westville), 42 Blenheim Road (Pinetown), 10 Delaware Avenue (Vir-

ginia), 24 Fraser Place (Greenwood Park), 611 Main Road (Northdene), 7 Majuba Street (Pinetown), 22 Netta Road (Rossburgh), 7 Sherwood Place (Escombe), State Aid Indian School (Pinetown), Stella Road (Hillary), 2 Table Mountain Street (Shallcross), Uvongo Town Board (Port Shepstone), 78 Wood Road (Pinetown);
Seamark (unpublished): Beneva Tunnel, Mooiplaas Plantation.

Family	**VESPERTILIONIDAE**
	Vesper bats
Subfamily	**VESPERTILIONINAE**
Genus	***Myotis*** **Kaup 1829**

KEY TO SPECIES (After Meester et al. 1986):

1. Size larger, forearm over 52 mm . . . .
    M. (Chrysopteron) welwitschii, p. 44
 − Size smaller, forearm under 52 mm . . . .
    M. (Selysius) tricolor, p. 44

Myotis welwitschii (Gray 1866)
Welwitsch's hairy bat
Welwitsch-langhaarvlermuis

Taxonomic status
No subspecies were recognised by Meester et al. (1986).

Distribution
A single specimen was collected from Craigieburn near Umkomaas in 1989 (Taylor 1991; Figure 38), and represents the southernmost limit of the known distribution. Welwitsch's hairy bat has a wide but scattered distribution in Africa, from Tanzania in the north to South Africa in the south. Only 21 museum specimens are known from southern Africa, including 10 from South Africa (eight from greater Transvaal and one each from the Free State and KwaZulu-Natal).

Protected areas
This species is not known to occur in any protected areas in KwaZulu-Natal.

Conservation status
Rare in KwaZulu-Natal. It is not clear whether the specimen collected represents a resident population, or whether it was a vagrant individual.

Habitat
The single specimen was collected roosting in daylight on a low bush in a narrow riverine strip of coastal forest bordered by sugar cane and open thornveld. The vegetation falls under Acocks (1988) Coastal Forest and Thornveld (Type 1a) veld type, or Phillips (1973) Coast hinterland.

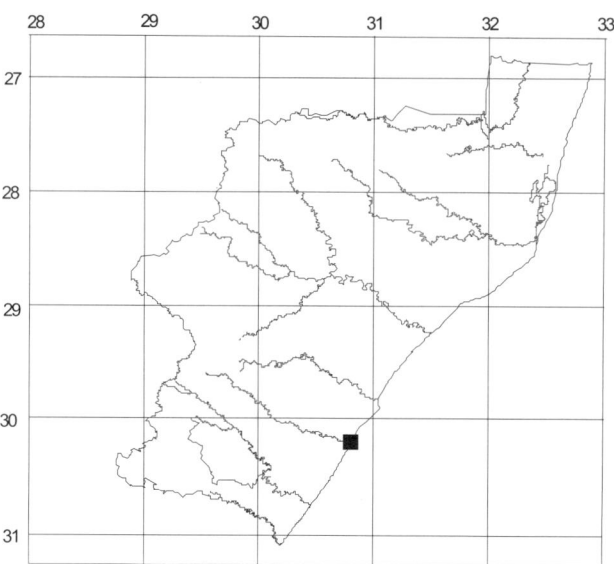

Figure 38

The species has elsewhere been recorded from a range of habitats, including a factory, a cave, a scrubby bush in the Lebombo Mountains, in rolled-up banana leaves and in savanna grassland (Rautenbach 1982; Smithers and Tello 1976; Smithers and Wilson 1979).

Habits
The species uses trees and bushes as day-roosting sites and enters houses at night when foraging (Skinner and Smithers 1990).

Breeding
No information available.

Linear measurements (mm) and mass (g)
The specimen from Craigieburn had a head and body length of 65 mm, tail length of 55 mm, hind foot of 11 *cu*, forearm length of 57 mm, and a mass of 17.2 g.

Records of occurrence
DM: Craigieburn.

Myotis tricolor (Temminck 1832)   PLATE 16
Temminck's hairy bat
Temmink-langhaarvlermuis

Taxonomic status
No subspecies were recognised by Meester et al. (1986).

Distribution
Temminck's hairy bat is widespread in KwaZulu-Natal (Figure 39). While the distribution map based on known specimen records indicates that it is absent from the western half of KwaZulu-Natal, including the Highland and Montane bioregions, the species has been collected from the Drakensberg in the eastern Free State and Lesotho (Watson 1990; Lynch 1994) and it is likely to occur in the

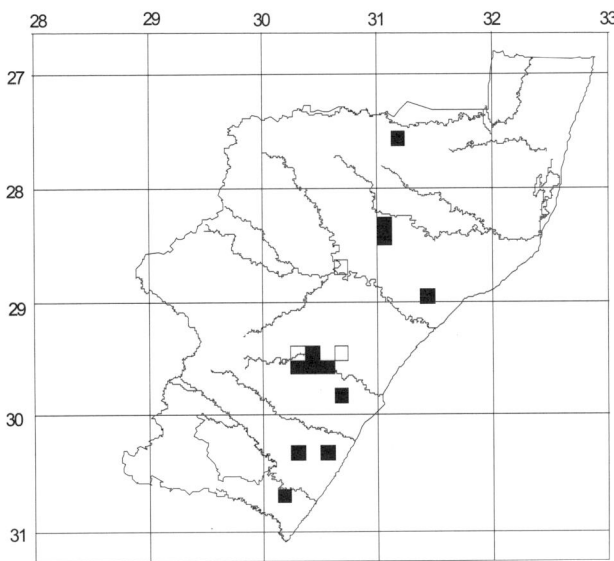

Figure 39

KwaZulu-Natal Drakensberg wherever suitable caves are present. Avery (1991) recorded *M. tricolor* from archaeological remains at Border Cave (upper Pleistocene).

Protected areas
Itala, Oribi Gorge, Vernon Crookes.

Conservation status
Fairly widespread in KwaZulu-Natal but localised to areas where suitable caves are present. Since it requires separate winter hibernation caves and summer maternity caves having different microclimatic requirements, the number of such sites is fairly limited, and any human disturbance of known maternity caves in the KwaZulu-Natal midlands (Laycock 1976) could have a severe impact on populations in KwaZulu-Natal.

Habitat
As noted elsewhere (Lynch 1994; Rautenbach 1982; Watson 1990), habitat seems to be governed by the availability of suitable caves or mine adits rather than particular vegetational associations.

Habits
Temminck's hairy bat is a social, cave-dwelling species. While colony size usually averages a few dozen individuals, colonies of up to 1 400 individuals have been recorded in the KwaZulu-Natal midlands (Laycock 1976). The species appears to undergo seasonal, local migrations in KwaZulu-Natal (see under **Breeding**). This species is often found roosting together with *Miniopterus schreibersii*, and clusters comprising both species are often seen. Individuals either hang by their hind claws from the cave roof, or cling onto the surface with the claws of their thumbs and hindfeet.

Breeding
In KwaZulu-Natal, young (one per female) are born between mid-November and mid-December after a gestation of around 63 days. Parturition (birth) is followed by a six-week period of lactation. Copulation occurs in mid-April, while ovulation and fertilisation occur in mid-September. Spermatozoa are stored in the uterine horns between mid-April and mid-September (Bernard 1982b).

Maternity colonies comprise only females and young. In Shongweni Dam Tunnel (and presumably elsewhere), females and volant young vacate the roost towards the end of January (Laycock 1976; personal observation). There is some uncertainty about the whereabouts and summer-time roosting behaviour of males. Bourquin and Mathias (1984) found a solitary adult male (**TM** 34344) roosting under a rock along a river in the Oribi Gorge Nature Reserve during January 1982, suggesting that males may become solitary during the breeding season. An adult male (**DM** 4873) and adult female (Seamark, unpublished) were collected with a harp trap in Dlinza Forest during March 1996, indicating that the sexes had re-united to mate prior to hibernation. A single adult male (**TM** 30147) was collected in Dumisa Mine in the Ixopo District during May 1979, while six females (**TM** 36253–36258) and a single adult male (**TM** 36259) were collected from the same site in September 1983. The above observations suggest that the species is capable of using the same roost for a summer maternity roost as well as a winter hibernaculum. However, all other roosts of this species recorded in the province appeared to be used exclusively as maternity roosts, with individuals being absent during winter. Major hibernacula have not been located in the province.

Linear measurements (mm)

	Males					Females				
	$\bar{x}$	s.d.	n	Min	Max	$\bar{x}$	s.d.	n	Min	Max
TL	102.7	9.1	9	85	108	–	–	1	107	107
HB	58.9	5.1	9	51	65	–	–	1	62	62
T	43.8	4.8	9	34	49	–	–	1	45	45
Hf	10.5	1.0	6	9	12	–	–	–	–	–
E	15.2	1.3	6	14	17	–	–	1	15	15
FA	50.1	5.2	6	41	54	–	–	1	51	51

Records of occurrence
SPECIMEN RECORDS:
BM: Town Bush;
DM: Dlinza Forest, Shongweni Dam;
KM: Doornhoek, Town Bush;
NM: Town Bush;
SI: Town Bush;
TM: Bishopstowe Cave, Dumisa Mine, Itala GR (Doornpan), Ferncliff, Goudhoek Farm, Oribi Gorge NR, Otto's Bluff, Town Bush, Vernon Crookes NR.

ADDITIONAL RECORDS:
Laycock (1976): Doornhoek, Laager Cave, Saxony Cave, Shongweni, Town Bush;
Bernard (1980): Babanango, Dumisa Mine, Ngubevu;
Avery (1991): Border Cave.

Genus	*Pipistrellus*	Kaup 1829

KEY TO SPECIES (After Meester et al. 1986):

1. Outer upper incisor more than half length
    of inner upper incisor; forehead strongly
    concave; upper canine and posterior
    upper premolar separated by a gap
    through which anterior upper premolar
    can be clearly seen rising above cingula
    of adjoining teeth; interdental palate
    clearly longer than broad . . . *P. (p.) nanus*, p. 48
—   Outer upper incisor less than half
    length of inner upper incisor; forehead
    usually flat or weakly concave; upper
    canine and posterior upper premolar
    in contact or barely separated,
    anterior upper premolar seen from
    the side only with difficulty; usually
    not rising above cingula of adjoining
    teeth; interdental palate not clearly
    longer than broad 2

2. Posterior upper incisor barely extending
    beyond cingulum of anterior upper incisor;
    anterior upper premolar a pointed tooth
    rising above cingulum of canine but not
    above congulum of posterior upper
    premolar; maxillary toothrow more than
    4.5 mm . . . . *P. (p.) kuhlii*, p. 46
—   Posterior upper incisor extending well
    beyond cingulum of anterior upper
    incisor; anterior upper premolar a flat-
    crowned tooth completely cingulum of
    adjoining teeth; maxillary toothrow less
    than 4.5 mm . . . . *P. (p.) anchietai*, p. 47

Pipistrellus kuhlii (Kuhl 1819)
P. k. broomi (Roberts 1948)
Kuhl's pipistrelle
Kuhl-vlermuis

Taxonomic status
Two southern African subspecies were recognised by
Meester et al. (1986), of which only the above is found in
KwaZulu-Natal.

Distribution
Distributed mainly in the eastern parts of the province, as
far south as Durban, with one isolated record from Rockcliff
Farm on the KwaZulu-Natal/Free State border (Figure 40).
Most specimen records originate from the bushveld regions
of northern Zululand, suggesting that the *P. kuhlii* is commoner in these areas.

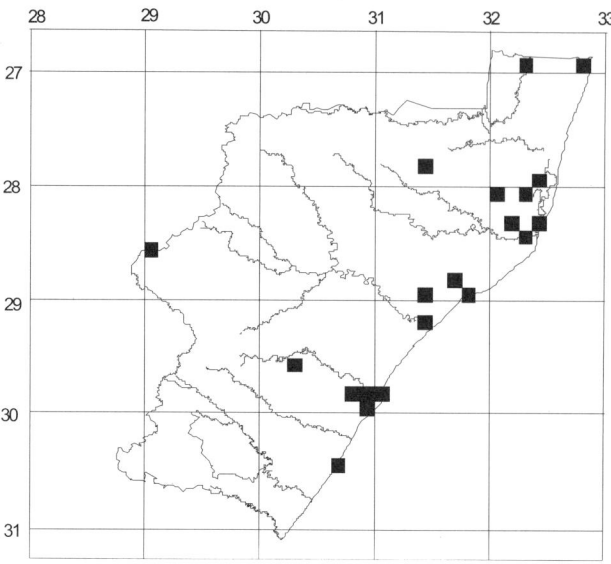

Figure 40

Protected areas
False Bay, Krantzkloof, Ndumu, St Lucia, Umlalazi.

Conservation status
Reasonably common.

Habitat
Kuhl's pipistrelle is distributed very widely throughout
Africa, Europe and the Middle East, where it occupies a
very wide range of habitats from deserts in the Middle East
to the Knysna Forest. However, it seems to be dependent,
in a given area, on a good water supply, plentiful food supply and suitable roosting spots (Skinner and Smithers
1990). In KwaZulu-Natal, Kuhl's pipistrelle has been recorded from most bioregions, being largely absent from
the Drier upland and Moist upland, Highland and Montane
regions.

Habits
Very little is known of their day-time roosting habits. All
specimens collected in KwaZulu-Natal were mistnetted at
dusk, usually in riverine situations in well wooded vegetation types, including Afromontane forest, coastal forest,
Terminalia-Acacia woodland, sand forest and riverine forest. In greater Transvaal and Botswana, they have been
collected under the bark of dead trees, as well as under the
roof of an old farm house (Smithers 1971; Rautenbach
1982).

Breeding
Pregnant females were collected during October at
Yellowwood Park (**DM** collection, two foetuses) and at
Dukuduku Forest (**TM** collection). In greater Transvaal,
pregnant females (with one foetus) have also been taken in
October (Rautenbach 1982).

Linear measurements (mm) and mass (g)

	Males					Females				
	x̄	s.d.	n	Min	Max	x̄	s.d.	n	Min	Max
TL	79.2	4.8	12	70	86	85.9	2.6	8	83	91
HB	47.2	3.8	12	38	53	50.4	2.3	8	47	55
T	31.8	2.8	12	25	35	35.5	0.9	8	34	37
Hf	6.5	1.0	10	5	7	6.4	0.5	8	6	7
E	10.9	0.7	12	10	12	11.6	0.5	8	11	12
FA	32.4	0.8	12	31	33	33.0	1.1	8	31	35
Mass	5.8	0.5	12	5.1	6.6	6.1	0.3	7	5.8	6.6

As also noted by Rautenbach (1982) in greater Transvaal, females tend to be larger and heavier than males. Specimens from KwaZulu-Natal are larger and heavier on average than specimens from greater Transvaal (compare mean total lengths of 79 mm and 86 mm above with values from the greater Transvaal of 76 mm and 80 mm for males and females respectively; mean masses of 5.8 g and 6.1 g above, compared with 4.6 g and 5.8 g for the greater Transvaal).

Records of occurrence

SPECIMEN RECORDS:
BM: Ngoye Hills;
CM: Pennington;
DM: Dlinza Forest NR, Durban, Eastern Shores NR (Iphiva Camp), False Bay Park, Harold Johnson NR, North Park NR, Pigeon Valley Reserve (Durban), Rossburgh (Durban), Yellowwood Park (Durban);
KM: Sweetwaters;
TM: Dukuduku Forest, False Bay Park (Dugundhlovu Camp), Hluhluwe GR (Research Camp), Kosi Lake (Department of Health Camp), Krantzkloof NR, Malvern, Ndumu GR, Ngome Forest Reserve, Rockcliff Farm, Umlalazi NR.

Pipistrellus anchietai (Seabra 1900)
Anchieta's pipistrelle
Anchieta-vlermuis

Taxonomic status

No subspecies were recognised by Meester et al. (1986). Kearney and Taylor (1997) pointed out the inadequacy of the current key for *P. kuhlii* and *P. anchietai* (Meester et al. 1986), based on dental characters which are strongly influenced by the degree of tooth wear. It is likely that many museum specimens of *P. anchietai* have in the past been misidentified as *P. kuhlii*. Bacular shape and karyotype (2n=26) provide a more objective basis for species identification, and these criteria were used to verify recent specimens in the Durban Natural Science Museum collection listed below (Kearney and Taylor 1997).

Distribution

Until recently, the only record of Anchieta's pipistrelle from South Africa was a specimen collected at Skukuza in the Kruger National Park (Skinner and Smithers 1990). Nine

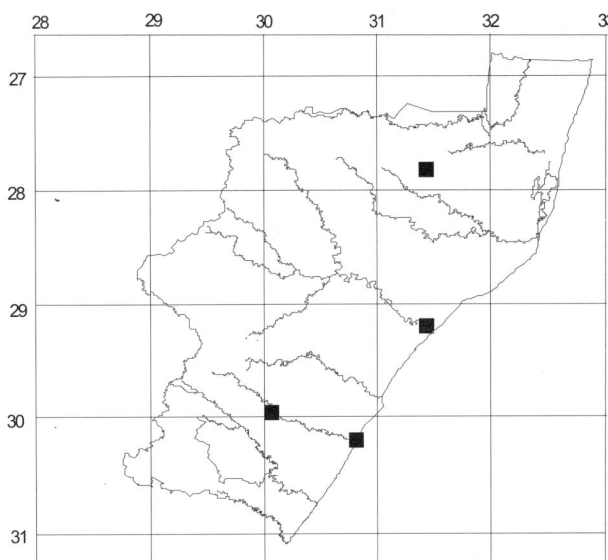

Figure 41

additional specimens from five localities have since been collected from KwaZulu-Natal, indicating a wide but scattered distribution in KwaZulu-Natal, from Ngome Forest Reserve in the north to Empisini Nature Reserve near Umkomaas in the south. (Figure 41).

Protected areas

Harold Johnson, Hella Hella, Krantzkloof.

Conservation status

Localised in distribution, although probably not as rare as previously thought.

Habitat, habits and breeding

Little is known about the biology of this species. Specimens collected in KwaZulu-Natal were either netted or collected in a harp trap in well wooded associations such as Afromontane forest, coastal forest, or bushveld. The specimen from Krantzkloof was netted over water at a height of 1.0–1.5 m.

A pregnant female with two foetuses was collected during October (**DM** collection).

Linear measurements (mm) and mass (g)

	Males					Females				
	x̄	s.d.	n	Min	Max	x̄	s.d.	n	Min	Max
TL	79.5	–	2	79	80	83.0	3.7	4	78	87
HB	45.5	–	2	45	46	47.0	1.8	4	45	49
T	34.0	–	2	34	34	36.0	3.2	4	32	39
Hf	6.0	–	2	6	6	5.2	0.5	4	5	6
E	10.5	–	2	10	11	11.5	1.0	4	11	13
FA	29.7	–	2	29	30	31.9	0.3	4	32	33
Mass	4.3	–	2	4.2	4.3	4.9	0.5	3	4.4	5.4

Females tend to be larger and heavier than males. This species is extremely similar in body size to the previous species.

Records of occurrence

SPECIMEN RECORDS:
DM: Empisini NR, Harold Johnson NR, Hella Hella GR;
TM: Krantzkloof NR, Ngome.

Pipistrellus nanus (Peters 1852)   PLATE 17
P. n. nanus (Peters 1852)
Banana bat
Piesangvlermuis

Taxonomic status

Although Meester et al. (1986) listed one subspecies from southern Africa, Skinner and Smithers (1990) recognise no subspecies on the basis that geographic variation has not been adequately assessed in the species and many of the described forms are likely to be synonyms.

Distribution

Widespread throughout the eastern parts of the province, occurring mostly in Coast lowlands, Coast hinterland, Lowveld, Valley bushveld and Mistbelt bioregions (Figure 42). The apparent gap associated with the Tugela Basin probably results from lack of collecting in this region.

Protected areas

Coastal Forest Reserve (=Kosi Bay), Eastern Shores, Hazelmere, Hluhluwe, Itala, Krantzkloof, Mkuzi, Ndumu, Oribi Gorge, Stainbank, Umlalazi.

Conservation status

Common in the eastern parts of KwaZulu-Natal, owing to their habit of roosting in banana plants in commercial banana plantations which are common on the South Coast of KwaZulu-Natal. They have also been recorded in suburban situations.

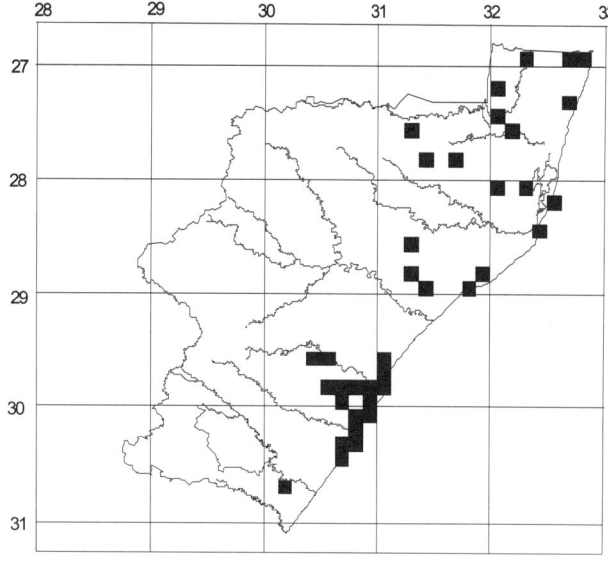

Figure 42

Habitat

They occur in a variety of vegetational associations in the province, from bushveld and coastal forest to mistbelt forest. They are absent from treeless grassland and montane habitats. Habitat requirements include a good water supply and availablity of their favourite day-time roosting sites, commercial banana or plantain plants, or naturally-occurring *Strelizia nicolai* plants.

Habits

Their favourite roosting sites are the unfurling, cylindrical-shaped, terminal leaves of banana plants, or inside the bunches of fruit. They roost in the terminal leaves, facing the narrow end, up to six or seven per leaf and one on top of the other. They are equipped with sucker pads on the wing and soles of the feet to assist with clambering up the smooth, slippery surfaces of banana leaves. Banana bats do not roost exclusively in banana plants, and have been recorded in roofs and in the thatch of rural huts (Skinner and Smithers 1990).

Breeding

In KwaZulu-Natal, reproduction is reported to be seasonal and monoestrus, with single or twin infants being born during November and December (La Val and La Val 1977). In Malawi, mating occurs in June and July, ovulation and fertilisation occurs in late August, and females invariably give birth to twins (Happold and Happold 1996; Bernard et al. 1997). Two pregnant females from Mkuzi Game Reserve were collected during September (**TM** collection).

Linear measurements (mm) and mass (g)

	Males					Females				
	x̄	s.d.	n	Min	Max	x̄	s.d.	n	Min	Max
TL	75.5	3.9	18	68	83	78.6	5.0	17	69	87
HB	43.2	4.8	21	36	53	43.6	3.2	18	37	49
T	32.8	3.3	18	23	37	35.2	2.8	17	30	38
Hf	5.1	1.0	8	3	6	5.8	0.7	7	5	7
E	9.0	1.9	14	6	12	9.2	1.4	11	6	12
FA	31.4	2.8	8	29	38	31.1	1.0	7	30	32.6
Mass	3.5	0.2	5	3.2	3.8	4.0	0.5	10	3.0	4.5

Data on mass obtained from Seamark and Kearney (personal communication)

Females appear to be slightly larger and heavier than males.

Records of occurrence

SPECIMEN RECORDS:
BM: Illovo, Malvern, Umkomaas;
CM: Pennington;
DM: Bonamanzi GR (Lalapanzi Pan), Empisini NR, Forest Hills (28 King George Avenue), Jozini Dam, Krantzkloof NR, Mkuzi GR, Mtunzini, New Germany (Rabe Street), Renishaw, Scottburgh, Shongweni Polo Club, Stainbank NR, Vuma Farm;
KM: Nagle Dam, Ndumu GR;
NM: Empangeni, Hluhluwe GR, Maputa (=Manguzi), Oribi Gorge, Scottsville, Umdoni Park, Umtwalume River (north bank);

SAM: Coastlands, Malvern;
SI: Maputa;
TM: Babanango, Eastern Shores NR (Lake Bangazi), Hazelmere Dam NR, Ingwavuma, Itala GR (Craigadam, Doornpan), Kosi Lake (Department of Health Camp), Kranzkloof NR, Lake Sibayi, Malvern, Mid-Illovo, Mkuzi GR, Ngome Forest Reserve, Nongoma, Rob Roy Hotel, St Lucia Estuary, Umbumbulu, Umdoni Park, Umhlanga Reserve.

ADDITIONAL RECORDS:
Sapsford (unpublished): Eshowe, 16 Everton Road (Gillitts), George Campbell Technical High School (Durban), 26 Latham Gardens (Umbilo), 59 Prince Street (Athlone Park), 51 Seventh Avenue (Pinetown), 6A Stockville Road (Gillitts), 5 Tritoma Road (Pinetown), 12 Warrior Road (Hillcrest).

Genus	*Chalinolobus*	Peters 1866

Chalinolobus variegatus (Tomes 1861)
C. v. variegatus (Tomes 1861)
Butterfly bat
Vlindervlermuis

Taxonomic status
Only the above-mentioned subspecies occurs in southern Africa. A further four extralimital subspecies are known (Meester et al. 1986).

Distribution
In KwaZulu-Natal, *Chalinolobus variegatus* is restricted to Lowveld and Coast lowlands of northern Zululand, occurring as far south as Futululu, immediately south of St Lucia Estuary (Figure 43). This species has been recorded in very late Holocene archaeological remains at Border Cave (Avery 1991).

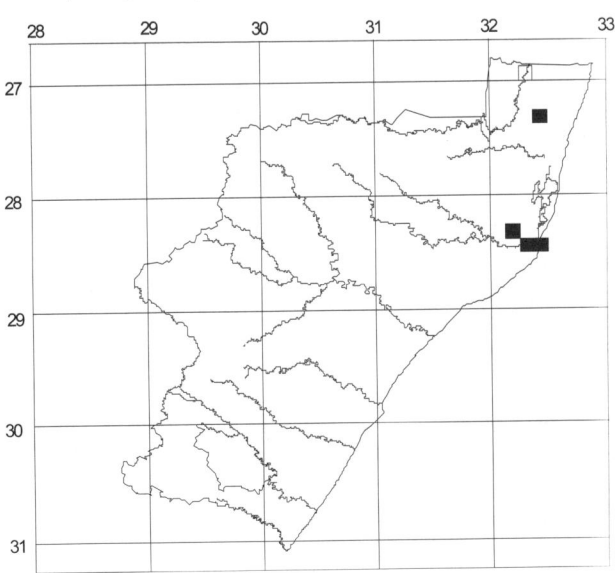

Figure 43

Protected areas
St Lucia, Ndumu.

Conservation status
While Bourquin (1988) reported only two old records for this species from the extreme north of KwaZulu-Natal (Dixon 1966; and a 1929 Transvaal Museum specimen), the collection of four specimens in the St Lucia area since 1980 (see under **Records of occurrence**) indicates that this species is probably less rare than was previously thought.

Habitat
Elsewhere they are apparently associated with open woodland, but not forest (Skinner and Smithers 1990). However, the specimen from Dukuduku was netted in the public picnic site within the densely forested Dukuduku Forest, and specimens from Futululu and St Lucia village were presumably also collected in forest.

Habits and breeding
Elsewhere, they have been found to roost in an old abandoned hut as well as between bunches of leaves. Nothing is known about their roosting habits in KwaZulu-Natal. No data on reproduction are available for southern Africa.

Linear measurements (mm) and mass (g)
The female specimen from Futululu had a head and body length of 51 mm, a tail of 44 mm, hind foot (*su*) of 8 mm, a forearm length of 44 mm and a mass of 14 g. The linear measurements given above fall below the range given for a series of *C. variegatus* from Zimbabwe (total length for females ranged from 106 to 109 mm (Smithers and Wilson 1979), compared with 95 mm for the Futululu specimen, suggesting that the latter was probably immature).

Records of occurrence
SPECIMEN RECORDS:
DM: Futululu Research Station;
KM: St Lucia Resort;
TM: Dukuduku Forest Station, Manaba, St Lucia Estuary.

ADDITIONAL RECORDS:
Avery (1991): Border Cave;
Dixon (1966): Ndumu.

Genus	*Laephotis*	Thomas 1901

Laephotis wintoni (Thomas 1901)
Winton's long-eared bat
Winton-langoorvlermuis

Taxonomic status
The validity of *L. wintoni* in South Africa is ambiguous. The species occurs in Kenya and Ethiopia, with isolated records from South Africa. Based on multivariate similar-

ity in cranial characters, specimens from the Cedarberg (Western Cape) (Rautenbach and Nel 1978), Lesotho and the Free State (Watson 1990), and KwaZulu-Natal (Kearney and Taylor 1997) have been referred to *L. wintoni*. Although Skinner and Smithers (1990) withdrew *L. wintoni* on the suggestion of Rautenbach (personal communication) that the specimens from the Western Cape called *L. wintoni* are more appropriately placed with *L. namibensis*, Meester et al. (1986) and Koopman (1993) retain *L. wintoni* in South Africa, and the validity of the species is provisionally upheld until more critical studies can be performed.

Distribution
So far the species has only been found at Game Valley Estates near Richmond (Figure 44).

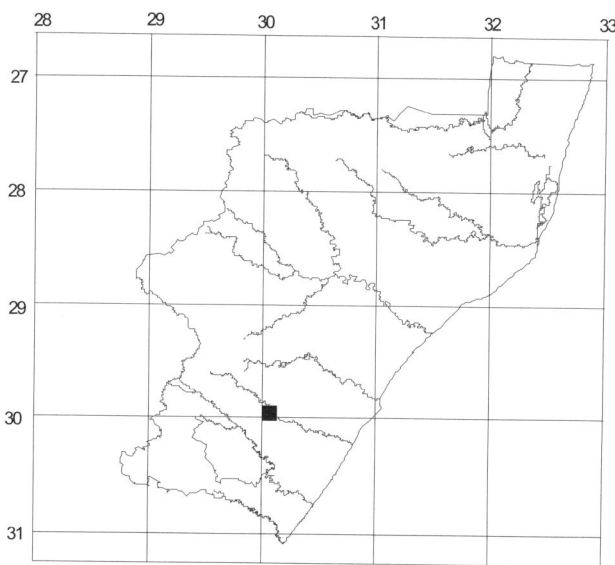

Figure 44

Protected areas
No formal protected areas.

Conservation status
Only known from one record in the province.

Habitat
The specimen from Game Valley Estates (Hella Hella) was recorded in Valley bushveld (northern variation Type 23a of Acocks 1988).

Habits and breeding
No information available for the subregion.

Linear measurements (mm) and mass (g)
The male specimen from Game Valley Estates (Hella Hella) had a head and body length of 51 mm, a tail of 43 mm, hind foot (*su*) of 6 mm, an ear length of 19 mm, forearm length of 36.9 mm and a mass of 6.6 g.

Records of occurrence
SPECIMEN RECORDS:
DM: Game Valley Estates (Hella Hella).

Genus	*Eptesicus*	Rafinesque 1820

KEY TO SPECIES (After Meester et al. 1986):

1. Wing membranes white or translucent . . . .
   E. rendalli, p. 50
   – Wing membranes dark 2

2. Size larger, forearm more than 40 mm, inner upper incisors unicuspid . . . .
   E. hottentotus, p. 51
   – Size smaller, forearm less than 40 mm; inner upper incisors bicuspid 3

3. Dorsal colour paler; wing membranes edged with white; dorsal fur up to 8 mm long; forearm 30–32 mm; condylobasal length usually less than 11.9 mm; skull lacking posterior occipital 'helmet'. . . . .
   E. zuluensis, p. 52
   – Dorsal colour darker; wing membranes not edged with white; dorsal fur not more than 5 mm long; forearm 29–35 mm; condylobasal length more than 11.9 mm; skull with posterior occipital 'helmet' . . . . *E. capensis*, p. 52

Eptesicus rendalli (Thomas 1889)
Rendall's serotine bat
Rendall-dakvlermuis

Taxonomic status
No subspecies recognised (Meester et al. 1986). Only recently recorded from KwaZulu-Natal and South Africa (Kearney and Taylor 1997).

Distribution
So far only recorded from Bonamanzi Game Reserve in Zululand (Figure 45).

Protected areas
No formal protected areas.

Conservation status
Because of its isolated occurrence in KwaZulu-Natal and South Africa, this species should in future be listed in the South African Red Data Book.

Habitat
Two specimens were mistnetted over a pan in Zululand Palm Veld (Type 1b; Acocks 1988).

Habits
The specimens from Bonamanzi were observed flying close to the ground (Kearney and Taylor 1997), as noted also by Skinner and Smithers (1990).

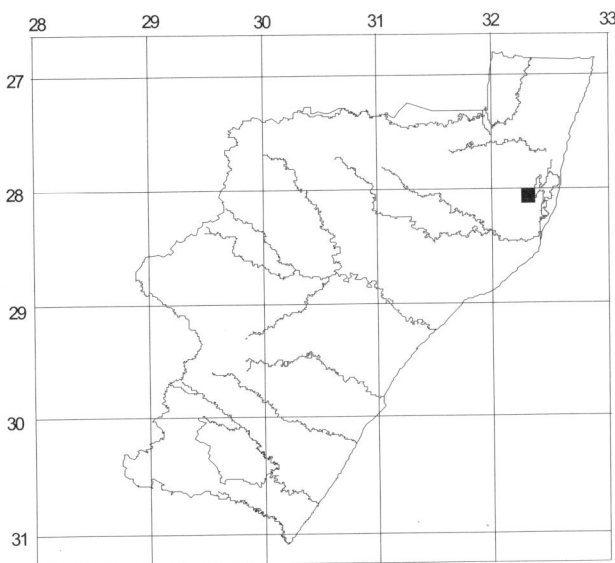

Figure 45

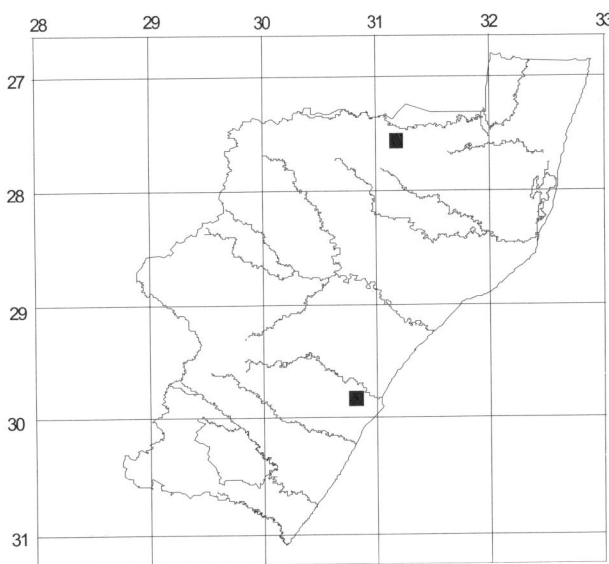

Figure 46

Breeding
A pregnant and lactating female with two foetuses was collected during December (**DM** collection).

Linear measurements (mm) and mass (g)

Cat. No.	Sex	TL	HB	T	Hf*su*	E	FA	Mass
DM5361	M	92	55	37	6	11	35.9	7.93[*]
DM5370	M	94	56	38	7	11	34.3	8.53[*]
DM5877	M	100	48	42	8	12	36.8	8.38
DM5878	F	99	49	41	7	12	37.9	–

[*] Bats kept in captivity for four weeks

Records of occurrence
SPECIMEN RECORDS:
DM: Bonamanzi GR (Lalapanzi Pan).

Eptesicus hottentotus (A. Smith 1833)
E. h. bensoni (Roberts 1946)
Long-tailed serotine bat
Langstretdakvlermuis

Taxonomic status
Three subspecies are recognised in southern Africa (Meester et al. 1986), of which only the above-mentioned occurs in KwaZulu-Natal.

Distribution
Known only from two specimens, one from Itala Game Reserve and one from Krantzkloof Nature Reserve (Figure 46). Although it has been reported to occur in the KwaZulu-Natal Drakensberg (Meester et al. 1986; Skinner and Smithers 1990), no specimens are available to substantiate this. While some 20 specimens of *E. capensis* have been collected by myself in the KwaZulu-Natal Drakensberg, no specimens of *E. hottentotus* have ever been collected. Avery (1991) recorded this species from Pleistocene deposits at Border Cave as well as (possibly) from late Holocene deposits at Nkupe in the upper Tugela Basin.

Protected areas
Itala, Krantzkloof.

Conservation status
Seldom collected and therefore probably uncommon in this province.

Habitat
Rautenbach et al. (1981) netted a single specimen over a river with well-wooded banks at Itala GR. The specimen from Krantzkloof was also probably netted in forest near water, although specific details are lacking. Judging from available evidence, the species is typically associated with more arid environments in the western regions of southern Africa.

Habits
In KwaZulu-Natal, nothing is known of their habits. Elsewhere, they have been found to occur in small groups of up to three or four, roosting in caves, rock hollows, or on the outside of buildings (Skinner and Smithers 1990).

Breeding
No data available.

Linear measurements (mm) and mass (g)
No data available for KwaZulu-Natal. They are the largest species of *Eptesicus* to occur in the subregion. In a sample from Zimbabwe, they had a mean total length of 115 mm, a tail of 47 mm and a mass of 16.6 g (Skinner and Smithers 1990).

Records of occurrence
SPECIMEN RECORDS:
TM: Itala GR (Doornkraal), Kranzkloof NR.

ADDITIONAL RECORDS:
Avery (1991): Border Cave, Nkupe.

Eptesicus zuluensis (Roberts 1924
Zulu serotine bat
Kaapse dakvlermuis

Taxonomic status
While *E. zuluensis* has in the past been regarded as a distinct species (e.g. Hayman and Hill 1971), Koopman (1975) regarded it to be a subspecies of *E. somalicus*, a view supported by Meester et al. (1986). However, based on chromosomal differences between *E. somalicus* and *E. zuluensis,* Rautenbach et al. (1993) regarded *E. zuluensis* to be a distinct species. This view is followed here. The type specimen of *E. s. zuluensis* was collected at Umfolozi Game Reserve. No subspecies are recognised.

Distribution
Known only from the Umfolozi and Mkuzi Game Reserves in northern Zululand (Figure 47).

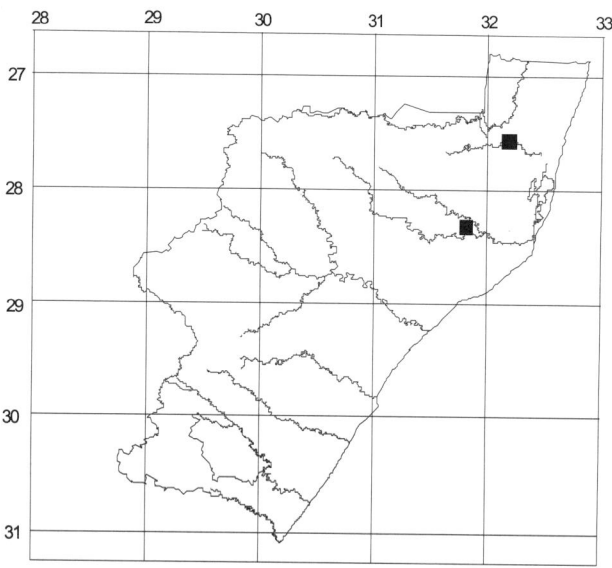

Figure 47

Protected areas
Mkuzi, Umfolozi.

Conservation status
Restricted distribution in KwaZulu-Natal.

Habitat
Associated with the Lowveld bioregion in northern Zululand.

Habits
Apparently found in association with the Cape serotine bat, *Eptesicus capensis* which is by the far the commoner species. Nothing is known of their day-time roosting habits.

Breeding
No data available for KwaZulu-Natal, although in southern Africa, generally, young are born during late November and early December (Skinner and Smithers 1990).

Linear measurements (mm)

Cat. No.	Sex	TL	HB	T	Hf	E	FL
DM5405	F	79	48	31	6	9	32.0

As indicated in the key above, this species tends to be smaller than *E. capensis*.

Records of occurrence
SPECIMEN RECORDS:
TM: Mkuzi GR (Malibali Pan), Umfolozi GR.

ADDITIONAL RECORDS:
Bourquin et al. (1971): Umfolozi GR (Thoboti outpost).

Eptesicus capensis (A. Smith 1829)   PLATE 18
E. c. capensis (A. Smith 1829)
Cape serotine bat
Kaapse dakvlermuis

Taxonomic status
Meester et al. (1986) recognised three subspecies, of which only the above occurs in southern Africa.

Distribution
Widespread in KwaZulu-Natal from sea level up to at least 1 600 m in the KwaZulu-Natal Drakensberg (Figure 48). This species also occurs very widely throughout sub-Saharan Africa excluding the Namib Desert. Avery (1991) recorded *E. capensis* from upper Pleistocene deposits at Border Cave.

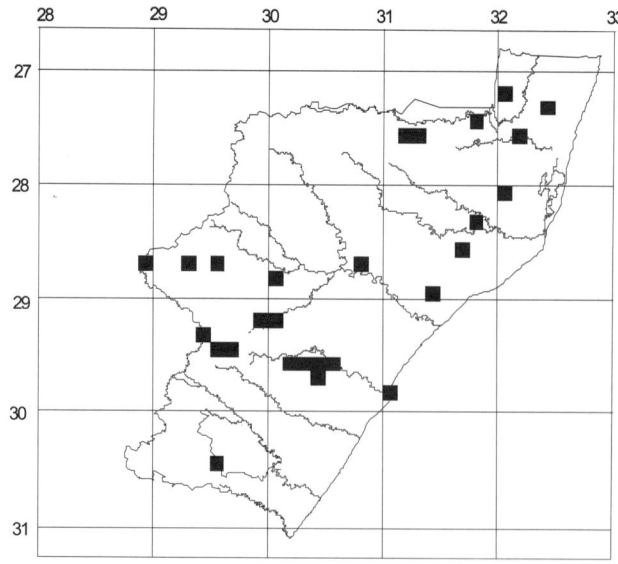

Figure 48

Protected areas
Dlinza Forest, Giant's Castle, Hluhluwe, Itala, Kamberg, Loteni, Mkuzi, Mount Currie, Queen Elizabeth Park, Royal Natal, Spoienkop, Umfolozi, Weenen.

Conservation status
Fairly common and widespread in KwaZulu-Natal.

Habitat
Occurs throughout all bioregions. Recently collected specimens (**DM** and **TM** collections) were netted in a variety of habitats including open Montane grassland, coastal forest, bushveld and *Acacia* woodland, usually in the vicinity of rivers, waterholes or pans.

Habits
Specimens from KwaZulu-Natal have been reported roosting in groups of between one and at least ten under the bark of trees, at the base of aloe leaves, in cracks in a dilapidated wall, and under the roofs of houses. They have been observed roosting under roofs of thatch, corrugated iron, or tiles. They emerge at dusk, and are initially slow and sluggish in flight, becoming more agile as they warm up. They follow specific flight paths under the canopy of trees, some 10–15 m above the ground, jinking and turning in a characteristic pattern as they search for insects (Skinner and Smithers 1990). According to Rautenbach (1982), Cape serotine bats have been reported to hunt insects in congregations.

Breeding
A pregnant female with two foetuses was collected at Itala Game Reserve during September (**TM** collection). From data elsewhere, females usually give birth to one or two young but pregnant females have been recorded having three or even four foetuses, an unusual occurrence in bats where a single young is the rule. The young are usually born around late November (Herselman and Norton 1985; Skinner and Smithers 1990; Lynch 1989).

Linear measurements (mm) and mass (g)

	Males					Females				
	x̄	s.d.	n	Min	Max	x̄	s.d.	n	Min	Max
TL	77.7	5.8	15	67	87	82.7	7.8	11	67	92
HB	49.1	4.6	15	42	58	52.6	5.5	11	40	58
T	28.6	2.9	16	22	33	30.1	3.5	11	22	35
Hf	6.2	0.8	13	5	8	7.1	0.9	5	6	8.5
E	11.4	1.5	12	9	14	12.0	0.8	4	11	13
FL	31.8	1.5	14	30	34	34.0	1.6	5	32	36
Mass	6.3	0.5	10	6	7	7.8	1.5	5	6	10

Females are distinctly larger than males, particularly with respect to mass, a result consistent with findings in studies in greater Transvaal (Rautenbach 1982), Free State (Lynch 1983), Eastern Cape (Lynch 1989) and Lesotho (Lynch 1994).

Records of occurrence
SPECIMEN RECORDS:
DM: Chase Valley Heights, Dlinza Forest NR, Itala GR, Kamberg NR (Glengarry), Loteni NR, Merrivale (Arden Estate), Mkuzi GR, Queen Elizabeth Park, Royal Natal NP, Spioenkop NR, Weenen NR;
NM: Firle Farm, Mfongosi;
SAM: Mfongosi;
TM: Ashburton, Cato Ridge (3 km north-east), Eshowe, Giant's Castle NR, Hluhluwe GR (Research Camp), Howick, Ingwavuma, Itala GR (Craigadam, Doornkraal, Doornpan), Kamberg NR, Leeuwspoor Farm, Manaba, Mkuzi GR (Bube Pan, Malibali Pan, Msinga Pan, Warden's house), Ntambanana, Mount Currie NR, Sea Cow Lake, Spioenkop NR, Umfolozi GR (Masinda Lodge), Weenen NR, Weston Agricultural College.

ADDITIONAL RECORDS:
Avery (1991): Border Cave.

| Genus | *Scotophilus* | Leach 1821 |

KEY TO SPECIES (After Meester et al. 1986):

1. Forearm 50–65 mm; skull length 20–23 mm . . . . *S. dinganii*, p. 53
– Forearm 43–50 mm; skull length 16–19 mm . . . . *S. viridis*, p. 55

Scotophilus dinganii (A. Smith 1833)   PLATE 19
S. d. dinganii (A. Smith 1833)
Yellow house bat
Geel dakvlermuis

Taxonomic status
There has been much confusion over the names given to the three species of *Scotophilus* in southern Africa. The name *nigrita* has been used widely in the literature to refer to the medium-sized species now known as *dinganii* but was shown by Robbins (1978) to refer to the largest of the three recognised southern African species. Meester et al. (1986) recognised three subspecies as occurring in southern Africa, of which only the nominate race is found in KwaZulu-Natal.

Distribution
The yellow house bat is widely distributed throughout the eastern parts of KwaZulu-Natal as far west as 30° longitude (Figure 49). Its association with houses accounts for the concentrated distribution of localities around the major urban centres of Durban and Pietermaritzburg. Avery (1991) recorded *S. dinganii* from upper Pleistocene deposits at Border Cave.

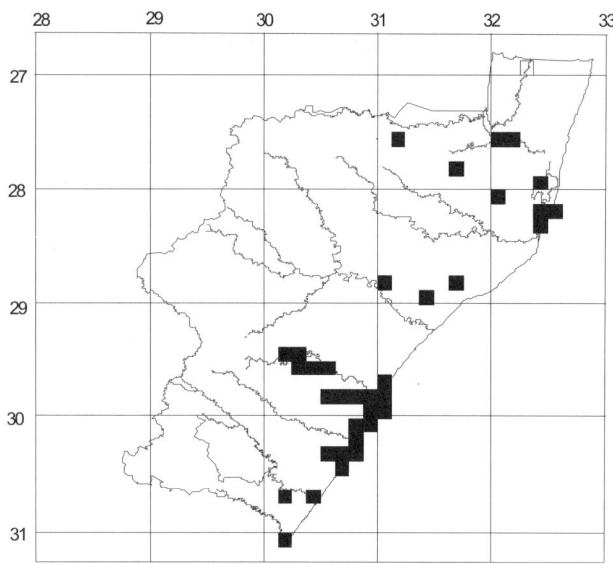

Figure 49

Protected areas
False Bay, Hluhluwe, Itala, Kenneth Stainbank, Krantzkloof, Mkuzi, Ndumu, Oribi Gorge, Queen Elizabeth Park, Umtamvuna, Vernon Crookes.

Conservation status
Along with *Chaerephon pumila,* this is the commonest bat species to be found occupying houses in built-up areas of eastern KwaZulu-Natal.

Habitat
Yellow house bats occur widely in Coast lowlands, Coast hinterland, Valley bushveld and Lowveld bioregions in KwaZulu-Natal, usually closely associated with houses or other buildings. While they are typically associated with savanna woodland and absent from forests (Skinner and Smithers 1990), specimens from KwaZulu-Natal have been collected from dense coastal or riparian forest.

Habits
As the name suggests they commonly roost under the roofs of houses and other buildings. In the Durban area they are frequently killed by domestic cats and dogs, and individuals are sometimes found drowned in swimming pools. They often fly into houses at night, a fact that frequently alarms members of the public. Group size of known roof-dwelling colonies in Durban and Pietermaritzburg varied from one to ten, with a mean of six (*n*=6). These colonies had been resident for between one and at least eight years. The bats typically roosted in a horizontal position between the top of the brickwork and the wooden beams and rafters, either at the level of the ceiling under the eaves or at the gable apex. Three specimens collected together at Ndumu Game Reserve were taken from the cavity wall of an outbuilding (Dixon 1966), and, in north-eastern Zululand, Roberts (1936) collected some specimens from a disused bird's nest.

Breeding
Pregnant females in the Durban Natural Science Museum

and Transvaal Museum collections were collected in September (*n*=2) and March (*n*=2): of these two had two foetuses and one had three. A lactating female with an unweaned juvenile clinging to it was collected in December. This pattern of parturition in early summer and frequent birth of twins has been confirmed by studies elsewhere in South Africa (Skinner and Smithers 1990). From captive data (*n*=7), births in the Durban and Pietermaritzburg areas occurred between early November and January, and young started flying when their forearms reached a length of 49 mm (E. Richardson, personal communication).

Linear measurements (mm) and mass (g)

	Males					Females				
	x̄	s.d.	n	Min	Max	x̄	s.d.	n	Min	Max
TL	121.3	10.2	12	104	137	122.4	6.6	9	110	130
HB	71.9	6.9	12	66	85	72.3	4.8	9	65	80
T	49.6	6.1	13	38	59	50.1	5.2	9	44	60
Hf	10.2	1.3	8	9	12	9.6	0.7	6	9	10.5
E	16.0	2.2	12	12	19.5	13.9	1.6	9	12	16
FA	54.1	2.3	9	51	58	52.7	1.7	7	50	55
Mass	24.8	1.3	4	23	25.7	–	–	–	–	–

Records of occurrence
SPECIMEN RECORDS:
BM: Ngoye Hills;
CM: Pennington;
DM: Bisley (Pietermaritzburg: 6 Holder Road), Cavendish, Durban, Eastern Shores (Cape Vidal), Empisini NR, Eshowe (4 Leigh Road, 5 Poynton Place), False Bay Park, Glenwood High School (Durban), Hillcrest, Hilton (12 Forest Lane), Howick, Moseley, Mount Moreland (12 William Street), Park Rynie, Pietermaritzburg Botanic Gardens, Pont Caravan Park, Queen Elizabeth Park, Shepstone Reserve, Umhlanga Rocks (2 Tibia Crescent), Umkomaas, Umpambinyoni, Waterfall (12 Lahle Crescent), Westville;
KM: Umdoni Park, World's View;
NM: Pietermaritzburg, Town Bush, Umdoni Park;
TM: Bluff (St Barnabas Church), Charter's Creek, Chase Valley, Dukuduku Forest Station, Durban, Eshowe, Hluhluwe GR (Egodeni), Itala GR (Doornkraal), Krantzkloof NR, Malvern, Mhlatuze River, Mission Rocks, Mkuzi GR (Bube Pan, Msinga Pan, Nhlonhlela Pan), Mkuzi River, Oribi Gorge NR, Pietermaritzburg, Queen Elizabeth Park, Ubombo (4 km south), Vernon Crookes NR.

ADDITIONAL RECORDS:
Avery (1991): Border Cave;
Dixon (1966): Ndumu GR;
Bourquin et al. (1971): Hluhluwe GR (Hilltop);
Rautenbach et al. 1981: Itala GR;
Sapsford (unpublished): 30 Abelia Road (Kloof), 33 Albers Road (Pinetown), 26 Atterbury Road (Rose Hill), Kearsney College (Botha's Hill), 44 Candella (Durban), 6 Chiselhirst Road (Westville), 21 Colinton Crescent (Pinetown), Colorado Guest Farm (Camperdown), 27 Cotham Road (Northdene), 100 Dan Pienaar Drive (Amanzimtoti), Dartnell Road (Amanzimtoti), 4 Dickens Road (Moseley), 49 Dilkoosh Road (Northdene), 20 Ebony Place (Glen

Anil), 2 Esser Road (Marianridge Park), 9 Finfoot Road (Woodhaven), 140 Glenardle Road (Bluff), Greenwich Farm (Umzinto), 59 Highlands Road (Pinetown), 17 Jan Smuts Avenue (Northdene), 78 Kingston Drive (Umhlanga), 633 Kingsway (Athlone Park), Knockinglan Farm (Eshowe), 92 Kiora Road (Bluff), 41 Leslie Drive (Pinetown), 144 Lewis Drive (Amanzimtoti), 20 Lewis Drive (Amanzimtoti), Macushla Drive (Glenhills), 531 Main Road (Escombe), 778 Main Road (Moseley), Marian Wood Park (Pinetown), 24 Matapan Drive (Westville), Mayville Terrace (Amanzimtoti: 89 Myland Court), 30 Middleton Road (Escombe), 330 Moore Road (Durban), 8 Nicolson Road (Pinetown), Old Main Road (Botha's Hill: Lot 70), 30 Park Drive (Umhlanga Rocks), 45 Parkview Drive (Pinetown), 4 Pearson Road (Everton), Phoenix (City Engineer's Department), Ridge Road (La Lucia), 40 Salisbury Avenue (Westville), Scottburgh Police Headquarters, 31 Sloan Road (Fynnlands), 12 Sunnyside Drive (Westville North), 4 Sylvania Avenue (Westville), 67 Thames Drive (Westville), Umtentweni (Port Shepstone), 43 Union Lane (Pinetown), University of Durban-Westville Campus (Westville), 57 Vause Road (Durban), Virginia Airport, 32 Westridge (Umhlanga Rocks), 25 Wheeler Road (Howardene), 39 Willington Avenue (Kloof), 90 Wood Road (Moseley Park);

Bat Interest Group files: 57 Dilkoosh Road (Northdene), 97 Frances Staniland Road (Montrose, Pietermarizburg), Old Main Road (Amanzimtoti: 4 Weiba Park).

Scotophilus viridis (Peters 1852)
Lesser yellow house bat
Klein geel dakvlermuis

Taxonomic status
This species was originally known as *S. viridis*, distinct from *S. borbonicus* which was originally described from Réunion. Meester et al. (1986) regarded *viridis* to be a subspecies of *S. borbonicus*. However, more recently, Koopman (1993) reverted to the name *viridis*, which is followed here.

Distribution
The lesser yellow house bat was previously known to occur as far south as Zululand (Bourquin 1988). However, new records from Garden Castle Nature Reserve and Pietermaritzburg extend the species' distribution as far south as 29°45´S (Taylor et al. 1994; Figure 50).

Protected areas
Eastern Shores, Garden Castle, Hluhluwe GR, Mkuzi, Ndumu, Sodwana Bay, St Lucia.

Conservation status
Fairly rare and localised in KwaZulu-Natal, although more widespread than previously thought.

Habitat
Mostly associated with bushveld habitats in Zululand, although the two specimens netted at Garden Castle were

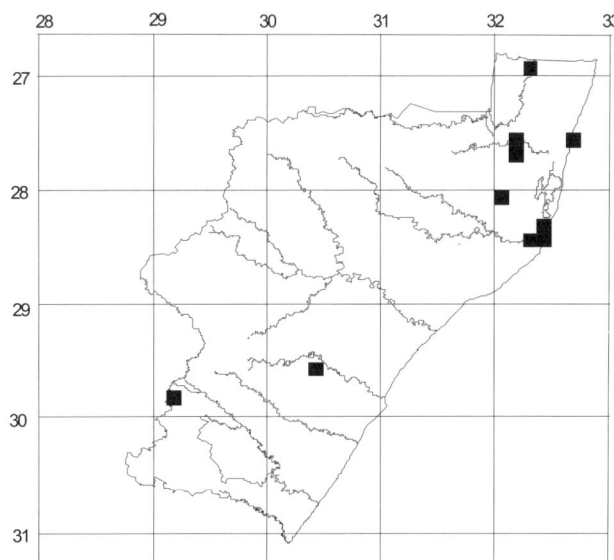

Figure 50

collected from the edge of a small wattle stand surrounded by open grassland at an altitude of 1 800 m. The specimen from Pietermaritzburg was collected near the town centre, which suggests it had been roosting in a building. A series of specimens collected at Mkuzi Game Reserve were all taken at pans, indicating a dependence on water.

Habits
Like the yellow house bat, they appear to roost commonly under the roofs of houses. Since most specimens from KwaZulu-Natal were mistnetted or found dead, little is known about their precise roosting habits in KwaZulu-Natal.

Breeding
A juvenile was collected in December at Hluhluwe Game Reserve. This is consistent with information from elsewhere which suggests that one or two young are born during November and December.

Linear measurements (mm) and mass (g)

	Males					Females				
	$\bar{x}$	s.d.	n	Min	Max	$\bar{x}$	s.d.	n	Min	Max
TL	105	9.6	4	97	116	121	–	1	121	121
HB	65	4.5	4	59	69	72	–	1	72	72
T	40	6.3	4	33	48	49	–	1	49	49
Hf	10	–	2	10	10	–	–	–	–	–
E	17	3.2	3	13	18	–	–	–	–	–
FA	49	3.0	4	45	52	46	–	1	46	46
Mass	18	–	2	18	18	24	–	1	24	24

Records of occurrence
SPECIMEN RECORDS:
DM: Futululu, Garden Castle NR, Hluhluwe GR (Egodeni), Pietermaritzburg (Berg Street), St Lucia village;
KM: Sodwana Bay (Jesser Point), St Lucia Forest Station;

TM: Dukuduku Forest, Mkuzi GR (Bube Pan, Malibali Pan, Msinga Pan, Nhlonhlela Pan), Ndumu GR, St Lucia village.

Genus	*Nycticeius*	Rafinesque 1819

Nycticeius schlieffenii (Peters 1859)
N. s. australis (Thomas and Wroughton 1908)
Schlieffen's bat
Schlieffen-vlermuis

Taxonomic status
Meester et al. (1986) listed two subspecies from the subregion, of which only the above occurs in KwaZulu-Natal.

Distribution
It has been recorded from three localities in north-eastern Zululand (Figure 51).

Protected areas
Mkuzi, Ndumu.

Conservation status
Because of its marginal occurrence in northern Zululand it is rare and localised in KwaZulu-Natal. However, this species is common throughout its wide African distribution which extends as far north as the Nile Delta in Egypt.

Habitat
Specimens from the Transvaal Museum and Durban Natural Science Museum were collected from pans and in riparian forest at Ndumu and Mkuzi Game Reserves. According to Rautenbach (1982) Shlieffen's bats have a very wide habitat tolerance, occurring in areas receiving as little as 250 mm and as much as 1 500 mm mean annual rainfall.

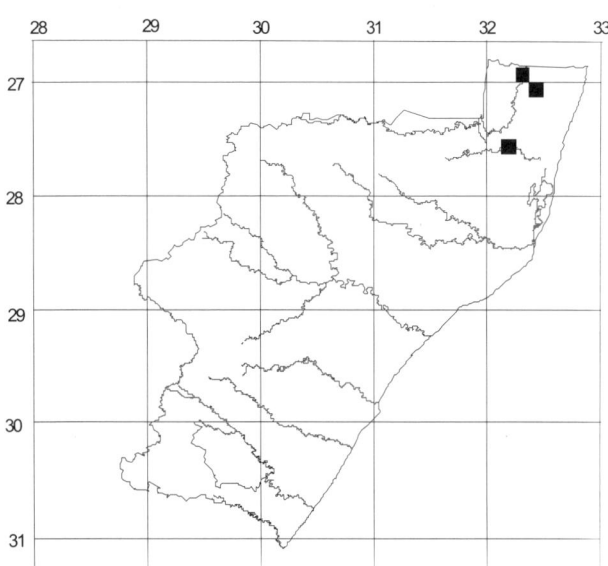

Figure 51

Habits
Nothing is known of its roosting habits in KwaZulu-Natal, although in greater Transvaal it is known to be a gregarious species which roosts in large numbers in buildings and crevices in rocks and trees (Rautenbach 1982).

Breeding
No data for KwaZulu-Natal, although in greater Transvaal most of the young are born during November. Up to three foetuses have been observed in pregnant females. Spermatozoa are stored in the female uterine horns from June (when copulation occurs) until September (when ovulation occurs) (Van der Merwe and Rautenbach 1986, 1987).

Linear measurements (mm) and mass (g)

Cat. No.	Sex	TL	HB	T	Hf	E	FA	Mass
DM5401	M	78	47	31	6	13	31.3	5.05

Schlieffen's bats have on average a total length of about 75 mm, with tails 30 mm in length, and a mass of 4.6 g (Skinner and Smithers 1990). The results for the single KwaZulu-Natal specimen are close to these values.

Records of occurrence
SPECIMEN RECORDS:
DM: Mkuzi GR;
TM: Mkuzi GR (Msinga Pan), Ndumu GR (Banzi Pan), Sihangwane.

Genus	*Scotoecus*	Thomas 1901

Scotoecus albofuscus (Thomas 1890)
S. a. woodi (Thomas 1917)
Thomas's house bat
Thomas-vlermuis

Taxonomic status
In revising the genus *Scotoecus*, Hill (1974) recognised two subspecies of *S. albofuscus* from Africa, of which only one is found in the subregion (Meester et al. 1986).

Distribution
The single record from St Lucia in KwaZulu-Natal (Figure 52) is the first record for both the province of KwaZulu-Natal and for South Africa, being known previously only as far south as the Zinave National Park in Mozambique (Kearney and Taylor 1997). This extension in the known range is not too surprising, given the distribution of at least two other vespertilionid bats, *Kerivoula argentata* and *Chalinolobus variegatus*, which extend into northern Zululand from Mozambique (Meester et al. 1986).

Plate 1
Myosorex varius
Forest shrew

J. Visser

Plate 2
Crocidura flavescens
Greater musk shrew

J. Visser

Plate 3
Crocidura cyanea
Reddish-grey musk shrew

J. Visser

Plate 4
Suncus infinitesimus
Least dwarf shrew

J. Visser

Plate 5
Amblysomus hottentotus
Hottentot golden mole

P. Taylor

Plate 6
Epomophorus wahlbergi
Wahlberg's epauletted fruit bat

H. Chittenden

Plate 7
Eidolon helvum
Straw-coloured
fruit bat

J. Visser

Plate 8
*Rousettus
aegyptiacus*
Egyptian fruit bat

J. Visser

Plate 9
*Taphozous
mauritianus*
Mauritian
tomb bat

D. Goode

Plate 10
Nycteris thebaica
Egyptian slit-faced bat

J. Visser

Plate 11
Rhinolophus clivosus
Geoffroy's horseshoe bat

J. Visser

Plate 12
Rhinolophus darlingi
Darling's horseshoe bat

P. Taylor

P. Taylor

Plate 13
Hipposideros caffer
Sundervall's leaf-nosed bat

P. Taylor

Plate 14
Cloeotis percivali
Short-eared trident bat

P. Royal

Plate 15
Miniopterus schreibersii
Schreiber's long-fingered bat

P. Royal

Plate 16
Myotis tricolor
Temminck's hairy bat

P. Taylor

Plate 17
Pipistrellus nanus
Banana bat

P. Taylor

Plate 18
Eptesicus capensis
Cape serotine bat

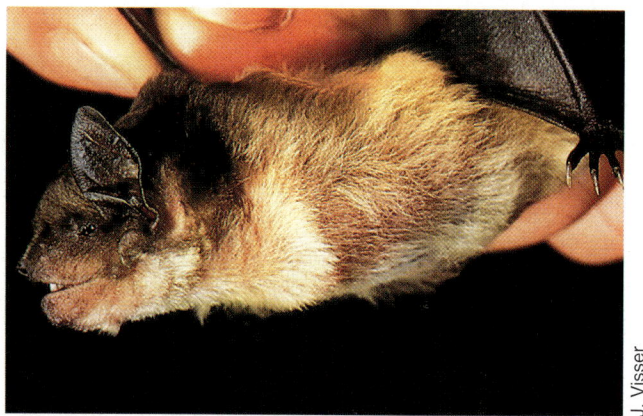

J. Visser

Plate 19
Scotophilus dinganii
Yellow house bat

E. Richardson

Plate 20
Otomops martiensseni
Large-eared free-tailed bat

J. Visser

Plate 21
Tadarida aegyptiaca
Egyptian free-tailed bat

J. Visser

Plate 22
Chaerephon pumila
Little free-tailed bat

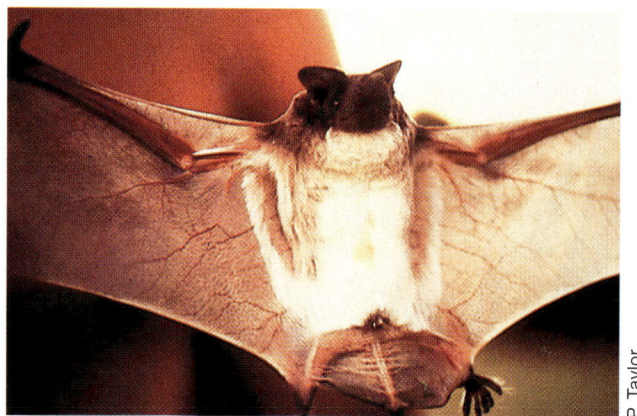

P. Taylor

Plate 23
Mops condylurus
Angola free-tailed bat

J. Visser

Plate 24
Papio hamadras
Chacma baboon

J. Kitchenside

Plate 25
Chlorocebus aethiops
Vervet monkey

Plate 26
Cercopithecus
mitis
Samango monkey

M. Macleod

J. Kitchenside

Plate 27
Procavia capensis
Rock dassie

J. Visser

Plate 28
Dendrohyrax arboreus
Tree dassie

J. Visser

Plate 29
Pedetes capensis
Springhare

J. Visser

Plate 30
Hystrix africaeaustralis
Porcupine

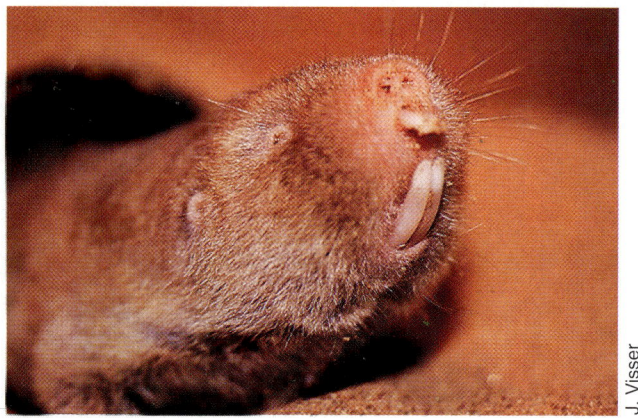

J. Visser

Plate 31
Cryptomys hottentotus
Common molerat

J. Visser

Plate 32
Georychus capensis
Cape molerat

A. Carter

Plate 33
Otomys angoniensis
Angoni vlei rat

Dr I. L. Rautenbach

Plate 34
Tatera leucogaster
Bushveld gerbil

J. Visser

Plate 35
Mystromys albicaudatus
White-tailed rat

J. Visser

Plate 36
Saccostomus campestris
Pouched mouse

Plate 37
*Dendromus
melanotis*
Grey climbing
mouse

J. Visser

Plate 38
Steatomys pratensis
Fat mouse

J. Visser

A. Carter

Plate 39
Lemniscomys rosalia
Single-striped mouse

A. Carter

Plate 40
Rhabdomys pumilio
Striped mouse

A. Carter

Plate 41
Dasymys incomtus
Water rat

A. Carter

Plate 42
Grammomys dolichurus
Woodland mouse

Plate 44
Mastomys natalensis
Natal multimammate mouse

A. Carter

Plate 43
Mus minutoides
Pygmy mouse

J. Visser

Plate 45
Thallomys paedulcus
Tree rat

Dr I. L. Rautenbach

Plate 46
Aethomys namaquensis
Namaqua rock mouse

J. Visser

Plate 47
Graphiurus murinus
Woodland dormouse

J. Visser

Plate 48
Petrodromus tetradactylus
Four-toed elephant shrew

J. Visser

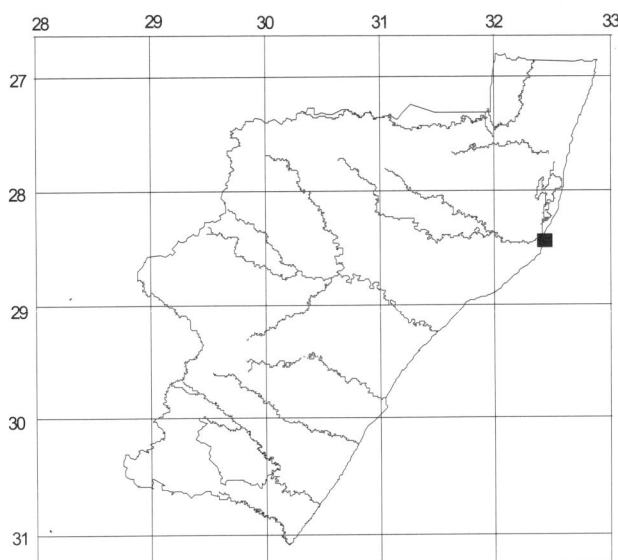

Figure 52

Protected areas
None.

Conservation status
Because of its isolated occurrence in KwaZulu-Natal and South Africa, this species should in future be listed in the South African Red Data Book.

Habitat, habits and breeding
Because it is collected so rarely, nothing is known of the biology of this species. The specimen collected from St Lucia on 30 November 1995 was a pregnant female which gave birth in captivity to twins.

Linear measurements (mm)

Cat. No.	Sex	TL	HB	T	Hf	E	FA
DM4885	F	85	53	32	8	9	31.0

This specimen is larger in body size and forearm length than a series of nine specimens measured by Hill (1974), and further collecting may show that it represents an undescribed species or subspecies.

Records of occurrence
SPECIMEN RECORDS:
DM: St Lucia (Perna Perna Resort).

Subfamily	**KERIVOULINAE**	
Genus	***Kerivoula***	**Gray 1842**

KEY TO SPECIES (After Meester et al. 1986):

1. Size larger, forearm 34–39 mm; skull >15 mm; colour above bright reddish-chestnut, below either whitish or buffy, never brown . . . . *K. argentata*, p. 57

—   Size smaller, forearm 30–36 mm; skull <13.5 mm, colour brownish or greyish-brown above, sometimes lightly grizzled with greyish-white or white, and brown or greyish-brown to greyish-white or white below . . . .
K. lanosa, p. 58

Kerivoula argentata (Tomes 1861)
K. a. zuluensis (Roberts 1924)
Damara woolly bat
Damara-wolhaarvlermuis

Taxonomic status
Meester et al. (1986) listed three subspecies from the subregion, of which only the above is found in KwaZulu-Natal.

Distribution
Localised to Maputaland and the Umfolozi Game Reserve (Figure 53). Avery (1991) recorded the possible occurrence of this species in late Holocene archaeological deposits at Nkupe in the upper Tugela Basin, some 200 km west of Umfolozi.

Protected areas
Umfolozi.

Conservation status
Localised to Maputaland and the Umfolozi Game Reserve. Bruton (1978) noted that they were common at Manzengwenya Forest north-east of Lake Sibayi.

Habitat
Bushveld habitats within the Lowveld bioregion.

Habits
The type specimen for *K. a. zuluensis* from Umfolozi was found roosting in the nest of the spectacled weaver, *Ploceus*

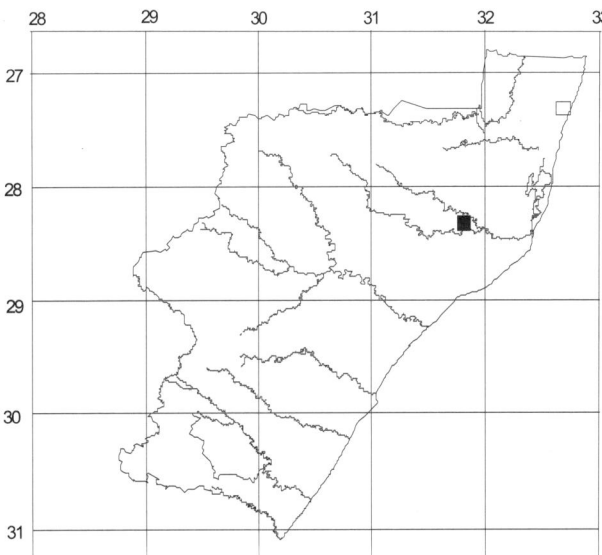

Figure 53

ocularis (see also Roberts 1951), as were two specimens collected by Austin Roberts from Msinyini Pan. Bruton (1978) noted that they were common at Manzengwenya Forest 'at forester's house and workshop', although it was not clear where specifically they were roosting in these buildings. Elsewhere, they have been found packed tightly in small groups of about four or five on the outer walls of rondavels, under the eaves (Skinner and Smithers 1990).

Breeding
No information available.

Linear measurements (mm) and mass (g)
No data available for KwaZulu-Natal. The Damara woolly bat is the larger of the two species of *Kerivoula* found in KwaZulu-Natal, with a total length of about 95 mm (including a tail of 47 mm) and a mass of between 6.0 and 9.0 g (Skinner and Smithers 1990).

Records of occurrence
SPECIMEN RECORDS:
TM: Msinyini Pan, Umfolozi GR.

ADDITIONAL RECORDS:
Bruton (1978): Manzengwenya.

Kerivoula lanosa (A. Smith 1847)
K. l. lucia (Hinton 1920)
Lesser woolly bat
Klein wolhaarvlermuis

Taxonomic status
Meester et al. (1986) listed two subspecies from the subregion, of which only the above is found in the province.

Distribution
Known from four localities in Maputaland and northern Zululand (Figure 54).

Protected areas
Coastal Forest Reserve (=Kosi Bay), Mkuzi.

Conservation status
Localised to Maputaland and northern Zululand in KwaZulu-Natal but occurring widely in Africa as far north as West Africa.

Habitat
Similar to *K. argentata*, i.e. well watered areas in dry woodland. Roberts (1951) recorded them from a forest in Zululand.

Habits
Like *K. argentata*, they often roost in abandoned bird nests during the day. In Zululand, Roberts (1951) recorded them from nests of spectacled weavers, *Ploceus ocularis*, three from one nest and two from another. Three specimens from

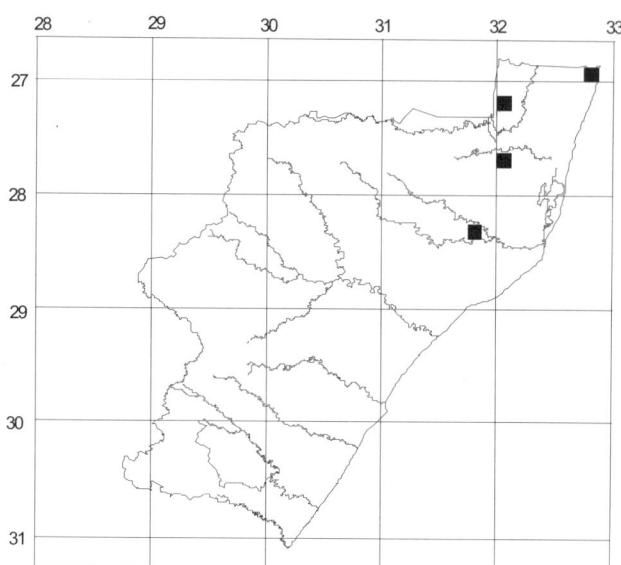

Figure 54

Ingwavuma in the Transvaal Museum collection were collected in a 'sunbird's' nest (species not given).

Breeding
No information available for the subregion.

Linear measurements (mm) and mass (g)
No information available for KwaZulu-Natal. As the name suggests, the lesser woolly bat is the smaller of the two species of *Kerivoula* found in KwaZulu-Natal, with a total length of 80 mm (including a tail of 36 mm), and a mass of between 6.0 and 8.0 g (Skinner and Smithers 1990).

Records of occurrence
SPECIMEN RECORDS:
NM: Mkuzi GR;
TM: Coastal Forest Reserve (=Kosi Lake), Ingwavuma, Umfolozi GR.

Family	MOLOSSIDAE	Free-tailed bats
Genus	*Otomops*	Thomas 1913

Otomops martiensseni (Matschie 1897)   PLATE 20
O. m. icarus (Chubb 1917)
Large-eared free-tailed bat
Bakoorlosstertvlermuis

Taxonomic status
Chubb (1917) described *Otomops icarus* from Durban as a distinct species. However, it is currently recognised as one of three recognised subspecies of *O. martiensseni* (Hayman and Hill 1971; Meester et al. 1986).

Distribution

The large-eared free-tailed bat has an unusual distribution. Although occurring widely in east and central Africa, sometimes in large cave colonies of hundreds of individuals (Mutere 1973), its southern African distribution is limited to two isolated records from the north of Zimbabwe, and the isolated population in the Durban area of KwaZulu-Natal, stretching from the coast as far inland as Hillcrest, and from Park Rynie in the south to Ballito in the north (Figure 55). Until recently very few specimens were available and the species was thought to be rare. Bourquin (1988) reported 16 specimens from four localities. Through specimens and roost sightings obtained by C. Sapsford during a bat rabies scare in Durban in 1980, as well as through the efforts of the Durban Bat Interest Group of the Durban Natural Science Museum since 1994, the present distribution is based on 54 specimens and some 14 additional verified roost sightings from some 38 localities (Richardson and Taylor 1995; Durban Bat Interest Group, unpublished data).

Protected areas

None.

Conservation status

Localised to the Durban area but fairly common locally, and not as rare as previously thought. However, two factors motivate for special conservation concern for this species: (1) it is not found in any protected areas, and (2) it occurs in the roofs of Durban houses which are prone to wood-borer attack of roof timbers, and therefore methyl bromide fumigation which kills resident bats. All 30 European species of microchiropteran bats have suffered declines, largely due to roof fumigation and the unsuitability of modern roofs for bat occupation (Tudge 1994). During 1996, this species was officially listed in the 'endangered mammal' category (Schedule 6 of the Natal Conservation Ordinance of 1974) by the provincial government of KwaZulu-Natal, affording it legal protection.

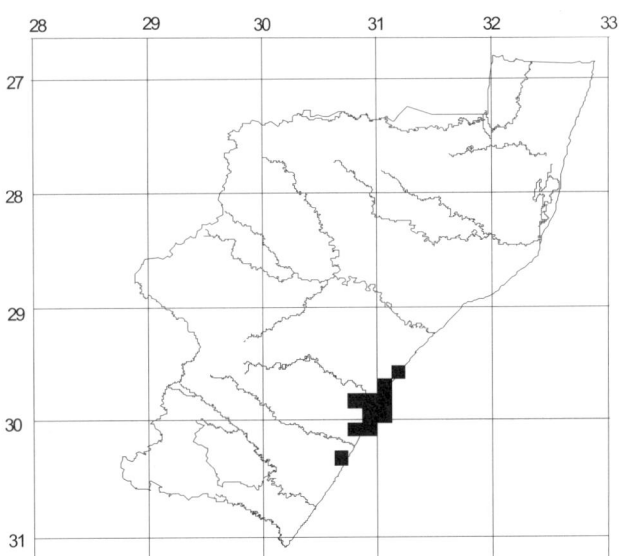

Figure 55

Habitat

It has a wide habitat tolerance in East Africa (Kingdon 1974), but in Durban it seems to be less dependent on habitat type than on the availability of suitable houses for roosting. Based on known colonies of *O. martiennseni* in the Durban area they prefer taller double or triple storey buildings constructed at least 30 years ago (Durban Bat Interest Group, unpublished data). Since roof timbers of pre-1950s houses were not treated with toxic wood preservatives, this may be a factor explaining the suitability of older houses for bat occupation. Older houses also tend to have a steeper roof pitch which provides a greater height of the gable apex from which bats usually exit (most molossid bats seem to require a large drop in order to become airborne). However, not all colonies of *O. martiensseni* were found in older double storey houses, with a few occupying single storey, modern houses.

Habits

In the Durban area, group sizes varied from one to 24, with a mean of 9.8 (*n*=13). In six smaller-sized colonies (group sizes of one to 10) where sex and age of individuals was determined, the following colony compositions were observed: single males, male-female pairs, and harems consisting of a single male and up to four females with young (Durban Bat Interest Group, unpublished data). Individuals hang upside down by the hind claws, with their ventral surface in contact with vertical surfaces such as a wall or rafter. They have been observed roosting on rafters close to the eaves, as well as on the inside of the gable wall, close to the gable apex (Richardson and Taylor 1995).

Otomops martiensseni has been found to feed almost exclusively on moths (97 per cent by volume), especially the larger species (Rydell and Yalden 1997).

Breeding

In a colony that was monitored in Carrington Heights, Durban (Richardson and Taylor 1995), young were born early in December. By the end of January, the young were probably still not weaned as the females were still lactating, while the young had reached 76 per cent of adult mass (mean=23.2 g; *n*=2), with forearm length 96 per cent (mean=60.6 mm; *n*=2) and total length 86 per cent (mean=110.9 mm; *n*=2) of the adult values. In additional colonies in the Durban area, juveniles were observed in October (*n*=1), November (*n*=1), January (*n*=1), February (*n*=1), March (*n*=2) and May (*n*=1) (Durban Bat Interest Group, unpublished data). A pregnant female with a single embryo was collected in September. Four pregnant females in the Durban Natural Science Museum collection were collected in April, suggesting an extended breeding season from September to at least April. These observations agree with Mutere's (1973) report of pregnancies in Kenyan *O. martiensseni* from October to January and in May and June. The young are initially a pink colour, but, by the second month they closely resemble the adult coloration (Richardson and Taylor 1995).

Linear measurements (mm) and mass (g)

	Males					Females				
	x̄	s.d.	n	Min	Max	x̄	s.d.	n	Min	Max
TL	136.3	5.8	3	133	143	129.3	4.2	7	125	136
HB	97.6	11.8	3	88	111	90.5	3.6	7	85	97
T	39.0	6.0	3	33	45	38.8	2.8	7	35	44
Hf	–	–	–	–	–	11	–	2	10	12
E	34.7	4.2	3	30	38	28.7	3.2	6	26	35
FA	66.2	–	2	66	66.3	62.8	0.9	4	61.9	64.1
Mass	32.5	–	1	32.5	32.5	39.9	2.1	4	28.5	33.0

In Kenya, Mutere (1973) recorded mean forearm lengths of approximately 70 mm for both males and females. In contrast, the limited data available for KwaZulu-Natal suggest that the sexes are dimorphic and that local populations exhibit smaller body size generally compared with those in East Africa.

Records of occurrence

SPECIMEN RECORDS:

BM: Durban (type of *O. m. icarus*);

DM: Bluff, Carrington Heights (33 Marshall Road), Clifton School (Durban), Durban, Hillcrest, Kingsburgh (14 St Boniface Maze), Morningside (Hime Road), Mount Edgecombe, Northdene (20 Jan Smuts Avenue), Park Rynie, Roseglen (25 Waller Crescent), Umhlanga (2 Tibia Crescent), Westbrooke, Westville;

KM: Warner Beach;

NM: Durban;

TM: Westville.

ADDITIONAL RECORDS:

Sapsford (unpublished): 30 Baily Road (Avoca), 10 Belgrave Crescent (Durban North), 17 Falmouth Avenue (Glenwood), 89 Fleming Johnson (Umbilo), 6 Glenmore Crescent (Durban North), 10 Haven Road (Westville), 37 Manor Drive (Durban), 62 Old Mill Way (Durban North), 5 Radbourne Road (Warner Beach), Salisbury Island, 14 Trent Place (Westville), 60 Wren Way (Yellowwood Park); Bat Interest Group files: 15 Beaumont Road (Ocean View), 85 Bowen Avenue, 4 Carnarvan Place (Durban North), 3 Chelsea Road (Athlone Park), Clifton School (Morningside), Dirkie Uys High School, 6 Doddington Crescent (Woodlands), Durban Technicon (Berea), 10 Haven Road (Westville), 1 Highgrove Close (Umgeni Park), 1 Hime Road (Morningside), 20 Jan Smuts Avenue (Northdene), 16 Marigny Road (Umbilo), 296 Marine Drive (Brighton Beach), 33 Marshall Grove (Glenmore), Park Rynie, 8 Portland Crescent (Durban North), 49 Prince Street (Amanzimtoti), 6 Stanley Road (Umbilo), 14 St Boniface Maze (Kingsburgh), 2 Tibia Crescent (Umhlanga), 34 Townsend Road (Ballito), 25 Waller Crescent (Morningside).

Genus	*Mormopterus*	Peters 1865

Mormopterus acetabulosus (Hermann 1804)
M. a. natalensis (A. Smith 1847)
Natal free-tailed bat
Natalse losstertvlermuis

Known on the mainland of Africa from a single specimen collected by Andrew Smith in 1833 on the edge of a forest near Durban, KwaZulu-Natal, as well as a further specimen collected from Ethiopia (Hayman and Hill 1971). The species is now almost certainly extinct on the mainland of Africa, occurring on the islands of Madagascar, Mauritius and Réunion. In spite of intensive collecting in the Durban area, no further specimens have been collected since 1833 (Figure 56).

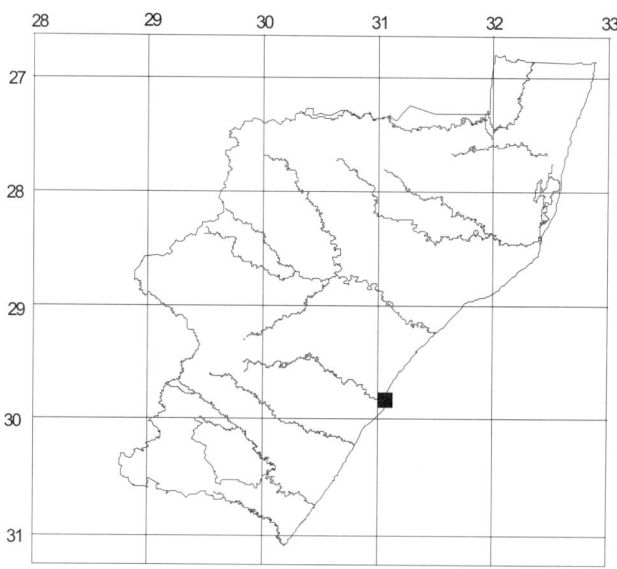

Figure 56

Genus	*Tadarida*	Rafinesque 1814

Tadarida aegyptiaca (E. Geoffroy 1818)   PLATE 21
T. a. aegyptiaca (E. Geoffroy 1818)
Egyptian free-tailed bat
Egiptiese losstertvlermuis

Taxonomic status

Two subspecies were recognised by Meester et al. (1986).

Distribution

The Egyptian free-tailed bat has a wide but scattered distribution in KwaZulu-Natal (Figure 57). While its range overlaps with that of *Chaerephon pumila* and *Mops condylurus*, it does not appear to be as common as the latter two species.

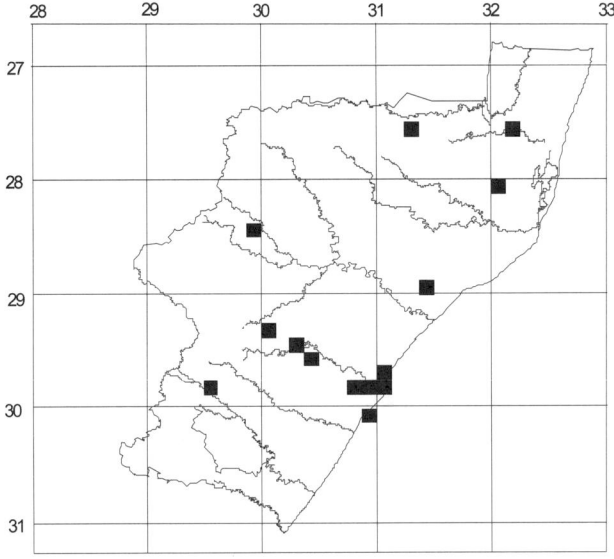

Figure 57

Protected areas

Hluhluwe, Itala, Mkuzi.

Conservation status

Widely distributed in KwaZulu-Natal.

Habitat

Occurs throughout a wide range of habitats and bioregions, from savanna and forest-dominated Valley bushveld, Lowveld and Coast lowlands to grassland-dominated Moist upland and Highland.

Habits

Observed roosting singly in a narrow crack in a sandstone boulder on a mountainside at Itala Game Reserve (Rautenbach et al. 1981), and together as a pair in a horizontal crevice in a small cliff near a stream bank at Hluhluwe Game Reserve (Bourquin et al. 1971). Around Durban, they are frequently found in roofs (though not nearly as commonly as *Chaerephon pumila*), often between the tiles and insulation or between the tiles and rafters, in groups of up to 50 or more. Several specimens have been collected from the central and beachfront residential areas of Durban where they presumably roost in the surrounding high-rise apartments. Although this species is said to prefer roosting alone (Herselman and Norton 1985), they were found on at least two occasions roosting together with *Chaerephon pumila* and on one occasion roosting together with *C. pumila* and *Otomops martiensseni*. Colonies of this species have a very pungent odour.

Breeding

Apart from a juvenile collected in Durban in March, no data on breeding are available for KwaZulu-Natal. From data elsewhere, it appears that a single young is born around November or December (Skinner and Smithers 1990). The species apparently establishes maternity colonies as all-female groups have been found in two colonies in the Western Cape during November (Herselman and Norton 1985).

Linear measurements (mm) and mass (g)

	Males					Females				
	$\bar{x}$	s.d.	n	Min	Max	$\bar{x}$	s.d.	n	Min	Max
TL	–	–	2	102	106	103.7	3.2	3	100	106
HB	–	–	2	70	70	66.3	1.5	3	65	68
T	–	–	2	32	36	37.3	4.7	3	32	41

Records of occurrence

SPECIMEN RECORDS:

DM: Addington (Durban), Durban North (15 Burleigh Crescent, 8 Portland Crescent), Gillitts, Hluhluwe GR (Mansiya Valley), Umhlanga Rocks (2 Tibia Crescent), University of Natal (Durban), Westville;
NM: Himeville, Spring Grove;
TM: Eshowe, Itala GR (Craigadam), Pietermaritzburg, University of Natal campus (Pietermaritzburg).

ADDITIONAL RECORDS:

Sapsford (unpublished): Avondale Road (Durban: 87 Avonmore Centre), Beach Road (Amanzimtoti: 2401 Sanlam Park), 208 Blackburn Road (Durban), 299 Blackburn Road (Durban), 6 Cox Place (Durban), 49 Gleneagles Drive (Durban North), 94 Hospital Road (Durban: Flat 2 Princeton), 23 James Henderson Crescent (Durban), 21 Maple Drive (Morningside), Marianhill (Pinetown), Marine Parade (Durban: 56 Grandborough Court), 47 Melbourne Road (Durban), 20 Morrison Street (Durban: Bosch Electrical), North Ridge Road (Durban: 2150 Kensington), Old Fort Road (Durban: Pavillion Hotel), Prince Street (Durban: 147 South Sands), Snell Parade (Durban: 1506 Grosvenor Court), 107 St Georges Street (Durban: 81 Arusha), Umhlanga Rocks (102 Villa Lax), 359 West Street (Durban: Eagle Building).
Durban Bat Interest Group files: Durban (Beares Furniture Shop: West Street), Durban North (15 Burleigh Crescent), Kloof (24 Surrey Lane), Glenmore (Convent Close), Nottingham Road (Clifton School), Pietermaritzburg (St Georges Church: Devonshire Road), Pinetown (Crompton Road), Westville (11 Beverley Drive).

Genus	*Chaerephon*	Dobson 1874

KEY TO SPECIES (Meester et al. 1986)

1. Anterior palatal emargination well developed, extending beyond upper incisors, and including area of incisive foramina; third commissure of last upper molar well developed, almost as long as second commissure . . . . . *C. ansorgei,* p. 62
—   Anterior palatal emargination greatly reduced, not extending behind upper incisors and separated from incisive foramina by a bony bar, third commissure much less developed than second . . . . *C. pumila,* p. 62

Chaerephon ansorgei (Thomas 1913)
Ansorge's free-tailed bat
Ansorge-losstertvlermuis

Taxonomic status
No subspecies were recognised by Meester et al. (1986). Although Hayman and Hill (1971) and Meester et al. (1986) include this species in the genus *Tadarida,* Freeman (1981) and Koopman (1993) are followed in including it in the genus *Chaerephon.*

Distribution
Only known in KwaZulu-Natal from a single specimen collected at Mkuzi Game Reserve in 1982 (Figure 58). This represents the southernmost distributional record for the species, which is widely though sparsely distributed in Africa as far north as Ethiopia and Central African Republic.

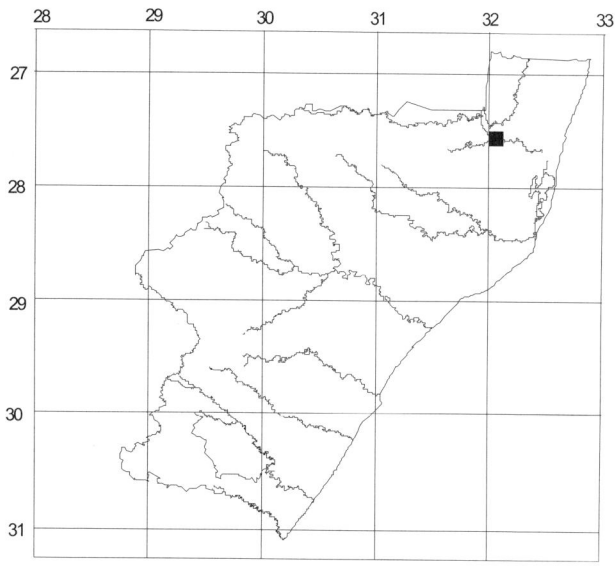

Figure 58

Protected areas
Mkuzi.

Conservation status
Marginal occurrence in KwaZulu-Natal.

Habitat
They apparently roost during the day in rock clefts, caves and mine adits.

Habits
Apparently a gregarious species, occurring in groups of up to hundreds, closely packed together (Skinner and Smithers 1990).

Breeding
No data available.

Linear measurements (mm)
A small species with a total length of about 106 mm and a tail of 36 mm (Skinner and Smithers 1990).

Records of occurrence
SPECIMEN RECORDS:
TM: Mkuzi GR.

PLATE 22
Chaerephon pumila (Cretzschmar 1830–1831)
Little free-tailed bat
Klein losstertvlermuis

Taxonomic status
Many forms of uncertain status have been assigned to this common and widespread species (Hayman and Hill 1971), and because of the confused state of its taxonomy, Meester et al. (1986) recognised no subspecies.

Distribution
Confined to Zululand and the coastal areas of southern KwaZulu-Natal (Figure 59). Skinner and Smithers (1990) gave the KwaZulu-Natal distribution as being confined to the northern parts. However, it has been recorded as far south as Oribi Gorge, and is extremely common in the Durban area. The species occurs continuously southwards down the coast as far south as the Eastern Cape, but an isolated population occurs in the south-west of the Western Cape.

Protected areas
Albert Falls, Coastal Forest Reserve, Eastern Shores, False Bay, Harold Johnson, Hazelmere, Kenneth Stainbank, Mkuzi, Ndumu, Oribi Gorge, St Lucia, Tembe Elephant Park, Umfolozi.

Conservation status
Fairly common throughout Zululand and the coastal areas of southern KwaZulu-Natal, becoming very common in the Durban area.

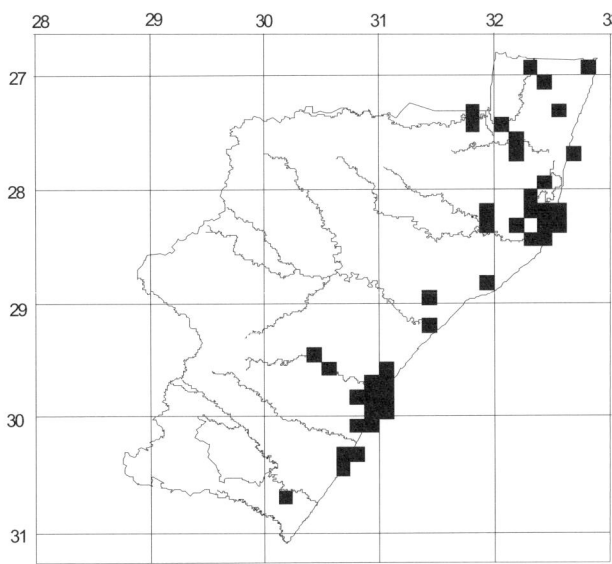

Figure 59

Linear measurements (mm) and mass (g)

	Males					Females				
	x̄	s.d.	n	Min	Max	x̄	s.d.	n	Min	Max
TL	89.7	4.2	15	82	98	85.8	4.7	9	77	93
HB	57.1	3.0	15	51	63	55.0	2.7	7	52	59
T	2.5	3.1	15	28	40	32.1	1.9	8	29	34
Hf*su*	6.6	1.1	5	5	7	5.9	3.4	5	7	9
E	13.8	2.8	10	10	18	13.8	1.2	6	12	15
FA	37.5	1.4	6	36	39	37.2	1.5	5	35	39
*Mass	11.1	0.6	4	10.5	12	11.8	1.7	4	10	14

* Masses from Seamark (personal communication)

Habitat

Restricted largely to the Valley bushveld, Lowveld and Coast lowland bioregions in KwaZulu-Natal (Figure 59). While they seem to prefer forest or savanna woodlands outside built-up areas, their presence in built-up areas is determined solely by the availability of suitable roofs for roosting.

Habits

Owing to the accumulation of guano in roofs, and the noise made by individuals in the roof, this species is frequently regarded as a nuisance by the public. Based on 31 colonies in KwaZulu-Natal excluded from roofs using humane methods by R. Sturgeon of Bat Man Bat Removals and Bat Proofing, colony size varied from around 20 to a maximum of 2 000 to 2 500 bats in buildings at a holiday resort in the St Lucia area. Based on information gathered by the Durban Bat Interest Group, preferred roosting sites of this species include the horizontal surfaces of rafters, between the rafters and brickwork, in the corrugations of corrugated roofs, between the roof tiles and the insulating material underneath (e.g. plastic, tar paper), as well as in air vents. Higher roofs appear to be favoured, and the access points used by little free-tailed bats are frequently the gable apices of roofs, at heights usually in excess of 5 m above the ground. Access points faced east, west or south, but never northwards.

Breeding

Juvenile little free-tailed bats have been collected during May and June in KwaZulu-Natal (**DM**), and pregnant females in December (**TM**) and October (E. Seamark, unpublished data). A sample of 20 individuals collected from a colony in Isipingo during March comprised 19 females and a single male; four of these females were lactating or post-lactating (personal observation), suggesting that the species forms female-dominated maternity colonies. Elsewhere, females of this species are known to give birth to a single young, and to have up to three births per year (early November, late January and early April), a fact attributed

to their ability to undergo post-partum oestrus. Females become sexually mature at the age of 5–12 months (Skinner and Smithers 1990).

Records of occurrence

SPECIMEN RECORDS:

BM: Berea (Durban);

CM: Pennington;

DM: Albert Falls NR, Amanzimtoti (8 Crawford Place), Bellair (80 Glendale Road), Berea, Chatsworth (51 Citizens Avenue), Clifton School (Durban), Durban, Durban North (15 Burleigh Crescent, 8 Portland Crescent), False Bay Park, Fynnland (40 Pitman Road), Glenhills (8 Wisteria Grove), Hluhluwe-Umfolozi Corridor (Masimba Camp), Isipingo (2 Sykes Road), Jozini Dam (Mountain Lake Adventures Campsite), Kenneth Stainbank NR, KwaSheleni, Morningside, Moseley, Renishaw (29 Maryland Road), Sea View (30 Cardiff Road), St Lucia, Tembe Elephant Park, Umbilo (15 Boxley Place), Westville (10 Haven Road);

SI: Makatini Flats, Maputa (Aerodrome, General Store, Police Station), Mkuzi GR (Headquarters);

KM: Ndumu GR, St Lucia, St Lucia Estuary, St Lucia Forest Station, St Lucia Resort, Umdoni Park, Umfolozi GR (Confluence of Black and White Umfolozi Rivers (13.5 km north; 21 km west);

NM: Coastal Forest Reserve (=Kosi Bay), Empangeni, Ndumu Police Station, Umdoni Park;

TM: Cape Vidal, Coastal Forest Reserve (=Kosi Bay), Charter's Creek, Dukuduku Forest Station, Eastern Shores NR (Cape Vidal, Lake Bangazi, Mission Rocks), Futululu Research Station, Jozini Dam, Jozini village, Leeuwspoor Farm, Mseleni, Ndumu GR, Ndumu Store, Nyalazi Forest, Oribi Gorge NR, Seacow Lake (=Northern Sewage Works), St Lucia, Umdoni Caravan Park, Umgeni Hatchery, Ziqhumene.

ADDITIONAL RECORDS:

Sapsford (unpublished): 106 Abrey Road (Pinetown), 4 Albert Road (Pinetown), 45 Albizia Place (Westville), Alexander Road (Pinetown), 5 Arundel Place (Pinetown), BR Construction Company (Inanda: Glen Anil), 2 Bells Avenue (Durban), 2 Bernadotte Street (Amanzimtoti), Botanic Gardens, 33 Canberra Avenue (Durban: Redhill), 3 Carlton Gardens (Durban), 1 Centre Street (Durban), 3 Chiral Drive (Westville), Colin Grove (Durban: Berea), 53 Dan Pienaar

Drive (Pinetown: Northdene), 11 Daventry Place (Durban), 8 Douglas Road (Durban: Redhill), East Street Primary School (New Germany), 55 Essenwood Road (Durban), Fenniscowles Road (Durban: 14 Latham Gardens), 21 Frere Road (Durban: Umbilo), 2 Fyfe Road (Durban: Puntans Hill), 108 Glendale Road (Durban), Goble Road (Durban: 32 Northgate), 4 Harcombe Road (Warner Beach), KwaMashu Police Station, 2 Limpus Road (Pinetown: Sarnia), 35 Linscott (Durban: Athlone Park), 46 Main Road (Durban: Malvern), St Mary's Hospital (Marianhill), 33 Marshall Grove (Durban: Glenmore), 81 Middleton Road (Durban: Malvern), Mount Edgecombe Police Station (Inanda), 543 Musgrave Road (Durban), 12 Oribi Crescent (Pinetown: Sarnia), 330 Park Station Road (Durban: Greenwood Park), 84 Perserverence Road (Durban: Bellair), Pickering Street (Durban: Lionel House), 40 Pidgeon Drive (Durban: Yellowwood Park), Povall Road (Durban: Berea: 3 Lyn Gate), 193 Queen Elizabeth Avenue (Durban: Manor Gardens), Ridge Road (Durban), Ridgeside Road (Durban: 12 Fairlawns), Riley Road (Durban: Overport: 10 Torbeth Lodge), 6 Ronan Road (Inanda: La Lucia), 21 Ryde Place (Durban North), Scottsburgh (Umzinto), 56 Seventh Avenue (Pinetown: Ashley), 5 Snundel Place (Pinetown), 21 Stellawood Road (Durban: Umbilo), Teignmouth Road (Durban: Umbilo: 31 Merilynn), 10 Tilbury Avenue (Durban: Rossborough), 892 Umgeni Road (Durban: Lion Match Company), 117 Underwood Road (Pinetown), 41 Villa Road (Durban: Sydenham: Flat 27), 801 West Street (Durban: Bales Building), 90 Wood Road (Pinetown: Moseley Park);

Seamark (unpublished): Harold Johnson NR.

Genus	*Mops*	Lesson 1842

Mops condylurus (A. Smith 1833)   PLATE 23
M. c. condylurus
Angola free-tailed bat
Angola-losstertvlermuis

Taxonomic status
Only the nominate subspecies of this species is known to occur in southern Africa (Meester et al. 1986).

Distribution
Zululand and Maputaland regions of northern KwaZulu-Natal, occurring within Valley bushveld, Lowveld and Coast lowland bioregions (Figure 60).

Protected areas
Coastal Forest Reserve (=Kosi Bay), Eastern Shores (Cape Vidal), Mkuzi, Ndumu, St Lucia, Umfolozi.

Conservation status
Common in Zululand; it is regarded as a pest in houses and other buildings in built up areas and tourist accommodation.

Habitat
Lowveld, Valley bushveld and Coast lowland bioregions.

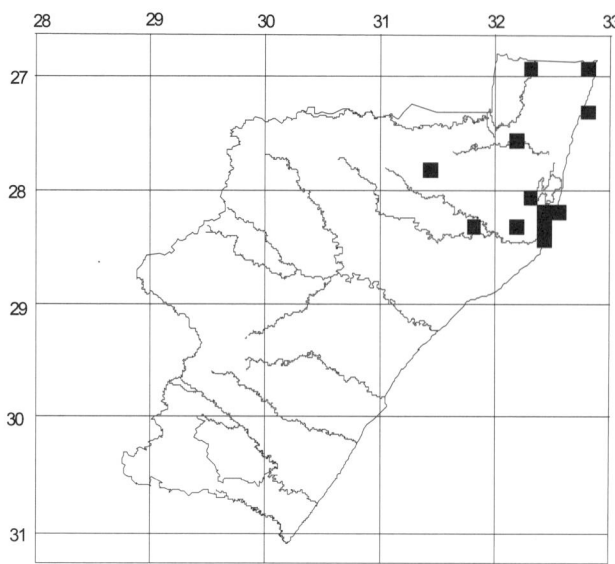

Figure 60

Habits
Very common in buildings in towns and hutted accommodation in nature reserves in Zululand, where their smell and the accumulation of guano may become a nuisance to residents. They roost in any available spaces provided by rafters, brickwork, timber poles, or roof thatch. A maximum of 26 bats were removed from a single roof in series collected in Zululand and Maputaland by I. L. Rautenbach (specimens in **TM**).

Breeding
Pregnant females were collected during January and December in KwaZulu-Natal (specimens in **TM**). In Mpumalanga, males are reproductively active for nine months; females undergo two reproductive cycles between September and early May. The first birth is followed by a post-partum oestrus during which females become pregnant again whilst still lactating. Gestation is 85 days (Vivier and Van der Merwe 1996, 1997).

Linear measurements (mm)

Cat. No.	Sex	TL	HB	T	Hf*su*	E	FA
DM1087	F	105	66	39	14	17	46
DM1096	F	101	66	35	11	18	48
DM3999	F	111	65	44	12	15	–
DM1098	M	114	82	32	18	21	49

Records of occurrence
SPECIMEN RECORDS:
DM: Bonamanzi GR, St Lucia Post Office;
KM: Ndumu, St Lucia Resort;
SI: Maputa;
TM: Dukuduku Forest, Eastern Shores NR (Cape Vidal, Mission Rocks), Kosi Lake (Department of Health Camp), Lake Sibayi Research Station, Madlangula, Mkuzi GR, Ndumu GR, Ndumu Store, Ngome Forest Reserve, Nyalazi Forest, St Lucia village, Umfolozi GR (Mpila Trails Store).

Order PRIMATES

KEY TO SUBORDERS AND GENERA (Meester et al. 1986)

1. Postorbital bar not forming a bony plate between orbit and temporal fossa (Suborder STREPSIRHINI) 2
   - Postorbital bar forming a bony plate between orbit and temporal fossa (Suborder HAPLORHINI) 3

2. Larger, head and body length more than 230 mm . . . . *Otolemur,* p. 65
   - Smaller, head and body length less than 230 mm . . . . *Galago,* p. 66

3. Size larger, greatest skull length greater than 132 mm; face very elongated, with terminal nostrils; very broad ischial callosities; tail shorter than head and body (Tribe Papionini) . . . . *Papio,* p. 66
   - Size smaller, greatest skull length less than 130 mm; face shorter, nostrils not terminal; ischial callosities small and rounded; tail as long as or longer than head and body (Tribe Cercopithecini) 4

4. Face pure black; upper parts grizzled; outer surface of arms not black . . . .
   Chlorocebus, p. 68
   - Face brownish; upper parts brownish on shoulders then posteriorly reddish-brown; outer surface of arms black . . . .
   Cercopithecus, p. 69

Suborder	STREPSIRHINI	Bushbabies
Family	LORISIDAE	
Genus	*Otolemur*	Coquerel 1859

Otolemur crassicaudatus (E. Geoffroy 1812)
O. c. crassicaudatus (E. Geoffroy 1812)
Thick-tailed bushbaby
Bosnagaap

Taxonomic status
Meester et al. (1986) recognised two subspecies in the subregion, of which only *O. c. crassicaudatus* is found in KwaZulu-Natal.

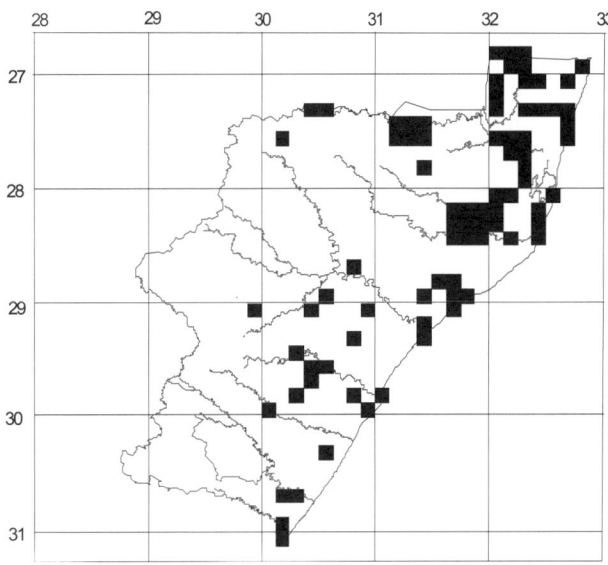

Figure 61

Distribution
Widespread in lower-lying Valley bushveld, Lowveld, Coast lowlands and Coast hinterland bioregions in Zululand, Maputaland and the coastal regions of KwaZulu-Natal south of Lake St Lucia, penetrating inland up major river valleys to about as far west as 30° east (Figure 61). Absent from higher-altitude Upland, Highland and Montane bioregions.

Protected areas
Cape Vidal, Dlinza, Eastern Shores, Enseleni, Entumeni, False Bay, Harold Johnson, Hluhluwe, Itala, Kenneth Stainbank, Mapelane, Mhlatuze, Mkuzi, Ndumu, Oribi Gorge, Pongola Bush, Sodwana Bay, St Lucia, Umfolozi, Umlalazi, Umtamvuna, Vernon Crookes.

Conservation status
Although the thick-tailed bushbaby appears to be widespread in low-lying, well-wooded areas in KwaZulu-Natal, Bourquin (1988) considered that fragmentation and shrinkage of its range must have resulted from the widespread destruction of its habitat, particularly in the coastal areas south of the Tugela River.

Habitat
They prefer low-lying, well-wooded habitats asssociated with the Valley bushveld, Lowveld, Coast lowlands (coastal forest) and Coast hinterland bioregions, and are absent from the grasslands associated with the Upland, Highland and Montane regions.

Habits
Thick-tailed bushbabies are nocturnal, resting in trees during the day, either singly, or in groups of two to six. Olfactory communication occurs through spreading urine on the hands, feet and substrate, as well as by rubbing their chests (on which glands are located) against trees and against

other individuals. While normally quiet, they are well known for their loud wailing cries which can carry over long distances. Some 18 calls have been recorded (Bearder 1997). They feed on fruit, invertebrates, small vertebrates, gum, soft leaves and flowers.

Breeding
A pregnant female with two foetuses was collected at Lake Sibayi in January (**TM** specimen). Thick-tailed bushbabies are seasonal breeders, and young (1–3) are born at the first rains, after a gestation period of 132 days (Bearder 1997).

Linear measurements (mm) and mass (g)

	||||| Males					Females				
	x̄	s.d.	n	Min	Max	x̄	s.d.	n	Min	Max
TL	681.4	23.9	10	640	710	619.0	25.4	3	590	637
HB	324.8	46.0	9	280	420	294.0	5.3	3	290	300
T	359.2	44.6	10	280	430	325.0	22.9	3	300	345
E	66.3	7.1	3	60	74	–	–	–	–	–

Males are somewhat larger than females, as was found in greater Transvaal (Rautenbach 1982). Both in this study and in that by Rautenbach (1982), twice as many males as females were represented in the sample, suggesting that females may be more shy (less frequently encountered) than males.

Records of occurrence
SPECIMEN RECORDS:
BM: Ngoye Hills;
DM: Berea (Durban), Durban, Harold Johnson NR, Mandini, Mfongosi, Umlalazi NR, Vernon Crookes;
NM: Ashburton Farm, Claridge (Pietermaritzburg), Gwaliweni Forest, Hluhluwe GR, Ingwavuma Gorge, Mseleni, Ntombeni, Otto's Bluff (Pietermaritzburg), Pietermaritzburg, Richmond, Seaforth Farm, Shemula's Pont (Pongola River), Table Mountain;
SAM: Mfongosi, Umfolozi Station;
TM: Bumbeni, Coastal Forest Reserve (=Kosi Bay), Hluhluwe GR, Ingwavuma, Ingwavuma River, Itala GR (Doornpan), Lake Sibayi Research Station, Lake St Lucia Estuary, Manaba, Ndumu GR, Ngoye Hills, Pongola River, Sihangwane, Sodwana Bay, Tete Pan, Ubombo, Umfolozi GR, White Umfolozi.

ADDITIONAL RECORDS:
Bruton (1978): Banda Banda Bay (east shore of Lake Sibayi), Dunule (east shore of Lake Sibayi), Mabibi, Mgobozeleni Lake;
Sight records (KwaZulu-Natal Nature Conservation Service): Aangelegen, Cape Vidal NR, Charter's Creek, Corridor between Umfolozi and Hluhluwe, Dlinza NR, Enseleni GR, Entumeni NR, False Bay NR, Harold Johnson NR, Hluhluwe GR, Itala GR, Kenneth Stainbank NR, Larne, Mapelane NR, Mkuzi GR, Mount Elias, Ndumu GR, North

Park, Oribi Gorge NR, Pongolabush NR, Saxony, Sodwana Bay NR, Springfield, The Start, Umfolozi GR, Umlalazi NR, Umtamvuna NR, Wendy Hill.

Genus	*Galago*	E. Geoffroy 1796

Galago moholi (A. Smith 1836)
Lesser bushbaby
Nagapie

In 1989 two *Galago moholi* were seen in a tree at Sihangwana in the garden of the Officer in Charge of Tembe Elephant Park in north-eastern Maputaland on the border between KwaZulu-Natal and Mozambique. They were subsequently seen and heard regularly over a period of several years. There have also been reports from residents of this species occurring in dry sand forests east of the Pongola River as far east as Muzi Swamp (Kyle 1996). Prior to these reports, there was no uniquivocal evidence of their ocurrence in KwaZulu-Natal, although it was suspected by Bourquin (1988).

The species is listed in the South African Red Data Book (Smithers 1986).

Suborder	HAPLORHINI	
Family	CERCOPITHECIDAE	
Tribe	PAPIONINI	Baboons
Genus	*Papio*	Erxleben 1777
Subgenus	*Papio*	Erxleben 1777

Papio hamadras (Linnaeus 1758)   PLATE 24
P. h. ursinus (Kerr 1792)
Chacma baboon
Kaapse bobbejan

Taxonomic status
Although previous authors have considered *ursinus* to be a distinct species (e.g. Meester et al. 1986), Groves (1993) is followed in including it as a subspecies of *P. hamadras*.

Distribution
Widespread in KwaZulu-Natal (Figure 62), with apparent gaps in the overall distribution corresponding either to populous urban areas (e.g. Durban and Pietermaritzburg) or areas of former KwaZulu where surveys or collecting have not occurred.

Protected areas
Cathedral Peak, Cobham, False Bay, Garden Castle, Giant's Castle, Highmoor, Hluhluwe, Itala, Kamberg, Loteni, Mkhomazi, Mkuzi, Monk's Cowl, Oribi Gorge, Pongola Bush, Royal Natal, Sodwana, Umfolozi, Umtamvuna, Vergelegen, Weenen.

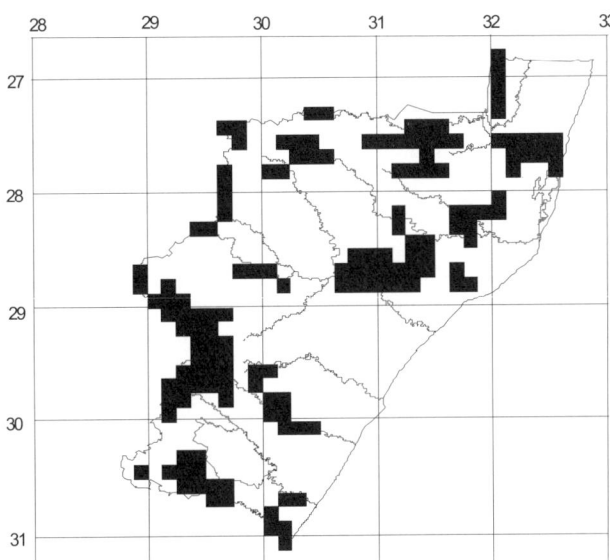

Figure 62

Conservation status

The Chacma baboon is adequately protected in reserves (Bourquin 1988; Figure 62; Table 2). However, because their presence is incompatible with most forms of farming, they are often shot on sight by landowners.

Habitat

They seem to occur extensively throughout both well-wooded and more open grassland and montane habitats, usually associated with mountainous or rocky situations.

Habits

Baboons are gregarious and live in troops which vary from a few animals to up to 130 in optimal situations. Troops studied in the KwaZulu-Natal Drakensberg had relatively smaller troop sizes (mean=22; n=61 troops; Henzi and Lycett 1995) and lower densities (home ranges of 10 km^2 to 18.9 km^2; Whiten, Byrne and Henzi 1987) than baboons from savanna regions. Troops consist of both sexes of all ages. A definite dominance hierarchy exists. No permanent pair bonds exist, and receptive females may be mated by all the males in the troops. A single female may be mated up to 100 times by different males during an oestrous cycle. A female in oestrus is mated first by the less dominant males and later, at the height of oestrus, by the dominant males. The oestrous cycle is clearly marked by swelling of the sexual skin around the ischial callosities. Baboons are omnivorous, with plant material (fruit, leaves, grass) making up the main component of their diet, and insects featuring throughout the year. They are known to eat smaller animals such as reptiles, birds, and small mammals. In the Drakensberg, baboons rely heavily on foraging for subterranean resources (bulbs and corms) to survive the period of food scarcity in winter (Whiten, Byrne and Henzi 1987).

Table 2. Population estimates of some KwaZulu-Natal baboon populations (NPB = former Natal Parks Board)

Population	Troop nos.	Ind. nos.	Data source
Lebombo Mts	5+	110	Pringle 1974, NPB 1986
Mkuzi GR	?	500	NPB 1995
Hluhluwe-Umfolozi	191	4 202	Pringle 1974
Ntambanana	1	22	Pringle 1974
Ngoye	2	44	NPB 1981
Nkandhla/Qudeni	40	880	NPB 1986
Melmoth/Ulundi	7	154	NPB 1986
Denny Dalton	4	88	NPB 1986
Magudu	6	132	NPB 1986
Morquenzon	1	22	NPB 1986
Magdalena/Ngome	3	66	Pringle 1974
Rendsburg	1	22	NPB 1981, 1986
Itala	?	100	NPB 1995
Dwaalhoek	2	44	NPB 1986
Balelesberg	9	198	NPB 1986
Pongola Bush	?	75	NPB 1995
Majuba	9	198	NPB 1986
Mullers Pass	9	198	NPB 1986
Nomandien	7	154	NPB 1986
De Beers Pass	1	22	NPB 1986
Drakensberg Park	335	7 540	Henzi and Lycett 1995
Tabamnyama	2	44	NPB 1986
Inhluzane	3	66	NPB 1986
Taylor's Ridge	1	22	NPB 1986
Amahwaqa	1	22	NPB 1986
Hella Hella	1	10	NPB 1985
Platt Estates	1	5	NPB 1985
Oribi Gorge	?	90	NPB 1986
Umtamvuna	?	200	NPB 1986
Ngeli/Weza	4	88	NPB 1986
Mount Currie	?	30	NPB 1986
Rusfontein	–	1	NPB 1986
Ncandu	?	100	NPB 1995

Breeding

Inter-birth interval is typically 24 months, up to 38 months in the Drakensberg population (Henzi, personal communication). Young (usually one per birth) are born throughout the year after a six-month gestation (Skinner and Smithers 1990).

Mass (kg)

Masses (kg) for a series from Clearwater Farm in the Umtamvuna Gorge on the southern border of KwaZulu-Natal are summarised below (this study), and compared with published data for three other localities in KwaZulu-Natal (Barrett and Henzi 1997).

Baboons from Umtamvuna Gorge and Giant's Castle Game Reserve have a lower mean mass than baboons from Zululand, and from Botswana (Smithers 1971), but are similar in mass to baboons from the greater Transvaal (Rautenbach 1982).

		Males					Females			
x̄	s.d.	n	Min	Max		x̄	s.d.	n	Min	Max
Umtamvuna:										
22.6	7.1	17	13.0	31.0		15.6	2.7	16	11.0	19.0
Umfolozi:										
28.8	–	2	26.2	31.5		16.3	–	2	15.1	17.5
Hluhluwe:										
30.3	–	2	27.0	33.5		15.5	–	3	14.5	17.1
Giant's Castle:										
23.0	–	6	20.5	26.4		15.9	–	1	15.9	15.9

Records of occurrence

SPECIMEN RECORDS:
DM: Clearwater Farm, Mkuzi Game Reserve;
SI: Mkuze River, Umsunduri River;
TM: Mkuzi River (near Ubombo).

ADDITIONAL RECORDS:
Pringle (1974): See Table 2;
Rautenbach et al. (1980): Gwaliweni, Ingwavuma Gorge, Mkuzi GR, Ubombo;
Sight records (KwaZulu-Natal Nature Conservation Service files): See Table 2.

Tribe	**CERCOPITHECINI**	Monkeys
Genus	***Chlorocebus***	**Gray 1870**

Chlorocebus aethiops (Linnaeus 1758)   PLATE 25
C. a. pygerythrus (F. Cuvier 1821)
C. a. cloeti Roberts. 1931
Vervet monkey
Blouaap

Taxonomic status

Of the two subspecies in KwaZulu-Natal, *C. a. cloeti* occurs in Zululand and *C. a. pygerythrus* occurs throughout the rest of the province (Meeeter et al. 1986).

Distribution

Vervet monkeys are widespread in KwaZulu-Natal wherever some form of tree cover is present, being largely absent from grassland-dominated habitats associated with the Drier upland, Highland, and Montane bioregions (Figure 63).

Protected areas

Figures in parentheses represent 1995 population estimates, courtesy of KwaZulu-Natal Nature Conservation Service.

Albert Falls (6), Beachwood Mangroves (20), Bluff (30), Cape Vidal, Chelmsford, Dlinza Forest (150), Doreen Clark (15), Eastern Shores, Enseleni (150), Entumeni (100), False Bay (200), Gxalingenwa (10), Harold Johnson (60), Hazelmere (360), Itala (400), Kenneth Stainbank (80), Krantzkloof (60), Mhlatuze, Mkuzi, Mpenjati (30), North Park (40), Oribi Gorge (150), Pongola Bush, Queen Elizabeth Park (10), Skyline (30), Soada Forest (150), Sodwana Bay, Sodwana State Forest, Umfolozi, Umhlanga Lagoon (50), Umlalazi (50), Umtamvuna (200), Vernon Crookes (50), Vryheid (30).

Conservation status

Fairly common and widespread in savanna and woodland habitats in the province.

Habitat

Wooded savanna to coastal and riverine forest, including suburban situations.

Habits

Vervet monkeys are diurnal and gregarious, occurring in heterosexual troops of up to 38 including more than one unrelated male in a troop. There is a clear dominance hierarchy within a troop with dominance being established by threat and aggression. Aggression, signalled by a variety of displays, includes head bobbing, specific postures, eyelid flashing and genital signalling. Grooming between individuals plays an important part in troop cohesion. The diet is primarily vegetarian, including wild fruits, flowers, leaves, seeds and seed pods, although they take a greater proportion of invertebrates than do samangos (Lawes, personal communication). They are known to raid farm crops and suburban gardens, and often become a public menace in built up areas such as Durban and Pietermaritzburg.

Breeding

No data for KwaZulu-Natal. Young (usually one but sometimes twins) are born throughout the year after a gestation of 150–160 days.

Linear measurements (mm)

		Males					Females			
	x̄	s.d.	n	Min	Max	x̄	s.d.	n	Min	Max
TL	1136.8	72.9	4	1055	1217	1000.7	102.7	6	914	1142
HB	512.5	38.4	4	470	545	464.3	69.5	6	386	558
T	624.2	44.2	4	565	672	536.3	43.4	6	470	584

Males are larger than females. The range of values recorded for vervet monkeys from KwaZulu-Natal falls within the range recorded for greater Transvaal (Rautenbach 1982).

Records of occurrence

SPECIMEN RECORDS:

DM: Belleview Farm (near Southbroom), Burman Bush (Durban), Hlatini Rose Farm, Link Road (Pietermaritzburg), Oribi Gorge NR, Richmond, Stellawood (Durban), Umgeni Valley, Umhlanga Bush;

NM: Eastern Shores, Elandslaagte, Glenbella, Mkuzi GR, Mount Edgecombe, Mseleni, Ngoye Forest, Richmond, Town Bush, Umtwalume, Wartburg;

SAM: Durban;

TM: Coastal Forest Reserve (=Kosi Bay), Ingwavuma, Lake Sibayi Research Station, Mkuzi River, Nondwana Falls (Durban), Pongola River, Tete Pan, Umfolozi GR, White Umfolozi River.

ADDITIONAL RECORDS:

Bourquin et al. (1971): Hluhluwe-Umfolozi Park;
Bourquin and Mathias (1984): Oribi Gorge NR;
Bourquin and Sowler (1980): Vernon Crookes NR;
Dixon (1966): Ndumu GR;
Rautenbach et al. (1981): Itala GR;
Sight records (KwaZulu-Natal Nature Conservation Service, excluding unnamed grid square records): Arthur's Seat, Baviaankloof, Blinkwater, Cape Vidal, Chelmsford Dam, Debeerspass, Dlinza NR, Eastern Shores, Entumeni NR, Estcourt, Gretna, Itala GR, Hazelmere NR, Lowlands, Mapelane, Soada Forest NR, Umtamvuna NR.

Tribe	**CERCOPITHECINI**	**Monkeys**
Genus	*Cercopithecus*	**Linnaeus 1758**

Cercopithecus mitis (Wolf 1822)   PLATE 26
C. m. labiatus (I. Geoffroy 1843)
C. m. erythrarchus (Peters 1852)
Samango monkey
Samango-aap

Taxonomic status

Two subspecies are recognised in KwaZulu-Natal, *C. m. labiatus* from the midlands, South Coast and parts of Zululand and *C. m. erythrarchus* from Zululand as far south as the Umfolozi River (Lawes 1990a; Meester et al. 1986). Lawes (1990a) suggested that *C. m. labiatus* represents an ancient radiation of the species associated with expansion of Afromontane forests and associated woodlands during wetter interglacial to pre-glacial times (10 000–20 000 years BP), while *C. m. erythrarchus* represents a much later radiation associated with the establishment of coastal lowland forests on the Mozambique coastal plain (6 000 years BP).

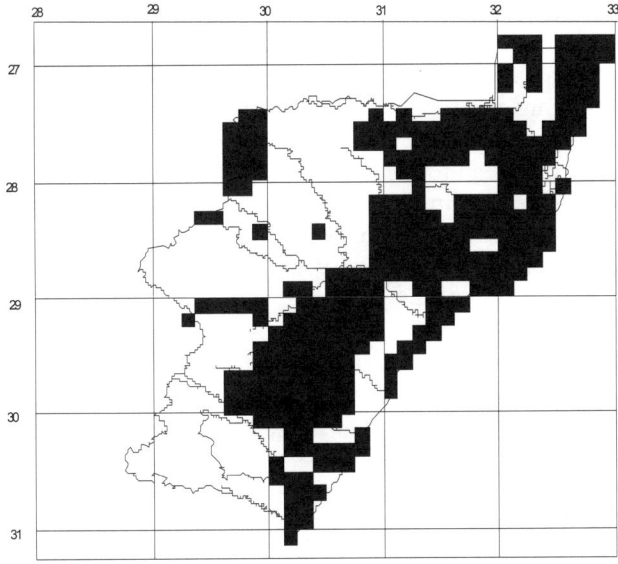

Figure 63

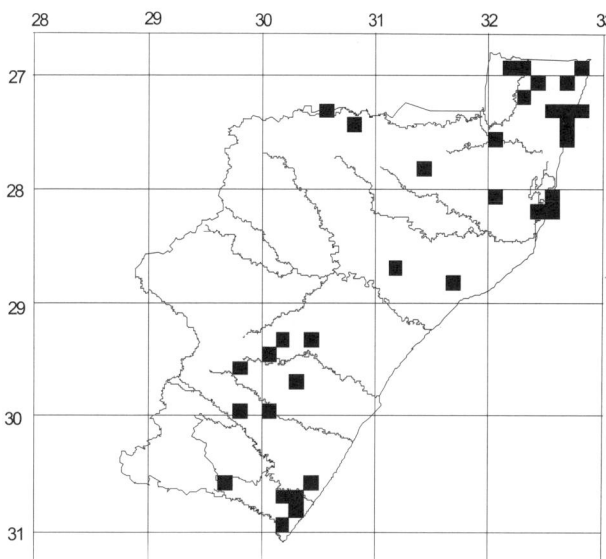

Figure 64

Distribution

Scattered distribution in the midlands, southern KwaZulu-Natal, Zululand, and northern KwaZulu-Natal, associated with the location of suitable forest habitats (Figure 64; Table 3; Lawes 1990b, 1992).

Protected areas

Eastern Shores, Hluhluwe, Itala, Karkloof, Mhlatuze, Mkuzi, Oribi Gorge, Pongola Bush, Soada Forest, Sodwana Bay, Sodwana State Forest, St Lucia Park, Umtamvuna.

Conservation status

According to Bourquin (1988) the samango is in a generally precarious position, relying on the continued existence of limited forest patches. Samangos are found in 54 indigenous forest patches totalling 51 647 ha, or 57 per cent of KwaZulu-Natal's indigenous forest area (Lawes 1992).

Habitat
Restricted to indigenous forests within Coast lowland, Coast hinterland, and Mistbelt bioregions, with *C. m. labiatus* occurring in Afromontane forest habitats and *C. m. erythrarchus* occurring in coastal forest habitats. The species exhibits a broad forest habitat tolerance, with viable populations occurring on all forest types from sand forest to mature *Podocarpus* forest at Pongola Bush and swamp forest at Kosi Bay (Lawes 1990b, 1992).

Habits
Samango monkeys are diurnal and gregarious, living in troops of between four and 30 individuals. Troops normally contain a single male. The size of four troops studied at Cape Vidal and Ngoye Forest in KwaZulu-Natal varied from 16 to 35 (Lawes, Henzi and Perrin 1990). As visual communication is difficult in forest habitat, they use a wide range of vocal signals. Adult males infrequently emit loud calls including low frequency, resonant 'boom' and sharp, hacking 'pyow'. Females and immature samango monkeys emit a variety of sounds including bird-like chirps and click alarm calls, soft grunts and trilling contact calls (Lawes 1997). The diet consists primarily of fruits (50 per cent), leaves (25 per cent); also flowers (13 per cent) and insects (6 per cent).

Breeding
Unlike baboons and vervet monkeys, samango monkeys are seasonal breeders. At Cape Vidal in KwaZulu-Natal, young were born between October and December (Henzi and Lawes 1987). Mating behaviour usually takes place over a 10-week period. Infants are usually born during the period of early summer rains (Swart and Lawes 1996).

Linear measurements (mm) and mass (g)

	Males					Females				
	$\bar{x}$	s.d.	n	Min	Max	$\bar{x}$	s.d.	n	Min	Max
This study:										
TL	1210.0	115.6	4	1053	1320	–	–	1	1016	1016
HB	546.8	73.7	4	445	600	–	–	1	444	444
T	663.2	45.8	4	608	720	–	–	1	572	572
Mass	6020	2389	4	2720	8180	–	–	1	4320	4320

Lawes unpublished (in Skinner and Smithers 1990), Cape Vidal:

HB	595	–	5	564	616	468	–	2	–	–
T	707	–	5	680	730	591	–	2	–	–
Mass	7160	–	5	6200	7800	4450	–	2	–	–

Records of occurrence
SPECIMEN RECORDS:
NM: Donnybrook, Eastern Shores, Hluhluwe GR, Lake Sibayi, Ngome Forest, Ngoye Forest, Ntombeni, Oribi Gorge NR, Pongola Bush;
TM: Bracco, Eastern Shores NR (Cape Vidal), Karkloof Forest, Lake Sibayi Research Station, Tete Pan.

ADDITIONAL RECORDS:
See Table 3 (Lawes 1990b, 1992).

Table 3.   Population estimates for some populations of samango monkeys in KwaZulu-Natal (Lawes 1990b, 1992) (NPB = former Natal Parks Board; KNDC = former KwaZulu Directorate of Nature Conservation)

Property	Owner	Lat/long	Population
Ngome State Forest	State	2752/3123	2700
Karkloof	Farm/NPB	2918/3014	300
New Forest	Farm	2929/2953	38
Pongola Bush	NPB	2719/3029	230
Cape Vidal	NPB	2808/3233	3700
Manguzi	KDNC	2658/3246	415
Dukuduku	NPB/State	2822/3221	3200
Hluhluwe	NPB	2805/3202	90
Umtamvuna	NPB	3102/3012	320
Oribi Gorge	NPB	3042/3014	580
Ngoye	KDNC	2848/3135	150
Kosi Bay	KDNC	2701/3246	1300
Sihangwane	KDNC	2700/3230	1060
Ndumu	KDNC	2654/3220	80
Mkuzi Gorge	NPB	2731/3202	20

Order   HYRACOIDEA

KEY TO GENERA (After Meester et al. 1986)

1. Molars hypsodont; P_1 sometimes absent; length of P^1– P^4 much less than length of M^1–M^3; mammae one pair pectoral, two pairs inguinal; temporal ridges in close contact or form a sagittal crest . . . .
    Procavia p. 70
—   Molars brachyodont; P_1 always present; length of P^1–P^4 exceeding length of M^1–M^3, usually one pair of inguinal mammae; skull concave dorsally, temporal ridges beaded and far apart . . . .
    Dendrohyrax, p. 72

Family	PROCAVIIDAE	Dassies
Genus	*Procavia*	Storr 1780

Procavia capensis (Pallas 1766)   PLATE 27
Procavia capensis capensis (Pallas 1766)
Rock dassie
Klipdas

Taxonomic status
Only one subspecies is currently recognised from southern Africa. There is much uncertainty about the number and

status of subspecies (Meester et al. 1986), with earlier authors recognising as many as five distinct species (Bothma 1971). Based on a series of 32 skins from KwaZulu-Natal, Pringle (1974) described a colour gradient, uncorrelated with distribution, from yellowish-grey to brownish-black and concluded that no subspeciation has occurred in KwaZulu-Natal.

Distribution
Widespread in KwaZulu-Natal, excluding north-eastern Zululand (Figure 65).

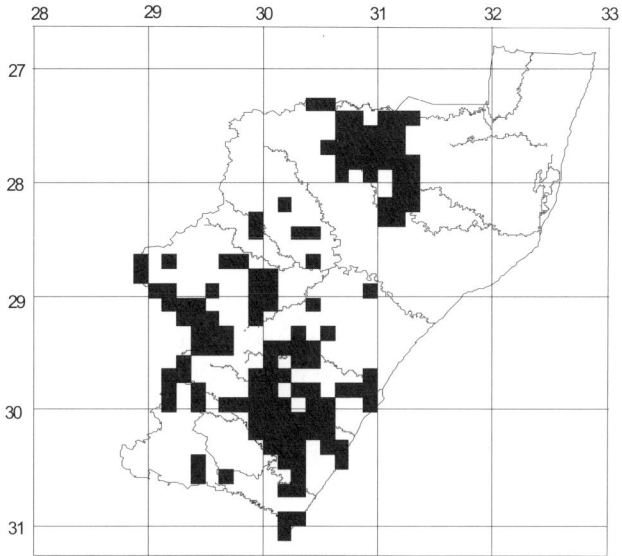

Figure 65

Protected areas
Albert Falls, Cathedral Peak, Cobham, Coleford, Garden Castle, Giant's Castle, Itala, Kamberg, Karkloof, Krantzkloof, Loteni, Mkhomazi, Monk's Cowl, Mount Currie, Oribi Gorge, Pongola Bush, Queen Elizabeth Park, Royal Natal, Soada Forest, Spioenkop, Umtamvuna, Vernon Crookes, Weenen.

Conservation status
Common and widespread in KwaZulu-Natal.

Habitat
Krantzes, cliffs, rocky outcrops and piles of rocks in association with bushes and trees and other food plants. They can colonise large donga formations or artificial structures such as culverts.

Habits
Diurnal and gregarious. They occur in harems of one male with up to 17 females. Colony size varies, according to the extent of rocky habitat available, from four to over 100. The species has a wide vocal repertoire, with 21 distinct vocal sounds recorded (Fourie 1977). Secretions of the dorsal gland play a role in reproductive behaviour and in mother-infant relationships. Rock dassies eat a variety of grasses, forbs and shrubs including species which are poisonous to other animals.

Breeding
Based on specimen records in the Natal Museum, six out of eight lactating females were collected in January and two in February. Juveniles were recorded in the months of October (one), November (one), December (one), January (two) and February (two). A pregnant female with two foetuses was collected in December, while one with three foetuses was collected in August. This evidence suggests a birth period between August and February. These data are consistent with data summarised by Skinner and Smithers (1990) indicating a birth season from September to October in the Western Cape, becoming progressively later in the year with decreasing latitude (October to November in the Eastern Cape, November to December in the Free State, December to February in greater Transvaal, and March to April in Zimbabwe). Gestation is 230 days, and litter size varies from one to six.

Linear measurements (mm) and mass (kg)

	Males					Females				
	$\bar{x}$	s.d.	n	Min	Max	$\bar{x}$	s.d.	n	Min	Max
HB	480.0	29.4	6	450	522	519.0	61.9	19	344	610
Hf	55.3	9.8	4	47	69	64.5	6.4	2	60	69
E	28.8	3.7	6	23	33	27.7	3.0	3	25	31
Mass	2.7	0.7	9	1.9	4.1	3.2	0.5	20	2.0	4.1

Females are slightly larger than males, contrary to evidence from the Free State (Lynch 1983) and greater Transvaal (Rautenbach 1982), where males average slightly larger than females.

Records of occurrence
BM: Estcourt;
DM: Belleview Farm, Inanda, Ketelfontein, Kloof, Malvern (Fifth Avenue), Pietermaritzburg, Roseleigh Farm, Royal Natal NP, Shongweni Dam, Vernon Crookes NR;
KM: Dundee, Giant's Castle GR;
NM: Carter's Nursery (Pietermaritzburg), Chase Valley (Pietermaritzburg), Dartford Farm, Eensaam, Geluk Farm, Giant's Castle GR, Hilton, Kambula, Kilgobbin Farm, Langewacht, Linwood, Loskop, Malden, Maritzdaal, Mike's Pass, Pietermaritzburg, Pinnacles (Hilton College), Redlands, Spring Grove, The Hoek, Town Bush;
TM: Goodhope Farm, Kilgobbin, Oribi Gorge NR, Royal Natal NP, Vernon Crookes NR.

ADDITIONAL RECORDS:
Rautenbach et al. 1981: Itala GR;
Sight records (KwaZulu-Natal Nature Conservation Service, excluding unnamed grid square records): Albert Falls, Ashley, Baynesfield Estate, Beinn Mheadon, Carlisle, Dolombe, Elands Hoek, Four Gates, Harden Heights, Hope Dale, Kai Ora, Kleinthal, Llangollen, Mount Currie NR, Naawpoort, Newton, Sevenfountains, Sunnyside, The Reef, The Rockeries, Umtamvuna NR, Weenen NR, Welverdiend.

Genus	*Dendrohyrax*	Gray 1868

Dendrohyrax arboreus (A. Smith 1827)   PLATE 28
D. a. arboreus (A. Smith 1827)
Tree dassie
Boomdas

Taxonomic status
Bothma (1971) recognised eight subspecies throughout Africa, of which only the nominate race occurs in southern Africa.

Distribution
Restricted to suitable forest patches in the midlands and South Coast regions to the Weza Forest (Figure 66).

Protected areas
Karkloof, Oribi Gorge and Vernon Crookes.

Conservation status
Although localised in distribution and generally uncommon, recent surveys employing tape recordings of tree dassie vocalisations to elicit responses has revealed their distribution to be somewhat wider than previously supposed (N. Langley, unpublished data; Figure 66).

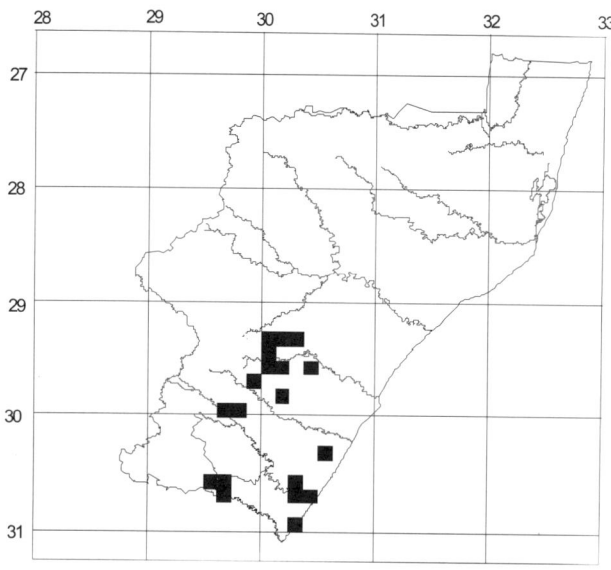

Figure 66

Habitat
Well-developed and undisturbed evergreen coastal forest and midland forests, of probably at least 300 ha in area (Bourquin 1988).

Habits
Unlike the rock dassie, they are solitary and nocturnal. Both sexes emit a loud vocalisation, starting off with a series of cackling barks followed by piercing screams which rise to a high crescendo.

Breeding
Two pregnant females collected at Karkloof during March and April had two and three foetuses respectively. No comparitive data are present from the subregion (Skinner and Smithers 1990). In Kenya the species apparently breeds throughout the year, and has a gestation of between seven and eight months (Kingdon 1971).

Linear measurements (mm)
Two males had head and body lengths of 400 mm and 440 mm, and a single female had a head and body length of 470 mm. These values fall within the lower range of head and body length given for the species (428–520 mm; Bothma 1967).

Records of occurrence
SPECIMEN RECORDS:
DM: Beacon Hill, Belleview Farm (near Southbroom), Ngeli Forest;
NM: Baynesfield, Colbourne, Karkloof, The Forest.

ADDITIONAL RECORDS:
Pringle 1974: Dargle Forest, Mpetyane Forest, Ngeli Forest; Sight records (KwaZulu-Natal Nature Conservation Service files): Belview Farm (Mbizane River), Elandshoek, Mount Shannon, Terlings.

ORDER	TUBULIDENTATA	
Family	ORYCTEROPODIDAE	
		Aardvark
Genus	*Orycteropus*	G. Cuvier 1798

Orycteropus afer (Pallas 1766)
O. afer afer (Pallas 1766)
Aardvark
Erdvark

Taxonomic status
Eighteen subspecies were listed for the whole of Africa by Meester (1971), only one of which is found in southern Africa. It is probable that far too many subspecies are currently recognised (Meester et al. 1986).

Distribution
Based on sight records and observations of occupied burrows, aardvarks are widely distributed in KwaZulu-Natal, with the exception of much of the coastal strip, the lower and middle reaches of the Tugela Basin, northern Maputaland and the extreme north-west (Figure 67).

Protected areas
Albert Falls, Blinkwater, Eastern Shores, Fort Nottingham, False Bay, Hluhluwe, Itala, Kamberg, Karkloof, Mhlatuze, Mkuzi, Sodwana, Spioenkop, St Lucia, Umfolozi, Weenen.

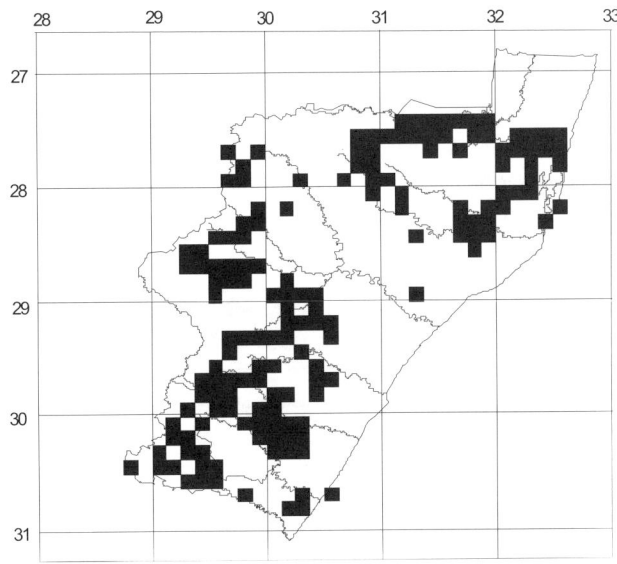

Figure 67

Conservation status

Widely but sparsely distributed in KwaZulu-Natal. Melton (*in litt.* cited in Smithers 1986) suggested that owing to hunting pressures in former KwaZulu, they may be extinct here. Pringle (1974) considered the species to be rare outside the Zululand game reserves. The snout and claws are in high demand for traditional medicine and witchcraft, and the flesh is highly palatable.

Habitat

They occur in all bioregions except Montane, and are mostly absent from Coast lowlands. According to farmers' reports they appear to be most common in open grassland in the midlands region (e.g. Mooi River District).

Habits

Solitary, nocturnal and burrow-dwelling. Temporary, foraging burrows and permament, living burrows are excavated with the aid of powerful limbs and sharp, long claws of the forefeet. They occupy a home range of some 3.5 km² and can travel up to 8 km or more a night in search of food (Melton 1976, Skinner and Smithers 1990). The diet consists primarily of formicid ants and termites – ants being commoner in the diet during winter and termites during summer (Melton and Daniels 1986).

Breeding

No data for KwaZulu-Natal. Based on evidence for the subregion as a whole, a single young is born after a gestation of seven months. Neonates or near-term foetuses have been recorded during July, September and November in Zimbabwe (Skinner and Smithers 1990).

Linear measurements (mm)

A single subadult female collected from Mkuzi GR had the following measurements:

Cat. No.	Sex	TL	HB	T	Hf	E
DM2426	F	1510	1083	427	20.1	15.5

Records of occurrence

SPECIMEN RECORDS:
DM: Mkuzi GR.

ADDITIONAL RECORDS:
Sight records (KwaZulu-Natal Nature Conservation Service files): numerous unnamed grid squares (Figure 67).

ORDER	PHOLIDOTA	Pangolin
Family	MANIDAE	
Genus	*Manis*	Linnaeus 1758
Subgenus	*Smutsia*	Gray 1865

Manis temminckii (Smuts 1832)
Pangolin
Ietermagog

Taxonomic status

No subspecies were recognised by Meester et al. (1986).

Distribution

Known from few records in Zululand and Maputaland and along the northern border between KwaZulu-Natal and the Mpumalanga Province (Figure 68).

Protected areas

False Bay, Hluhluwe, Mkuzi, Ndumu.

Conservation status

This species is classed as specially protected game in KwaZulu-Natal. It is listed in the South African Red Data Book as Vulnerable. Because of the heavy use made of this species for traditional medicine and witchcraft, Bourquin (1988) considered this species to be on the verge of extinction both inside and outside reserves in KwaZulu-Natal. The situation is exacerbated by the fact that the pangolin

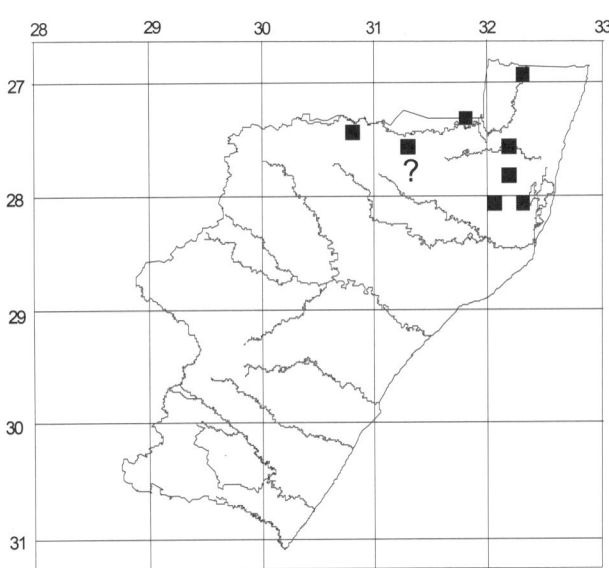

Figure 68

is at the limit of its distribution in KwaZulu-Natal, and therefore probably occurring in sub-optimal habitats.

Habitat
Woodland, grassland and savanna within the Valley bushveld and Lowveld bioregions of Zululand.

Habits
Solitary and predominantly nocturnal. They occupy a home range of between 1 km² and over 8 km², and will travel up to 6 km or more in a night whilst foraging. Pangolins are specialist feeders on formicid ants and termites, the former being better represented in stomach samples studied to date (Skinner and Smithers 1990).

Breeding
No data for KwaZulu-Natal. Based on limited data for the subregion, young appear to be born during the colder, drier months of the year (Skinner and Smithers 1990).

Linear measurements (mm) and mass (g)
No data for KwaZulu-Natal. Elsewhere, total length varies from 702 mm to 1 049 mm (mean of 809 mm) and mass varies from approximately 7 kg (Zimbabwe) to approximately 11 kg (Skinner and Smithers 1990).

Records of occurrence
ADDITIONAL RECORDS:
Dixon (1964): Mkuzi GR;
Pringle (1974): Biala (=Bayala?), Candover, False Bay Park, Hluhluwe GR, Ndumu GR, Paulpietersburg (near); Rautenbach et al. (1981): Itala GR (dubious record, not presently occurring; Rowe-Rowe, personal communication).

Order   RODENTIA

KEY TO SUBORDERS AND FAMILIES (After Meester et al. 1986)

1. Lower jaw with angular process distorted outwards by a limb of lateral superficial masseter muscle ('SUBORDER HYSTRICOMORPHA') 2
—   Lower jaw with angular process not so distorted 4

2. Infraorbital foramen small, not transmitting muscle; fibula reduced and fused with tibia; cheekteeth 4/4, occlusal surfaces simple, without enamel infoldings or islands; eyes suppressed, ear pinnae absent; pelage soft; animals adapted to fossorial life ....
        Family BATHYERGIDAE, p. 78

—   Infraorbital foramen much enlarged for muscle transmission; fibula well developed, not fused with tibia; cheekteeth 4/4, occlusal surfaces with enamel infoldings or islands; eyes well developed, ear pinnae present; pelage modified to quills, bristles or springy hair 3

3. Size large (head and body >600 mm); body covered with long spines; skull with dorsal profile oval, orbits placed far back, nasals very broad; cheekteeth 4/4, with wavy enamel patterning, occlusal surfaces flat; upper incisors smooth, ungrooved ....
        Family HYSTRICIDAE, p. 76
—   Size smaller (HB<200 mm); body without spines ....
        Family THRYONOMIDAE, p. 77

4. Infraorbital foramen not, or scarcely, transmitting muscle; skull with post-orbital processes ('SUBORDER SCIUROMORPHA', in part) ....
        Family SCIURIDAE, p. 75
—   Infraorbital foramen enlarged for muscle transmission; skull without postorbital processes 5

5. Zygomatic plate narrow, situated below much enlarged infraorbital foramen; cheekteeth 4/4, rootless, premolars normally as large as molars; molars with simplified occlusal pattern; body modified for bipedal and saltatorial way of life, with long hind limbs and tail ('SUBORDER SCIUROMORPHA', in part) ....
        Family PEDETIDAE, p. 76
—   Zygomatic plate broadened; infra-orbital foramen well developed; cheekteeth rooted and complex, either 3/3 or 4/4, premolars tend to be smaller than molars ('SUBORDER MYOMORPHA') 6

6. Zygomatic plate not tilted upwards, infra-orbital foramen not very large; cheekteeth 4/4; bullae large; jugal usually large; tail densely furred, bushy ....
        Family GLIRIDAE, p. 112
—   Infraorbital foramen flattened by zygomatic plate, the latter tilted upwards to a greater or lesser degree; cheekteeth not exceeding 3/3; jugal usually short; tail usually not bushy .... Family MURIDAE, p. 81

Suborder	SCIUROMORPHA	
Family	SCIURIDAE	Squirrels
Genus	*Paraxerus*	Forsyth Major 1893

Paraxerus palliatus (Peters 1852)
P. palliatus ornatus (Gray 1864)
P. palliatus tongensis (Roberts 1931)
Red squirrel
Rooi-eekhoring

Taxonomic status

Paraxerus palliatus ornatus was described from the Ngoye Forest in Zululand, and *P. p. tongensis* from Manguzi Forest in north-eastern Zululand. Ngoye Forest represents a relic lowland-evergreen forest island which has been isolated (apart from brief interludes) since the early Pleistocene from the savanna woodland-dominated coastal plains of Zululand and southern Mozambique, thus accounting for differentiation between the two subspecies of red squirrels in KwaZulu-Natal (Viljoen 1989). Viljoen (1989) therefore considered both these subspecies to be valid, but regarded *P. p. tongensis* to be a synonym of *P. p. bridgemani* Dollman, 1914, from southern Mozambique. Nevertheless, pending further studies, Meester et al. (1986) is followed in retaining *tongensis*.

Distribution

Paraxerus palliatus ornatus is restricted to the Ngoye Forest in Zululand, whereas *P. p. tongensis* is more widely distributed through north-eastern Zululand and Maputaland (Figure 69).

Protected areas

Cape Vidal, Eastern Shores, False Bay, Mkuzi, Ndumu, Sodwana Bay, St Lucia Park.

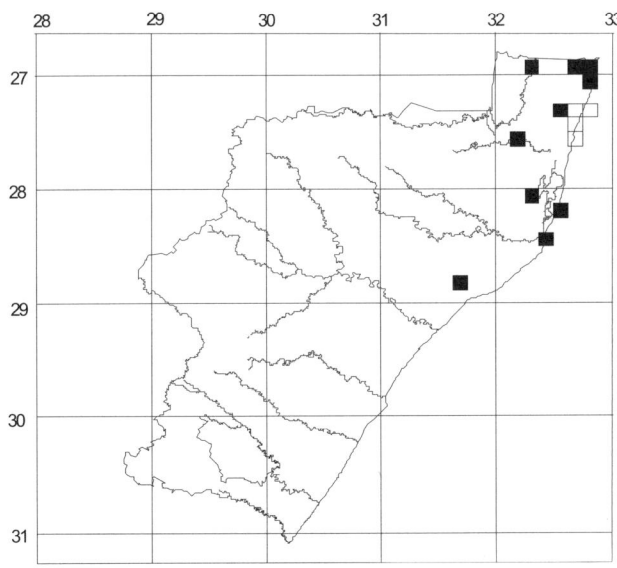

Figure 69

Conservation status

Localised in distribution and considered rare in some areas (e.g. Mkuzi), but fairly common (Ndumu) and common (Eastern Shores and Sodwana) in other areas (Bourquin 1988).

Habitat

Coastal forest, Afromontane evergreen forest and riverine thickets within the Lowveld and Coast lowlands bioregions. In the Mount Silinda Forest in Zimbabwe they are associated with forest edges and clearings where sufficient light penetrates to encourage a dense understorey of shrubs, small trees and climbers (Skinner and Smithers 1990).

Habits

Red squirrels are diurnal and solitary. Viljoen (1980) compared the habits of the two subspecies in KwaZulu-Natal and found *P. p. tongensis* to be more aggressive in behaviour than *P. p. ornatus*. Males generally have larger home ranges (3.2–4.2 ha) and range further (up to approximately 1 000 m) than females (home range 1.2–2.2 ha; maximum distance moved approximately 700 m). Communication occurs by tail-fluffing and flicking, foot-stomping, and scent-marking by urine-dribbling or anal-dragging.

Breeding

A pregnant and lactating female was collected in May from coastal forest at Cape Vidal. Red squirrels from Zululand are known to give birth to litters of one or two young (after a gestation of 60 to 65 days) between the months of August and March (Viljoen 1980). In captivity, red squirrels may have several litters but it is currently thought that they have only one litter in the wild (Skinner and Smithers 1990). The collection of a pregnant, lactating female in May implies that they may have more than one litter a year, and that in Zululand they may give birth during the cooler winter months.

Linear measurements (mm)

	Males					Females				
	x̄	s.d.	n	Min	Max	x̄	s.d.	n	Min	Max
TL	367.7	14.3	3	352	280	365.2	13.3	4	349	380
HB	182.7	9.3	3	175	193	192.8	11.8	4	183	210
T	185.0	13.0	3	177	200	172.5	9.7	4	160	182
Hf	43.3	0.6	3	43	44	44.0	2.0	3	42	46
E	18.3	1.5	3	18	20	19.0	0.0	4	19	19

The above measurements for *P. palliatus tongensis* match those given by Viljoen (1989), with females being slightly larger than males in head and body length. This subspecies is somewhat smaller in size (and paler in coloration) than *P. palliatus ornatus* from Ngoye Forest which averages head and body length of 219 mm for males and 225 mm for females (Viljoen 1989).

Records of occurrence

SPECIMEN RECORDS:
DM: Eastern Shores NR (Lake Bangazi), St Lucia village;

KM: St Lucia Resort;
NM: Coastal Forest Reserve (=Kosi Bay), Eastern Shores NR, Mseleni (Lake Sibayi), Ngoye Forest;
SAM: Ngoye Hill, Ngoye Forest;
TM: Coastal Forest Reserve (=Kosi Bay), Manguzi Forest, Maputa, Ngoye Forest.

ADDITIONAL RECORDS:
Bruton (1978): Dumile, Lake Sibayi Research Station, Mbibi, Nyanene Forest, Sodwana Bay NP;
Dixon (1964): Mkuzi GR;
Dixon (1966): Ndumu GR;
Sight records (KwaZulu-Natal Nature Conservation Service files): False Bay Park, Mkuzi GR, Zinave Game Ranch.

Family	PEDETIDAE	Springhares
Genus	*Pedetes*	Illiger 1811

Pedetes capensis (Forster 1778)   PLATE 29
P. capensis capensis (Forster 1778)
Springhare
Springhaas

Taxonomic status
Only one subspecies is currently recognised from southern Africa (Meester et al. 1986).

Distribution
Restricted to the extreme north-west of KwaZulu-Natal (Figure 70).

Protected areas
None.

Conservation status
Although only marginally present in KwaZulu-Natal, it is generally common throughout its wider range.

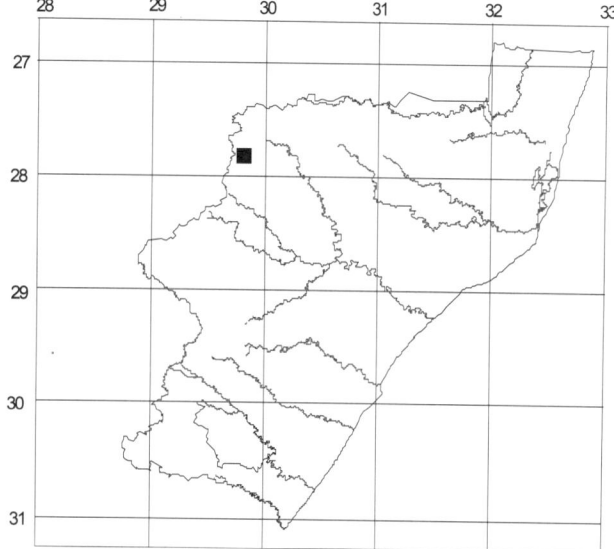

Figure 70

Habitat
Open sand country in lightly wooded, short grassland. An important habitat requirement is a substratum of compacted sandy soil in which to dig burrows.

Habits
Springhares are nocturnal. Single individuals excavate and occupy their own burrow system, usually containing a sloping entrance tunnel marked by a crescent-shaped mound of sand, as well as vertical escape tunnels. Earthen plugs are often found near the tunnel entrances to deter predators. Springhares are selective grazers, preferring certain grasses such as couch grass *Cynodon dactylon*, which is widespread in KwaZulu-Natal (Tainton et al. 1976). They can become a pest in agricultural areas (Skinner and Smithers 1990).

Breeding
A pregnant female in the Natal Museum collection was collected during September. Elsewhere, reproduction may be seasonal or it may continue throughout the year (with over three pregnancies per female) depending on local environmental conditions (see studies cited in Skinner and Smithers 1990). A single young (or rarely twins) is born after a gestation of between 72–82 days (Skinner and Smithers 1990).

Linear measurements (mm) and mass (g)

Cat. No.	Sex	TL	HB	T
NM1913	F	700	380	320
NM1914	M	680	340	340

Elsewhere (see Skinner and Smithers 1990), head and body length varies from 335 mm to 420 mm with a mean of approximately 390 mm, and mass varies from 2.4 kg to 3.9 kg with a mean of approximately 3 kg.

Records of occurrence
SPECIMEN RECORDS:
NM: Kranskop (east of).

Suborder	HYSTRICOMORPHA	
Family	HYSTRICIDAE	Porcupines
Genus	*Hystrix*	Linnaeus 1758

Hystrix africaeaustralis (Peters 1852)   PLATE 30
Porcupine
Ystervark

Taxonomic status
No subspecies are recognised (Meester et al. 1986).

Distribution
Porcupines are widely distributed throughout KwaZulu-Natal, except for the Drier uplands and Moist uplands in

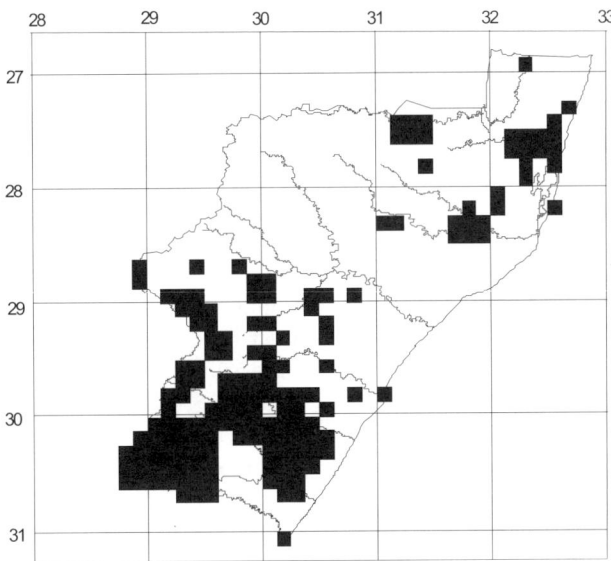

Figure 71

the north-west of the province and the Coast lowlands between St Lucia and Durban (Figure 71). Their apparent absence from much of the Tugela Basin may be due to poor sampling in this area of former KwaZulu.

Protected areas
Cathedral Peak, Cobham, Eastern Shores, False Bay, Garden Castle, Giant's Castle, Highmoor, Hluhluwe, Itala, Kamberg, Karkloof, Loteni, Midmar, Mkhomazi, Mkuzi, Monk's Cowl, Ndumu, Oribi Gorge, Royal Natal, Spioenkop, St Lucia, Sodwana, Umfolozi, Vernon Crookes, Weenen.

Conservation status
Widespread and fairly common, often becoming a pest in agricultural areas.

Habitat
They use a wide range of habitats wherever sufficient rock or scrub cover is present, but are absent from the interior of forests. Altitudinal range varies from sea level to over 2 000 m. Caves, rock crevices and holes in the ground are important for day-time shelter, and porcupines tend to be commoner in areas having broken terrain.

Habits
Porcupines are nocturnal, living in extended family groups (adult pair with variable number of offspring from consecutive litters). They forage either singly or in groups of two or three (usually an adult pair and offspring or adult male and offspring). The species is predominantly vegetarian and relies on bulbs, tubers and roots which are dug up from the ground, as well as on fallen fruits (Skinner and Smithers 1990).

Breeding
No data available for KwaZulu-Natal. Porcupines are monogamous and mating occurs throughout the year. In areas subject to seasonal climates reproduction is seasonal, but is continuous throughout the year in captive populations. Females usually conceive once in a year, and give birth to between one and three young (Skinner and Smithers 1990, and references within).

Linear measurements (mm) and mass (g)
No data available for KwaZulu-Natal. Elsewhere (see Skinner and Smithers 1990) they can reach a mass of 12 to 19 kg when adult; the sexes are of similar size.

Records of occurrence
SPECIMEN RECORDS:
DM: Durban, Mooi River, Umtamvuna NR;
KM: Doornhoek Mine Tunnel;
NM: Langewacht;
TM: Ngome Forest Reserve, Umfolozi GR.

ADDITIONAL RECORDS:
Sight records (KwaZulu-Natal Nature Conservation Service files, excluding unnamed grid squares): Baynesfield Estate, Boston House, Carthorpe, Connemara, Doornhoek, Fairfields, Four Gates, Gracelands, Inhlamvunkulu, Kleinthal, Kloof (Ronalds Road), Kutani turnoff from Lower Mkuzi Road, Llangollen, Mount Ashley, Mount Currie NR, Naauwpoort, Mount Shannon, Priscilla Vale, Rosebank, Southdown, The Rockeries, Sevontein, Spioenkop Resort, Tugela Drift NR, Welverdiend, Weston Agricultural College, Winterhaven, Woolstone;
Bourquin and Mathias (1984): Oribi Gorge NR;
Bourquin and Sowler (1980): Vernon Crookes NR;
Bourquin et al. (1971): Hluhluwe GR, Umfolozi GR;
Bruton (1978): Manzengwenya, Mbazwane;
Dixon (1964): Mkuzi GR;
Dixon (1966): Ndumu GR;
Mentis (1972): Giant's Castle GR;
Oatley (1972): Kilgobbin;
Rautenbach et al. (1981): Itala GR.

Family	THRYONOMYIDAE	Cane-rats
Genus	*Thryonomys*	Fitzinger 1867

Thryonomys swinderianus (Temminck 1827)
Greater cane-rat
Grootrietrot

Taxonomic status
No subspecies are recognised (Meester et al. 1986).

Distribution
Widespread throughout the north-eastern, eastern and southern parts of the province, being absent from the Drier uplands in the north-west (Figure 72). There is evidence that the range of this species is expanding westwards as greater cane-rats have recently been recorded on several occasions in the Free State (J. Eckstein, personal communication).

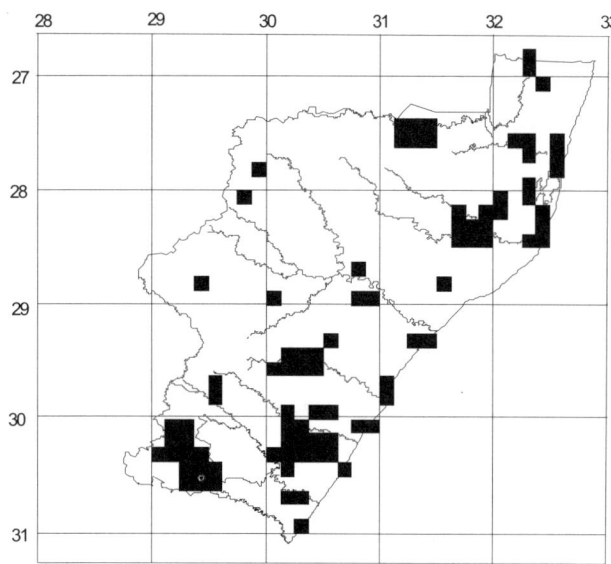

Figure 72

Protected areas
Albert Falls, Chelmsford, Eastern Shores, False Bay, Hluhluwe, Itala, Mhlatuze, Midmar, Mkuzi, Mount Currie, Ndumu, Oribi Gorge, Queen Elizabeth Park, Spioenkop, St Lucia, Sodwana, Umfolozi, Vernon Crookes.

Conservation status
Widespread and common wherever suitable conditions are found. They benefit from agricultural practices, and can become pests on sugar farms.

Habitat
Reed beds or areas of dense, tall grass in the vicinity of lakes, swamps or rivers; including sugar cane near water.

Habits
Nocturnal and crepuscular, as well as essentially solitary, although feeding parties of several adults have been observed (Skinner and Smithers 1990). Their presence is marked by distinct runways and piles of cut grass and reeds as well as faeces. They will use existing holes or cover of dense vegetation for resting places. They are highly prized in rural areas as a food source.

Breeding
No data for KwaZulu-Natal. Reproductive biology has been reviewed by Schröder and Mensah 1987: between one and eight (mean=4) young are born during and just before the warm wet spring months, after a gestation of 152 days.

Linear measurements (mm) and mass (kg)

	Males					Females				
	x̄	s.d.	n	Min	Max	x̄	s.d.	n	Min	Max
TL	641.7	24.7	3	625	670	601.7	57.5	6	530	662
HB	450.0	43.6	3	420	500	440.7	43.1	6	415	494
T	171.7	22.5	3	150	195	161.0	31.3	6	120	198
Hf	76.0	5.3	3	72	82	73.7	5.5	3	70	80
E	30.0	1.0	3	29	31	27.8	5.0	6	20	33
Mass	2.3	–	1	–	–	1.5	–	1	–	–

Records of occurrence
SPECIMEN RECORDS:
BM: Hlatwa District, Umfolozi;
DM: Amanzimtoti, Belleview, Durban, Chelmsford Dam NR;
KM: Kearsney College;
NM: Hilton, Ndumu GR, Pietermaritzburg, Umdoni Park;
NMB: Mount Edgecombe;
SAM: Durban, Mfongosi;
TM: Itala GR (Craigadam), Maurann (Umfolozi Flats), Ndumu GR, Pietermaritzburg Golf Course, Queen Elizabeth Park, St Lucia Estuary, Umfolozi GR, Zululand.

ADDITIONAL RECORDS:
Bourquin and Matthias 1984: Oribi Gorge NR;
Bourquin and Sowler 1980: Vernon Crookes NR;
Dixon 1964: Mkuzi GR;
Rautenbach et al. 1980: Lake Sibayi, Mkuzi GR, Ndumu GR, Sihangwane;
KwaZulu-Natal Nature Conservation Service files (excluding unnamed grid squares in Figure 72): Albert Falls Dam, Ashley, Carthorpe, Chelmsford Public Resort, Clovelly Farm, False Bay NR, Himeville (3 km north), Howick North, Lincoln Haven, Mapelane Public Resort, Maxwilton, Mount Currie NR, Ntunjambili, Tweedie, Weenen NR, Welverdiend, Wetherley Farm, Wonderfontein.

Family	**BATHYERGIDAE**	Molerats

KEY TO GENERA (After Meester et al. 1986)

1. Cheekteeth simple, ring-shaped in adults; posterior tooth erupts early in life; jugal fitting into long groove on zygoma; face not contrastingly marked . . . . *Cryptomys*, p. 78
–   Cheekteeth retaining one inner, one outer fold to old age; posterior tooth erupts late in life; jugal fitting dove-tail fashion into zygoma; face contrastingly marked: black cap on head, white ring around ear, cheeks black, nose white . . . .
    Georychus, p. 80

Genus	*Cryptomys*	Gray 1864

Cryptomys hottentotus (Lesson 1826)   PLATE 31
C. h. natalensis (Roberts 1913)
Common molerat
Knaagdiermol

Taxonomic status
Recent cyto-systematic and biochemical systematic studies of *C. hottentotus sensu lato* have resulted in the elevation of certain subspecies recognised by Meester et al.

(1986) to full specific rank (e.g. *C. damarensis*: Nevo et al. 1987; Honeycutt et al. 1987). Although *C. h. natalensis* shares the same diploid number with *C. h. hottentotus*, allozyme (Nevo et al. 1986, 1987) and mitochondrial DNA (Honeycutt et al. 1987), studies suggest that the two populations are as genetically divergent as are different genera of Bathyergidae. Thus, while *C. h. natalensis* is provisionally retained as a subspecies, it is likely that future taxonomic treatments will treat it as a species distinct from *C. hottentotus*.

Distribution

Fairly widespread in the more mesic Highland, Mistbelt, Coast hinterland and Coast lowlands bioregions; largely absent from drier habitats represented in the Drier upland, Moist upland, Lowveld and Valley bushveld bioregions (Figure 73). The apparent distributional gap in central KwaZulu-Natal may be partly explained by biases in previous collecting, although there may be a geological basis to this distribution (see under **Habitat**). The common molerat is well represented in archaeological remains from a number of sites throughout KwaZulu-Natal, from the upper Pleistocene to the very late Holocene (Avery 1991; see under **Records of occurrence**).

Protected areas

Cathedral Peak, Eastern Shores, Garden Castle, Giant's Castle, Hazelmere Dam, Ian Ellis, Itala, Kamberg, Loteni, Mgeni Vlei, Mkuzi, Ndumu, Oribi Gorge, Royal Natal, St Lucia, Umtamvuna, Vernon Crookes.

Conservation status

They are common wherever suitable soils are present and can become pests in suburban gardens where they eat the bulbs of a large variety of plants.

Habitat

In KwaZulu-Natal they have been collected in a variety of habitats from short, mesic grassland (in Montane, Highland, Mistbelt and Coast hinterland bioregions) to dense

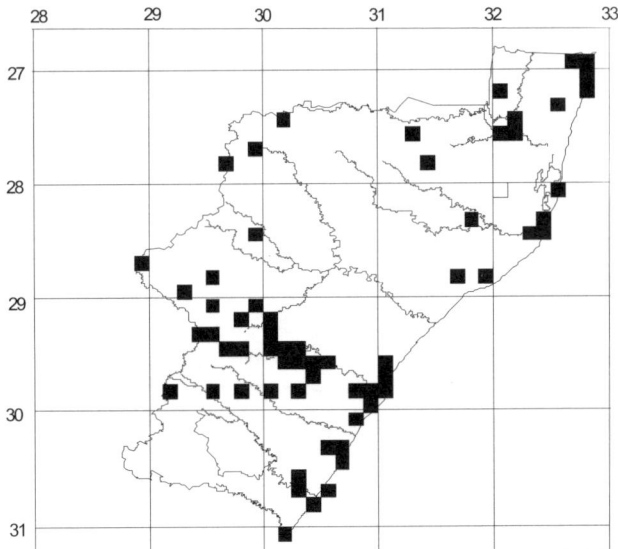

Figure 73

coastal forest (Coast lowlands bioregion) to a variety of man-made habitats including lawns, golfcourses, cultivated gardens and vegetable plots. While they seem to prefer deep sandier soils (such as are found on granites and on Natal Group Sandstone in KwaZulu-Natal), they have been taken from stony soils on hillslopes in the Drakensberg. They apparently avoid heavy, compacted clay soils such as Mopane woodlands in Zimbabwe. Their apparent absence (or at least scarcity) throughout much of central and western KwaZulu-Natal may be owing to the predominance of Karoo Sequence ECCA shales in these regions (Department of Mineral and Energy Affairs 1984). Such shales are known to give rise to hard compacted clay soils which would be unsuitable for burrowing by molerats. In fact the distribution shown in Figure 73 correlates fairly closely with the distribution of Natal Group Sandstone as well as Karoo Sequence dolerites, mudstones and sandstones in the south and with the distribution of alluvial soils in northeastern KwaZulu-Natal (Department of Mineral and Energy Affairs 1984).

Habits

Cryptomys hottentotus natalensis is a diurnal, social, subterranean rodent occurring in small family groups of two to six individuals (Hickman 1979, 1982). Reproduction is usually limited to a dominant pair. The common molerat excavates extensive tunnel systems of up to 340 m in length, comprising long, linear, shallow (200 mm) foraging tunnels and a deeper (30 cm) nest area. Short lateral branches are used to push up excavated soil into mounds (Hickman 1979). Molerats are particularly quick to plug any openings in their burrow system (probably to deter predators such as the mole snake, *Pseudaspis cana*). Molerats are extremely aggressive and will threaten intruders with a widely gaping mouth, revealing their sharp, protruding teeth, and a loud squeaking noise. When trapping molerats it is important to position traps well within the most recent side tunnel (determined by the position of the most recent mound), with minimal disturbance to the tunnel system (i.e. no loose dirt), and to close up the hole with a suitable plug of soil or grass. Failure to set traps carefully will result in animals plugging their burrows before reaching the trap. Their diet of bulbs, fleshy roots, tubers and underground stolons of grasses have made them a major pest to many gardeners (Willan 1992; Taylor 1994). A survey to assess the effectiveness of 13 different methods of molerat control by 44 Durban gardeners (including traps, repellents and poisons, smoking, flooding and excavating burrows, protective covering of bulbs, hunting dogs, and the use of ground vibrations) demonstrated that the use of non-toxic chemicals and traps, as well as protective covering of bulbs, was at least as effective (and less environmentally hazardous) than the use of toxins such as phostoxin and disulphide (Taylor 1994).

Breeding

Two pregnant females with three foetuses were collected in December; a pregnant and lactating female was collected in November; and a juvenile was collected in July (**DM** collection). Indications elsewhere are that between one and three young may be born at any time throughout the year, after a gestation of some 98 days (Rautenbach 1982; Skinner and Smithers 1990, and references within)

Linear measurements (mm) and mass (g)

	Males					Females				
	$\bar{x}$	s.d.	n	Min	Max	$\bar{x}$	s.d.	n	Min	Max
TL	149.3	14.1	44	115	178	138.7	15.2	47	114	194
HB	134.6	13.0	44	101	160	124.7	14.5	47	101	174
T	14.7	3.8	44	8	22	14.4	3.3	47	7	23
Hf	22.2	1.8	10	18	24	20.6	1.5	17	18	24
Mass	86.4	26.8	6	48	122	68.4	14.4	14	48	89

Males are larger than females in all measurements. Based on measurements given above, *C. h. natalensis* from KwaZulu-Natal is similar in size to *Cryptomys* (including *C. h. hottentotus* and *C. h. natalensis*) from greater Transvaal (Rautenbach 1982), but larger than *C. h. hottentotus* from the Free State (Lynch 1983) and Western Cape (Bennett 1989). The extreme variation in adult mass (range = 48–122 g; CV = 31 per cent and 21 per cent for males and females respectively) is also seen in samples from greater Transvaal (Rautenbach 1982: 28–181 g) and the Free State (Lynch 1983: 44–125 g), and is a reflection of the effect of dominant status on body size, with subordinate males in a colony remaining small for years (Bennett 1989).

Records of occurrence

SPECIMEN RECORDS:

BM: Estcourt, Illovo, Umfolozi;

CM: Pennington;

DM: Bosch Hoek Farm, Botanic Gardens (Durban), Durban, Garden Castle, Glenmore (52 Bowen Avenue), Hazelmere, Highmoor, Hillcrest, Hilton (12 Forest Lane), Ian Ellis NR, Kamberg NR, Malvern (7 Houghton Road), Merrivale (24 Geekie Road), Ncandu NR, Newcastle (26 Gasel Avenue), Ngome, Pietermaritzburg, Royal Natal NP, Sarnia Road (Durban), St Lucia NR, St Lucia NR (Iphiva Campsite), St Winifred's (Kingsburgh), Umbilo (Stella Road), Umgeni Park (Marine View Avenue), Umhlanga Rocks (60 Hilken Drive), Umtamvuna NR, University of Natal (Pietermaritzburg), Warner Beach, Waterfall Farm, Westville (111 Blair Atholl Drive, 55a Jan Hofmeyer Road, 37 Springdale Road), Winterton;

KM: Hillcrest, Mkuzi GR, Winterton;

NM: Alexander Road (Pietermaritzburg), Balgowan, Botanic Gardens (Pietermaritzburg), Bulwer, Carter's Nursery (Pietermaritzburg), Castle View, Empangeni, Esperanza, Giant's Castle NR, Manderston, Maputa (=Manguzi), Mason's Mill (Pietermaritzburg), Merrivale, Mseleni, Nagle Dam, Oribi Gorge, Pietermaritzburg, Richmond, Royal Natal NP (Tendele Camp), Spring Grove, Tabamhlope, Town Bush, Umdoni Park, Umtwalume, Underberg, Zwartkop Valley;

SAM: Coastlands, Howick, Estcourt;

SI: Coastal Forest Reserve (=Kosi Bay), Drakensberg Gardens Hotel (3 km north, 3 km west), Groenkloof Farm, Kilgobbin Farm, Makatini Flats;

TM: Berea, Bluff NR, Blackridge, Botanic Gardens (Durban), Cathedral Peak NR (research area), Clarendon, Dargle, Dukuduku Forest Station, Durban, Durban Country Club, Eastern Shores NR, Hazelmere Dam NR, Hilton Road, Iron Watch, Itala GR (Craigadam), Kamberg NR, Karkloof (Type specimen), Kilgobbin, Kloof, Malvern, Manzengwenya Forest Station (Lalanek Inspection Quarters), Maputa, Mgeni Vlei NR, Ngome Forest Reserve, Pietermaritzburg, Pinetown, Ronan Egg Farm, Southport, Ubombo, Umtamvuna NR, St Lucia Estuary (Type specimen), Vernon Crookes NR.

ADDITIONAL RECORDS:

Avery (1991): Border Cave, Clarke's, Collingham, Diamond I, Gehle, Mbabane, Mgede, Mhlwazini, Nkupe, Umhlatuzana;

Bourquin et al. (1971): Hluhluwe GR.

Genus	*Georychus*	Illiger 1811

Georychus capensis (Pallas 1779)   PLATE 32
Cape molerat
Blesmol

Taxonomic status

Meester et al. (1986) did not recognise any subspecies. Based on extreme allozyme (Nevo et al. 1987) and mitochondrial DNA (Honeycutt et al. 1987) differences, the isolated KwaZulu-Natal population may prove to be a different species.

Distribution

A relic population of this species occurs in south-western KwaZulu-Natal, from Bulwer in the south to Nottingham Road in the north and to Giant's Castle Game Reserve in the west (Figure 74). The species occurs continuously in the Western and Eastern Cape. Apart from in KwaZulu-Natal, a further relic population is found in the vicinity of Belfast and Ermelo in the Mpumalanga Province (Skinner and Smithers 1990). That this species has occupied a wider distribution in the recent past is evidenced by its occurrence in early (10 000–7 000 years BP) and late (4 000–1 000 years BP) Holocene deposits at Umhlatuzana outside Durban (Avery 1991), indicating that it must have occurred along the coast of KwaZulu-Natal as is presently the case in the Western Cape and Eastern Cape. The present relic population has been in place at least since the middle Holocene (7 000–4 000 years BP) as indicated by its occurrence in middle and late Holocene deposits at Collingham (grid reference 2929DBB) and Gehle (2929BBD) respectively (Avery 1991; see under **Records of occurrence**).

Protected areas

Giant's Castle? One specimen was apparently collected at Witteberg Gate, Giant's Castle, but subsequently lost (D. Rowe-Rowe, personal communication).

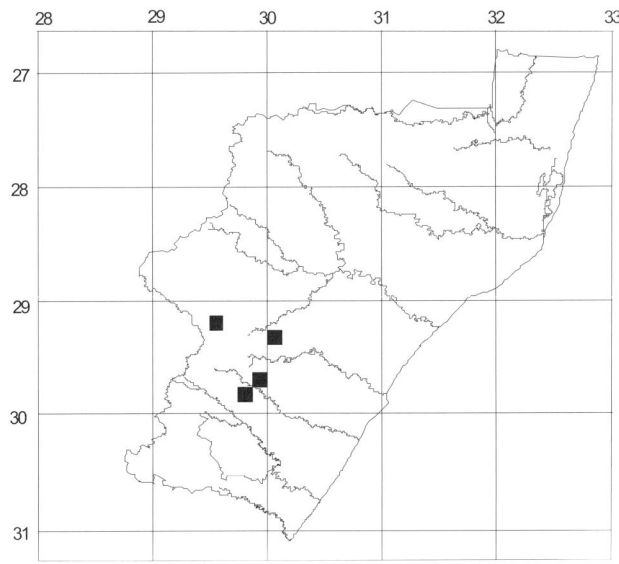

Figure 74

Conservation status
Uncommon in KwaZulu-Natal, although this may reflect both lack of collecting and the localised nature of the species' distribution in the province.

Habitat
In KwaZulu-Natal the species is associated with mesic grasslands on sandy soils within the Highland and Mistbelt bioregions of the KwaZulu-Natal midlands. In the Western Cape and Eastern Cape they are associated with coastal sand dunes, sandy alluvium along river valleys, and montane regions (Skinner and Smithers 1990).

Habits
They are solitary animals which, like the common molerat, use their teeth to excavate burrows marked by above-ground mounds. They feed on roots, tubers and bulbs which are stored in underground storage chambers. Entire plants are undermined and pulled into their burrows (Skinner and Smithers 1990).

Breeding
No data for KwaZulu-Natal. In the Western Cape, young are born betweeen September and December (Taylor et al. 1985). Up to two litters may be born in a year. Litter size varies from one to 10 with a mean of six.

Linear measurements (mm) and mass (g)
A single male examined from Virginia Farm in the Impendle District had a head and body length of 140 mm and a tail of 18 mm. Elsewhere, head and body length may reach around 200 mm (mean of around 160 mm for both sexes) while mass may exceed 320 g (mean of around 180 g for both sexes). The tail is approximately 20 mm in length (Taylor et al. 1985).

Records of occurrence
SPECIMEN RECORDS:
DM: Virginia Farm;
KM: Foxtail Farm (Giant's Castle GR);
NM: Bulwer, Nottingham Road.

ADDITIONAL RECORDS:
Avery (1991): Collingham, Gehle, Umhlatuzana.
Unpublished specimen record (Rowe-Rowe, 1983): Giant's Castle NR.

Suborder	**MYOMORPHA**	
Family	**MURIDAE**	**Mice and rats**

KEY TO SUBFAMILIES (After Meester et al. 1986)

1. M^3 the largest tooth; cheekteeth compactly laminate . . . . Subfamily OTOMYINAE, p. 82
   - M^1 the largest tooth; cheekteeth not compactly laminate 2

2. M^1 with three cusps in anterior row . . . . Subfamily MURINAE, p. 97
   - M^1 with only two cusps in anterior row 3

3. Bullae much enlarged, at least 25 per cent of greatest skull length; upper cheekteeth with cusps in two longitudinal rows, weakly laminate when worn . . . . Subfamily GERBILLINAE, p. 87
   - Bullae less enlarged, less than 25 per cent of greatest skull length 4

4. Soles of hind feet partly haired; upper cheekteeth with cusps in two longitudinal rows, weakly laminate when worn . . . . Subfamily CRICETINAE, p. 89
   - Soles of hind feet naked; upper cheekteeth with cusps in three longitudinal rows 5

5. With cheek pouches; upper incisors ungrooved; head and body length more than 110 mm . . . . Subfamily CRICETOMYINAE, p. 90
   - Without cheek pouches; upper incisors grooved or ungrooved; head and body length less than 110 mm . . . . Subfamily DENDROMURINAE, p. 91

Subfamily	OTOMYINAE	Laminate-toothed rats
Genus	*Otomys*	F. Cuvier 1824

KEY TO SPECIES (Meester et al. 1986)

1. One deep outer and sometimes also a
   shallow inner groove on lower incisors 2
 — Lower incisors ungrooved . . . *O. sloggetti*, p. 85

2. M³ with 9–10 laminae, M₁ with 6–7 . . . .
   . . . . *O. laminatus*, p. 82
 — M³ with 4–7 laminae, M₁ with 4 3

3. Posterior petrotympanic foramen
   (on postero-internal surface of bulla)
   slit-like . . . . *O. angoniensis*, p. 83
 — Posterior petrotympanic foramen
   round . . . . *O. irroratus*, p. 84

Otomys laminatus (Thomas and Schwann 1905)
O. l. laminatus (Thomas and Schwann 1905)
O. l. fannini (Roberts 1951)
Laminate vlei rat
Bergvleirot

Taxonomic status
The type locality of the species is Sibhudeni in the Nkandhla Forest (altitude = 1 050 m) just north of Eshowe in northern KwaZulu-Natal (Meester et al. 1986; Figure 75). Roberts (1951) recognised a second KwaZulu-Natal subspecies, *O. l. fannini*, from the Dargle area in the midlands. However, Meester et al. (1986) cautioned that *fannini* may be synonymous with the nominate race.

Distribution
This species has a patchy, scattered distribution in KwaZulu-Natal. Although De Graaff (1981) showed the species to be widespread in the province, he did not list localities, specimens examined or sources of reference. Bourquin (1988) listed a single known specimen record from Oribi Gorge. Subsequently, specimens have been collected from several new localities from throughout KwaZulu-Natal from Dukuduku Forest in the north to the Umtamvuna Gorge in the south, and to Royal Natal National Park in the west (Taylor et al. 1994, Figure 75). Most of the specimens collected to date originate from the midlands (Kilgobbin Farm and Karkloof) and the South Coast (Oribi Gorge and Clearwater Farm). Avery (1991) recorded this species from Quaternary sites at Border Cave on the Swaziland border, Umhlatuzana near Durban, and from several sites in the upper Tugela Basin and Drakensberg foothills (see under **Records of occurrence**).

Protected areas
Oribi Gorge, Royal Natal.

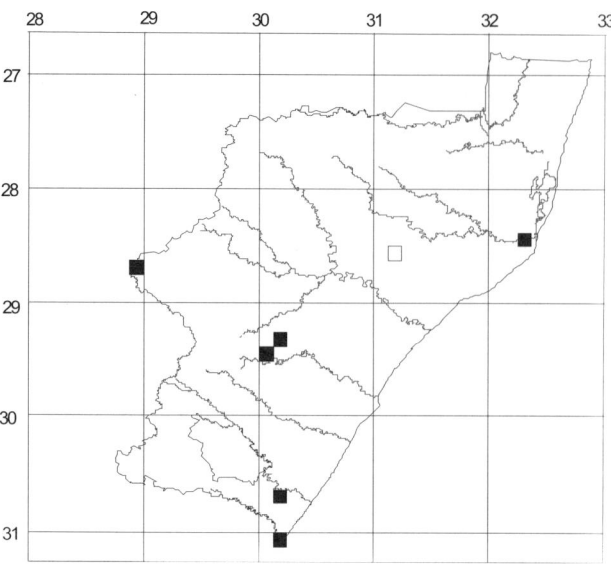

Figure 75

Conservation status
Widespread although much less common in trap captures than the vlei rat *O. irroratus* with which it co-occurs.

Habitat
They have been collected mostly in well-vegetated, grass-dominated wetland and plateau situations in the Coast hinterland and Mistbelt bioregions, occurring both sympatrically and syntopically (in the same habitat) with the more numerous *O. irroratus*. However, they have also been collected in coastal forest at Dukuduku Forest in the Coast lowlands bioregion where they are sympatric with both *O. irroratus* and *O. angoniensis* (Taylor et al. 1994).

Habits
No information is available; presumably their habits are similar to those of *O. irroratus*.

Breeding
No data for KwaZulu-Natal.

Linear measurements (mm) and mass (g)
An adult female from Karkloof had a head and body length of 184 mm and a mass of 140 g; an adult male from Oribi Gorge had a head and body length of 189 mm and a tail length of 111 mm.

Records of occurrence
SPECIMEN RECORDS:
DM: Clearwater Farm, Karkloof, Royal Natal NP;
NM: Oribi Gorge NR;
TM: Kilgobbin Farm (Dargle), Oribi Gorge NR;
Private collection of T. Bodbijl (No. M9): Dukuduku Forest.

ADDITIONAL RECORDS:
Avery (1991): Border Cave, Clarke's, Collingham, Diamond I, Mhlwazini, Umhlatuzana;
Bourquin and Mathias (1984): Oribi Gorge NR.

Otomys angoniensis (Wroughton 1906)   PLATE 33
O. a. tugelensis (Roberts 1929)
Angoni vlei rat
Angoni-vleirot

Taxonomic status
Meester at al. (1986) recognised three subspecies from southern Africa, of which only *O. a. tugelensis* is found in KwaZulu-Natal.

Distribution
Widespread throughout the lower-lying (0–1 000 m above sea level), drier Coast lowlands, Lowveld, Valley bushveld, Moist upland and Drier upland bioregions (Figure 76). De Graaff (1981) gave the distribution in KwaZulu-Natal as being central and northern, while Roberts (1951) considered the Tugela River to be the southernmost limit of *O. tugelensis tugelensis* (= *O. angoniensis tugelensis*). The present data extend the southernmost limit of the species to 30°46′ South in the Port Shepstone area (Figure 76). The species has been recorded in Quaternary archaeological remains at Border Cave and Nkupe in the north and north-west of KwaZulu-Natal (Avery 1991; see under **Records of occurrence**).

Protected areas
Albert Falls, Beachwood Mangroves, Bluff, Chelmsford, Coastal Forest Reserve (=Kosi Bay), False Bay Park, Hazelmere, Hluhluwe, Ian Ellis, Itala, Kenneth Stainbank, Midmar, Moor Park, Spioenkop, Umfolozi, Vernon Crookes, Vryheid, Weenen.

Conservation status
Common.

Habitat
Open *Acacia* woodland and grassland associated with coastal forest and bushveld habitats, usually in the vicinity of water (rivers, dams and vleis), although they have been collected kilometres from the nearest water source. They prefer habitats containing dense, broad-leafed, palatable grass clumps which provide suitable cover for their characteristic runways. They have been collected in gardens, often associated with grasses such as *Panicum spp.*, and are often brought in by cats in gardens.

Habits
Solitary and crepuscular, mostly diurnal although daily activity patterns seem to vary both geographically and seasonally (Bronner and Meester 1988). At Albert Falls Nature Reserve, over six trap-nights, 12 animals were collected in early morning checks (07:00–08:00) while two animals were collected in late afternoon (15:00–16:00) checks. Elsewhere in KwaZulu-Natal, I have collected them only during morning checks, suggesting that in KwaZulu-Natal they are predominantly crepuscular or nocturnal, with minimal diurnal activity. They use intricate, well established runways among grass tussocks, and these are often marked by fresh droppings and piles of discarded grass clippings. They are strictly herbivorous, feeding mostly on a variety of grasses, newly sprouting reeds and the rhizomes and roots of reeds (Bronner and Meester 1988).

Breeding
Pregnant females were recorded in July (three foetuses) and November (two foetuses). The female collected in July was also lactating. Elsewhere, the species appears to be seasonally polyoestrous with pregnant and lactating females having been recorded between the months of August and May (Bronner and Meester 1988). Two to five semi-precocial young (mean of three) are born after a gestation of 37 days. Minimum age at sexual maturity is five weeks in females and eight weeks in males (Pillay, personal communication).

Linear measurements (mm) and mass (g)

	Males					Females				
	$\bar{x}$	s.d.	n	Min	Max	$\bar{x}$	s.d.	n	Min	Max
TL	236.0	26.1	35	176	292	238.4	19.9	20	208	292
HB	153.2	18.6	38	113	192	157.4	17.7	25	127	190
T	83.9	11.4	34	63	102	82.1	8.6	20	65	93
Hf	26.1	1.5	12	24	29	25.5	1.0	11	24	27
E	19.0	2.9	22	15	28	20.6	2.0	18	18	25
Mass	121.2	35.3	12	71	176	122.6	28.0	11	78	182

Records of occurrence
SPECIMEN RECORDS:
DM: Albert Falls NR, Beachwood NR, Bluff NR, Bosch Hoek (SANDF base), Burman Bush, Chase Valley Heights, Chelmsford NR, Escombe (2 Dale Place), Eston, False Bay Park, Game Valley Estates (Hella Hella), Glenmore, Greenwood Park, Hazelmere Dam NR, Hillcrest, Hlambanyati-Mkuze, Ian Ellis NR, Itala GR, Johnson Farm (Hillcrest), Kenneth Stainbank NR, Lake Eteza, Malvern (1 Rindel Road), Mbumbazi NR, Midmar Dam NR, Moor Park NR, Mount Moreland, Nkonyeni Hospital, Pietermaritzburg Golf Course, Pigeon Valley (Durban), Shongweni,

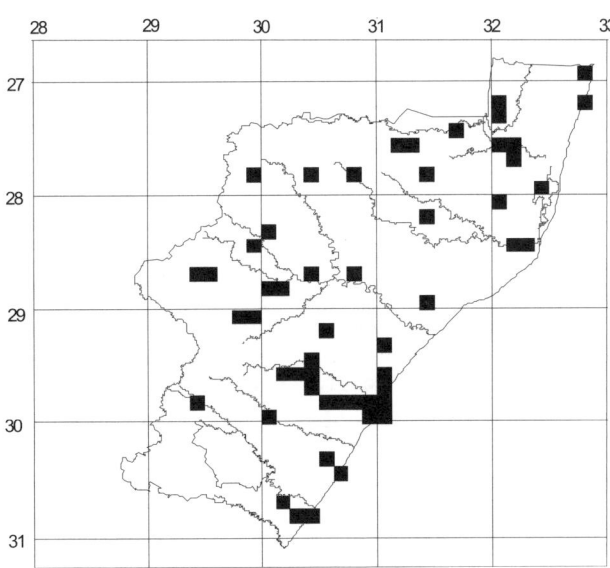

Figure 76

Spioenkop NR, Umgeni Park (144 Marine View Avenue), Umhlanga Rocks, Umhlanga Rocks (Hilken Drive), University of Natal (Durban), Vryheid NR, Weenen NR, Westville (111 Blair Atholl Road);

KM: Vergeval Farm;

NM: Belfort, Burman Bush, Carter's Nursery, Elandslaagte, Gwaliweni, Hluhluwe GR, Merrivale, Mfongosi, Scottsville, Sobantu village, Umdoni Park, Umtwalume;

TM: Ashburton, Bluff NR, Coastal Forest Reserve (=Kosi Bay), Darvill Sewage Works (Pietermaritzburg), Eshowe, Hluhluwe GR (Egodeni), Ingwavuma, Itala GR (Craigadam, Warden's Outpost), Kenneth Stainbank NR, Klipspruit, Lincolnmead, Malvern, Manzengwenya Forest Station (Lalanek Inspection Quarters), Midmar Dam NR, Mkuzi River, Monzi, Moor Park NR, Mount Edgecombe, Mvoti Vlei NR, Ngome Forest Reserve, Oribi Gorge NR, Pietermaritzburg (Armstrong Drive), Pietermaritzburg Golf Course, Pigeon Valley, Pinetown, Pomeroy (16 km from, on road to Greytown), Queen Elizabeth Park, Spioenkop Dam NR, Stanger (16 km from, on Glen Dali Road), Tugela Estates, Ubombo, University of Natal (Pietermaritzburg), Vernon Crookes NR, Woodhouse Road (Pietermaritzburg).

ADDITIONAL RECORDS:
Avery (1991): Border Cave, Nkupe;
Bourquin and Sowler (1980): Vernon Crookes NR;
Bronner and Meester (1987): Bluff NR;
Rautenbach et al. (1981): Itala GR.

Otomys irroratus (Brants 1827)
O. i. natalensis (Roberts 1929)
O. i. orientalis (Roberts 1946)
Vlei rat
Vlei-rot

Taxonomic status
Eight subspecies are recognised in southern Africa; two in KwaZulu-Natal, although Meester et al. (1986) cautioned that too many subspecies are probably recognised. Recent cytogenetic studies (Contrafatto et al. 1992a, b; Contrafatto 1996) have indicated the existence of a number of chromosomal races associated with climatic regions (Taylor et al. 1994), of which at least three (A, A1 and A2) are found in KwaZulu-Natal. The A1 race is defined by the presence of a tandem fusion re-arrangement of chromosomes 7 and 12, and it is restricted to the Highland and Montane bioregions (altitudes exceeding 1 400 m) in KwaZulu-Natal. The A race has been recorded from the southern region of KwaZulu-Natal, but no chromosomal data are as yet available for the northern regions of KwaZulu-Natal. The A2 race, found in the Midlands, contains two additional chromosome pairs.

Distribution
The vlei rat occurs widely throughout the higher-lying or moister Montane, Highland, Mistbelt and Coast hinterland bioregions, being absent from or extending marginally into the Drier upland, Moist upland, Coast lowlands, Valley

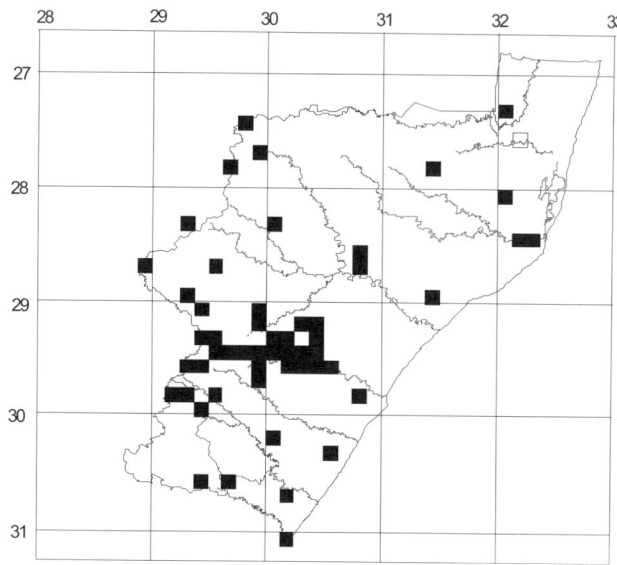

Figure 77

bushveld and Lowveld bioregions (Figure 77). Since *O. angoniensis* is largely associated with the latter drier habitats, the ranges of the two species in KwaZulu-Natal are largely parapatric. However, there are several localities, usually near the border between bioregions, where the two species are sympatric, e.g. Hluhluwe Game Reserve and the Dukuduku Forest in Zululand, Hillcrest, Vernon Crookes Nature Reserve, the Estcourt region, Midmar Dam and Itala Game Reserve. The species has been recorded from late Quaternary archaeological sites at Umhlatuzana and Collingham (see under **Records of occurrence**).

Protected areas
Cathedral Peak, Coleford, Fort Nottingham, Garden Castle, Giant's Castle, Highmoor, Itala, Kamberg, Karkloof, Loteni, Mgeni Vlei, Midmar, Monk's Cowl, Mkuzi, Ncandu, Oribi Gorge, Royal Natal, Umgeni Vlei, Umtamvuna, Mvoti Vlei, Vernon Crookes.

Conservation status
Common and widespread in KwaZulu-Natal, although generally restricted to mesic habitats.

Habitat
Typically collected in dense grassland (as well as grassland mixed with forbs, bracken and other ferns) in close proximity to vleis, rivers and dams. However, in high-altitude, high-rainfall areas such as the Drakensberg, vlei rats occur in sparser montane grasslands, often on rocky hillslopes kilometres from the nearest water source (Rowe-Rowe and Meester 1982). Where *O. irroratus* is found in sympatry with *O. angoniensis*, the former tends to occupy more mesic habitats than the latter (Davis 1973). Where extreme altitudinal variation is found at localities where the two species are sympatric (e.g. Itala Game Reserve and Vernon Crookes Nature Reserve), it is highly probable that *O. angoniensis* would occupy lower-lying areas and *O. irroratus* higher-lying areas.

Habits

They are active in bouts throughout the day and night. Vlei rats are shy, aggressive and solitary; they are more vocal than Angoni vlei rats and give a loud metallic 'chit' call as a threat to other individuals. They use well-worn runways in dense grass clumps and shelter in saucer-shaped grass nests in grass clumps above the water level. In montane areas where grass clumps are more sparse, they do not make runways and their characteristic piles of discarded grass clippings and droppings can be seen on open patches between grass clumps. Vlei rats are known to be herbivores, feeding almost exclusively on grasses (Davis 1973). Rowe-Rowe (1986) examined 17 stomachs collected at Giant's Castle Game Reserve: all contained green plant material, while 24 per cent contained bark.

Breeding

Of four pregnant females collected in KwaZulu-Natal (**DM** collection), three were collected in January and one in November. The number of embryos recorded in pregnant females was two ($n=2$) or three ($n=1$). Single lactating females were collected in November, December and January. At Giant's Castle Game Reserve, Rowe-Rowe and Meester (1982b) reported pregnant females during October, November and January, and a lactating female in May. Mean number of foetuses was 1.6 ($n=5$, range 1–2; Rowe-Rowe and Meester 1982b).

Laboratory breeding studies of two chromosomally distinct KwaZulu-Natal populations, Kamberg Nature Reserve (A1 race possessing the 7/12 tandem fusion) and Karkloof (A2 race lacking the tandem fusion, but with two additional chromosomal pairs), revealed mean litter sizes of 2.5 and 2.3 respectively. Litters were produced throughout the year at intervals of just under 50 days (Pillay 1990; Pillay et al. 1995). However, hybrids from cross-breeding between the two populations experienced reduced litter sizes, increased pre-weaning mortality (26.0–37.5 per cent compared with 3.9–6.2 per cent in intrapopulation crosses) and almost complete sterility (only one young produced from 48 backcross attempts; Pillay et al. 1995). In greater Transvaal (Davis and Meester 1981) and the Eastern Cape (Perrin 1980), the species has an extended birth season (from August through to May in greater Transvaal and throughout the year in the Eastern Cape). Between one and four young were produced per litter (mean of 2.33; Davis and Meester 1981).

Linear measurements (mm) and mass (g)

	Males					Females				
	x̄	s.d.	n	Min	Max	x̄	s.d.	n	Min	Max
TL	257.2	39.3	58	158	320	249.1	29.0	74	180	320
HB	171.9	29.2	59	107	222	165.7	25.0	78	100	245
T	88.4	13.7	54	51	117	84.3	9.6	75	48	110
Hf	28.1	2.7	29	24	34	26.2	2.7	38	17	32
E	22.2	25	51	15	26.5	22.4	1.9	68	17.5	25.5
Mass	162.8	63.8	41	58	281	138.3	45.5	51	55	30.5

Pillay (1993) noted sexual dimorphism and increased body size in a population studied at Kamberg NR in the Drakensberg, compared with absence of sexual dimorphism and smaller body size at Karkloof in the midlands. The mean and maximum masses recorded for males in the present study (162 and 281 g respectively) are considerably larger than recorded elsewhere in South Africa (De Graaff 1981; Lynch 1983, 1989, 1994; Rautenbach 1982). The maximum value of 281 g was obtained for an individual from Kamberg NR. The large masses in this study may be partly due to the fact that some individuals in the sample were kept for varying periods in captivity. Although no captive-born individuals were included, growth rates in captive animals exceed those in the wild (Davis and Meester 1981).

Records of occurrence

SPECIMEN RECORDS:

DM: Blinkwater NR, Cathedral Peak, Clearwater Farm, Coleford NR, Eshowe, Fort Nottingham NR, Garden Castle NR, Giant's Castle GR, Highmoor NR, Hillcrest, Hluhluwe GR, Kamberg NR, Karkloof (Clan Syndicate Forest, Merensky Trust Forest), Karkloof NR, Kokstad, Lidgetton, Loteni NR, Mgeni Vlei NR, Monk's Cowl Forest Station, Mtubatuba (St Lucia Road), Ncandu NR, Ngome, Nottingham Road, Royal Natal NP, Umgeni Valley Game Ranch, Underberg, Vernon Crookes NR, Vergelegen NR;

NM: Boston, Castle View, Champagne Castle, Dukuduku, Giant's Castle GR, Gwaliweni, Hilton, Hluhluwe GR, Mooi River, Moor Park GR, Ngome Forest Reserve, Pietermaritzburg, Town Bush, Tretower Farm, Treverton, Umgeni Valley Game Ranch, Umsindusi River (Pietermaritzburg), Underberg Trout Hatchery;

SAM: Estcourt, Mfongosi, Rietvlei;

SI: Dargle Station, Drakensberg Gardens Hotel (3 km north, 3 km west), Sani Pass (Makhake Store);

TM: Cathedral Peak NR, Dargle, Eersteling Farm, Hilton Road (Pietermaritzburg), Hluhluwe GR, Inhluzani Mountain, Itala GR (Breda, Craigadam), Kamberg NR, Karkloof Forest, Kilgobbin Farm, Ladysmith (50 km from, on Van Reenen road), Loteni NR, Midmar Dam NR, Monzi, Mtubatuba, Ngome Forest Reserve, Oribi Gorge NR, Pottershill Farm, Qudeni, Royal Natal NP, Umtamvuna NR.

ADDITIONAL RECORDS:

Avery (1991): Collingham, Umhlatuzana;
Bourquin et al. (1971): Hluhluwe;
Dixon (1964): Mkuzi;
Rowe-Rowe and Meester (1982b): Cathedral Peak NR, Giant's Castle GR.

Otomys sloggetti (Thomas 1902)
O. s. robertsi (Hewitt 1927)
Sloggett's rat
Sloggett-rot

Taxonomic status

Five subspecies were recognised by Meester et al. (1986). The subspecies *O. s. robertsi* occupies extreme altitudes, up to 3 283 m along the KwaZulu-Natal Drakensberg, and is also known as the ice rat.

Distribution
Occurs on the highest Drakensberg peaks and plateaux along the KwaZulu-Natal–Lesotho border, as well as along the KwaZulu-Natal–Mpumalanga border in the Wakkerstroom area (Figure 78). At Sani Pass in the Natal Drakensberg Park, Rowe-Rowe and Meester (1982a) found it to occur at altitudes above 2 750 m (alpine zone of Killick 1978), whereas elsewhere, in the Free State, Eastern Cape and Lesotho it is confined mainly to the subalpine and alpine ecological belts (Killick 1978) above 2 000 m; occasionally occurring as low as 1 600 m (Lynch 1983, 1989; Lynch and Watson 1992). Avery (1991) recorded *O.* cf. *sloggetti* from very late Holocene (<1000 years BP) deposits at two sites in the Valley bushveld bioregion in the upper Tugela Basin. If these records are correct they imply that the species once occupied a wider range of altitudes and habitats than present.

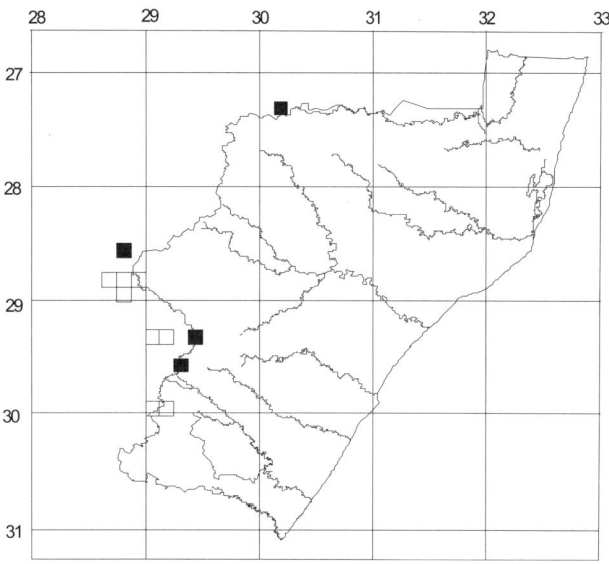

Figure 78

Protected areas
Cathedral Peak, Cobham, Giant's Castle, Royal Natal.

Conservation status
Although restricted in their occurrence in KwaZulu-Natal they tend to be common wherever they occur (densities of up to 100 per hectare were reported by Willan 1990).

Habitat
Otomys sloggetti is associated with *Erica-Helichrysum* alpine heath, occupying rocky areas on more xeric soils as well as raised molerat-excavated hummocks in mesic bog areas (Lynch and Watson 1992). Above the escarpment, all along the KwaZulu-Natal Drakensberg Park, they were observed next to burrows excavated in alluvial soils along river banks (Rowe-Rowe, personal communication). They are particularly common in man-made rocky structures such as stone walls, roadside cuttings and kraals where abundant potential nesting sites occur.

Habits
They are diurnal, and emerge from their shallow (200–300 mm) burrows to bask during the warmest hours of the day. Burrows are occupied by family groups within which only a single pair breeds (Willan 1990). Details of an excavated burrow are provided by Lynch and Watson (1992). When temperatures are very low they may remain in their burrows for several days. On the Mont-aux-Sources plateau at 3 000 m, one was seen (Rowe-Rowe, personal communication) emerging from its burrow at night into the bell of a tent, and to feed on underground portions of plants. Air temperature was -5°C and the ground was frozen. *Otomys sloggetti* is vegetarian and food is carried back to the burrows where it is hoarded. Lynch and Watson (1992) list 11 species of plants eaten by this species. Food plants may be manipulated with the forepaws whilst sitting up on the haunches; alternatively they may graze without the use of the forepaws.

Breeding
No data for KwaZulu-Natal. From captive breeding (Willan 1990) and field-based (Lynch and Watson 1992) studies on populations in Lesotho, the breeding season is from October to March; litter size varied from one to four ($\bar{x}$ = 2.2) in wild caught pregnant females and from one to two ($\bar{x}$ = 1.4) in captive animals. The young nipple-cling for two to three days, are weaned at 16 days and first mate at 18 weeks (Willan 1990).

Linear measurements (mm) and mass (g)
A single female collected in KwaZulu-Natal had the following measurements:

Cat. No.	Sex	TL	HB	T	Hf*cu*	E
DM260	F	223	163	60	27	18

Based on data from the Eastern Cape and Lesotho (Lynch and Watson 1992), females are larger (mean mass of 90.0–94.4 g; mean TL of 186–196 mm) than males (mean mass of 71.5–73.1 g; mean TL of 171–178 mm).

Records of occurrence
SPECIMEN RECORDS:
DM: Near Giant's Castle on KwaZulu-Natal–Lesotho border;
TM: Mount Ziesbok, Sani Pass (Black Range), Wakkerstroom.
Sight records: D. Rowe-Rowe, KwaZulu-Natal Nature Conservation Service: Kubelu-Kubedu Rivers, Mont-Aux-Sources plateau, Mont-Aux-Sources slope (bank of Tugela River), various other localities above the escarpment in KwaZulu-Natal Drakensberg Park.
Note: Records from the Transvaal Museum fall just outside the province but are included anyway. Similarly, records from Willan (1990) and Lynch and Watson (1992)

representing 1/64th-degree grid squares falling on or near the KwaZulu-Natal border are included in Figure 78.

<table>
<tr><td>Subfamily</td><td>GERBILLINAE</td><td>Gerbils</td></tr>
<tr><td>Tribe</td><td>TATERILLINI</td><td></td></tr>
<tr><td>Genus</td><td>*Tatera*</td><td>Lataste 1882</td></tr>
</table>

KEY TO SPECIES (After Meester et al. 1986)

1. Colour brighter, texture of fur sleek; sharp line of demarcation between flanks and belly; belly pure white; mammary formula 2 + 2 = 8; tail longer than head and body, well haired, particularly towards tip, never white tipped; pads of feet normally dark, feet narrower; upper incisors grooved, moderately to strongly opisthodont; molars relatively narrow, lightly built (*robusta* group) . . . . *T. leucogaster*, p. 87

— Colour duller, texture of fur fluffy or somewhat harsh; sharp line of demarcation between flanks and belly often indistinct; belly white, buffy or grey; mammary formula 1 + 2 = 6 or 2 + 2 = 8; tail shorter than or subequal to head and body, fairly evenly haired throughout, with short hairs, sometimes white tipped; pads of feet normally more light-coloured, feet broader; upper incisors grooved or plain, moderately opisthodont to orthodont; molars relatively broad, more heavily built (*afra* group) . . . .

T. brantsii, p. 88

Tatera leucogaster (Peters 1852)   PLATE 34

Bushveld gerbil

Bosveldnagmuis

Taxonomic status

Classified in the *robusta* group by Davis (1975). No subspecies are recognised (Meester et al. 1986).

Distribution

Confined to the Lowveld and Coast lowlands bioregions of Maputaland (Figure 79). Avery (1991) recorded a species of *Tatera* (probably *T. leucogaster*) from the upper Pleistocene and very late Holocene at Border Cave (see under **Records of occurrence**).

Protected areas

Coastal Forest Reserve (=Kosi Bay), Mkuzi, Ndumu, Tembe.

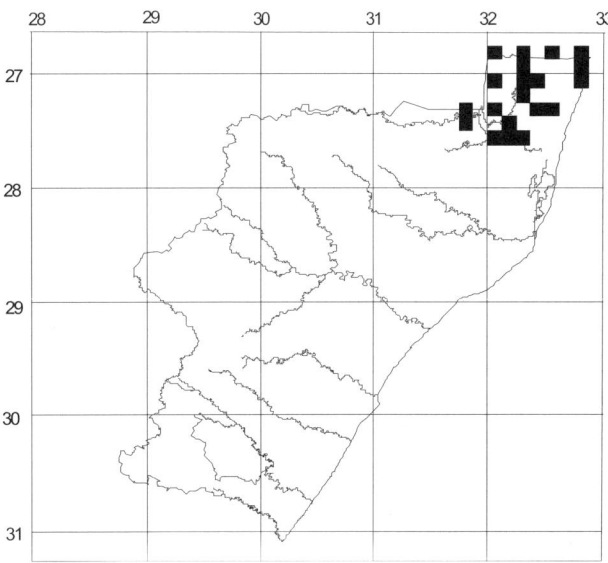

Figure 79

Conservation status

Commonly trapped in bushveld habitats wherever it occurs.

Habitat

Sandy soils in open woodland with more than 250 mm annual rainfall. During a trapping exercise at Mkuzi GR they were collected exclusively in sand forest and not in other sampled habitats which included: tall, dense *Panicum sp.* grassland in open thornveld, disturbed grassland associated with alluvial soils, and open *Acacia nilotica* woodland on stony hill slopes. At Tembe Elephant Park they were collected in termitaria veld on sandy soils, on the ecotone between a vlei and a swamp, as well as in closed woodland on sandy soils. A number of specimens in the Ndumu area were collected in the vicinity of buildings including schools, stores and residential compounds. They are elsewhere reported to occur in a variety of vegetational associations from grassland to savanna woodland (Skinner and Smithers 1990).

Habits

They are nocturnal. Burrows of some 40 mm diameter are occupied by a pair of individuals. Burrows are excavated in sandy soils, usually with the openings at the base of a bush or grass clump and marked by the presence of a conspicuous ramp of loose sand in the mornings. They eat almost equal amounts of insects, seeds and herbage (Perrin and Swanepoel 1987, cited in Skinner and Smithers 1990).

Breeding

No data for KwaZulu-Natal. They have a distinct summer breeding season in greater Transvaal and a more extended year-round season in Zimbabwe and Botswana. Between two and nine young are born, usually four or five (Skinner and Smithers 1990 and references within).

Linear measurements (mm) and mass (g)

	Males					Females				
	x̄	s.d.	n	Min	Max	x̄	s.d.	n	Min	Max
TL	284.5	20.1	6	246	301	299.6	13.9	9	279	320
HB	137.8	13.2	6	113	150	143.8	10.9	9	126	152
T	146.7	8.6	6	133	153	154.4	12.7	9	130	170
Hf	–	–	2	33	35	33.5	1.7	4	32	35
E	20.9	2.4	6	18	25	21.2	1.4	8	18	23
Mass	–	–	2	76.0	81.2	78.8	7.8	4	70.0	86.8

Records of occurrence

SPECIMEN RECORDS:

DM: Mkuze, Mkuzi GR, Tembe Elephant Park;

NM: Mseleni, Ndumu GR;

SI (species uncertain): Coastal Forest Reserve (=Kosi Bay), Maputa, Ndumu GR;

TM: Coastal Forest Reserve (=Kosi Bay), Candover (6 km from, on road to Collel), Ingwavuma River, Ingwavuma River (6.4 km north of), Leeuwspoor Farm (No. 647), Manaba, Manguzi Forest, Maputa, Maputa (3.2 km from, on Ubombo road), Maputa Store, Maputa Store (4 km from, on Ndumu road), Mkuzi GR (Bube Pan), Ndumu GR, Ndumu Police Station, Ndumu Store (1.6 km from, on Abercorn road), Otobothini, Shongwe Store (10.6 km from, on Ubombo road), Sihangwane, Sihangwane (6 km from, on road to Maputa), Tete Pan Camp, Tete Pan School.

ADDITIONAL RECORDS:

Avery (1991): Border Cave.

Tatera brantsii (Smith 1836)
T. b. brantsii (Smith 1836)
T. b. ruddi (Wroughton 1906)
Highveld gerbil
Hoëveldse nagmuis

Taxonomic status

Classified in the *afra* group by Davis (1975). Two of the three southern African subspecies are found in KwaZulu-Natal, with *T. b. brantsii* reported to occur in the midlands and *T. b. ruddi* reported to occur from Richards Bay to Kosi Bay (Meester et al. 1986).

Distribution

Western and north-eastern KwaZulu-Natal (Figure 80). Drier upland and Highland bioregions of western KwaZulu-Natal (*T. b. brantsii*), as well as the Coast lowlands bioregion of Zululand, from Richards Bay in the south to Maputa in the north (*T. b. ruddi*). A *Tatera* species (probably *brantsii* based on modern distribution) was recorded from middle and late Holocene deposits at two localities in the upper Tugela and Umgeni Basins (Avery 1991: see under **Records of occurrence**).

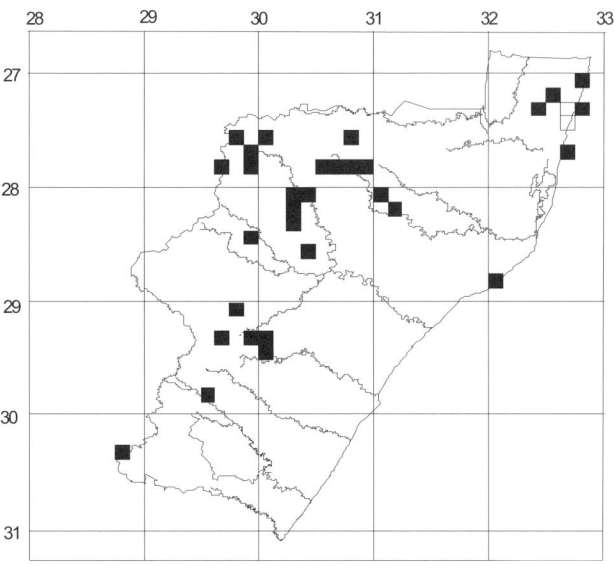

Figure 80

Protected areas

Eastern Shores, Ncandu, Sodwana State Forest, Umfolozi.

Conservation status

Locally common within their restricted distribution in KwaZulu-Natal.

Habitat

In KwaZulu-Natal they are associated with grasslands on sandy soils. They can tolerate drier conditions than can *T. leucogaster* (De Graaff 1981). According to De Graaff (1981) they prefer sandy peaty soils around vleis and marshes. They have been collected on agricultural lands with sparse, short grass cover and can become pests on agricultural lands as well as on golf courses. During the 1991/1992 drought, the species seems to have expanded its range eastwards into the Mistbelt bioregion of KwaZulu-Natal as evidenced by an invasion of these gerbils onto a golfcourse at Bosch Hoek Farm in the Balgowan area where they had not previously been recorded. Judging from their distribution in KwaZulu-Natal they seem to avoid bushveld habitats, and do not therefore occur in broad sympatry with *T. leucogaster* as suggested by Skinner and Smithers (1990). The distributions of the two KwaZulu-Natal sub-species of *T. brantsii* appear to be geographically separated by the Lowveld bioregion in Zululand.

Habits

Similar to *T. leucogaster*. Their warrens may cover an area as great as 65–70 ha; home ranges of males and females are 0.49 ha and 0.19 ha respectively. They are omnivorous, eating grass, seeds, termites and small insects (Skinner and Smithers 1990).

Breeding

No data for KwaZulu-Natal. From studies elsewhere, between one and five young (mean = 3.3) are born after a gestation of 22.5 days (Measroch 1954). Most studies indicate a year-round breeding season (Skinner and Smithers 1990) although De Moor (1969) found them to be seasonal breeders on the South African highveld.

Linear measurements (mm) and mass (g)

	Males					Females				
	x̄	s.d.	n	Min	Max	x̄	s.d.	n	Min	Max
TL	279.4	22.8	8	255	309	270.3	30.8	6	226	320
HB	138.6	11.6	8	122	160	128.7	17.5	6	106	159
T	140.8	14.8	8	122	165	141.7	16.0	6	120	161
E	–	–	–	–	–	–	–	1	17	17

Records of occurrence
SPECIMEN RECORDS:

DM: Dundee, Ncandu NR;

KM: Matatiele, Umfolozi GR;

NM: Nottingham Road, Matatiele, Castle View, Spring Grove;

TM: Buckland Downs, Buffalo River Bridge, Buffels River Bridge (Ingogo–Wakkerstroom road, Utrecht–Newcastle road), Bulwer (23 km west), Dundee (18 km from; 8 km from, on Pomeroy road), Estcourt (16 km from, on Nottingham Road road), Hazeldene, Kilgobbin Farm, Lake Bangazi, Lake Sibayi (Bandabanda Bay, Research Station), Manaba, Maputa, Maputa Aerodrome, Maputa Store (3 km from), Matatiele, Newcastle (3 km from, on Dundee road; 10 km from, on Utrecht road), Ngutshane Dip, Nottingham Road (4 km from, on Estcourt road), Piet Retief (32 km from, on Paulpietersburg road), Pomeroy (4 km north, on Dundee road), Richards Bay, Richards Bay (11 km west, on Empangeni road), Richards Bay Hotel (7 km from), Swartberg, Tayside, Underberg (3 km west, on Drakensberg Gardens road), Utrecht (32 km from, on Blood River road), Vryheid (19 km east), Vryheid (10 km from, on Louwsberg road), Vryheid (10 km from, on Magudu road).

ADDITIONAL RECORDS:

Avery (1991): Collingham, Nkupe.

Sight records (PJT): Hlatikulu Vlei, Bosch Hoek Farm.

Subfamily	CRICETINAE	
Genus	*Mystromys*	**Wagner 1841**

Mystromys albicaudatus (A. Smith 1834)   PLATE 35
White-tailed rat
Witstertrot

Taxonomic status
No subspecies were recognised by Meester et al. (1986).

Distribution
It occurs at low densities throughout the higher-lying, grassland-dominated Highland, Montane, Drier upland and Moist upland bioregions of KwaZulu-Natal (Figure 81). Records of occurrence of this species in the Richards Bay and Pongolapoort areas (De Graaff 1981) are unsubstantiated by known specimen records. Avery (1991) recorded this species from upper Pleistocene and very late Holocene deposits at Border Cave as well as from the middle Holocene (7 000–4 000 years BP) onwards at five sites in the Tugela Basin (see under **Records of occurrence**).

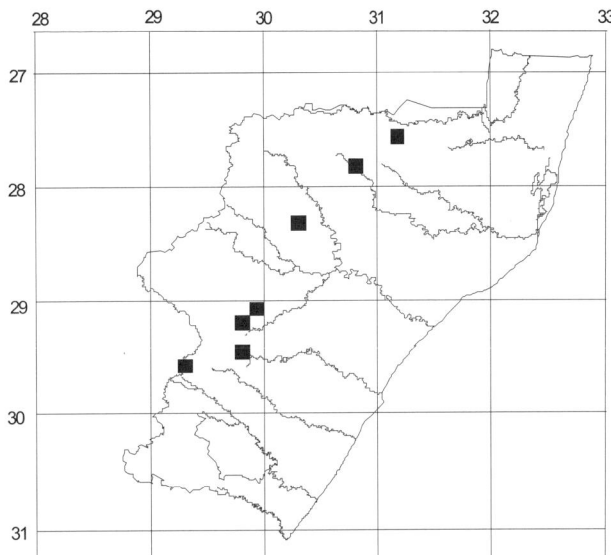

Figure 81

Protected areas
Giant's Castle, Itala, Mgeni Vlei.

Conservation status
Recorded in the South African Red Data Book (Smithers 1986) as Vulnerable. The white-tailed rat is endemic to greater South Africa (i.e. including Swaziland). Owing to low population densities and its K-selected reproductive strategy, Dean (1978) motivated for stricter conservation measures for this species. Few specimens have been collected in KwaZulu-Natal, although Lynch (1994) found it to be common in Lesotho.

Habitat
Occurring mainly within the Southern Savanna Grassland Biome, *Mystromys albicaudatus* has been said to prefer particularly dense grasslands on sandy soils in greater Transvaal (Rautenbach 1982), whereas in Lesotho (Lynch 1994), it favours rocky habitats near rivers, often associated with sparse vegetation. At Mgeni Vlei in KwaZulu-Natal they were caught in habitat resembling the latter, i.e. short, sparse grassland on a gentle stony slope on the margin of the Mgeni Vlei itself.

Habits
Nocturnal, living either in cracks in the ground or in burrows constructed themselves or by other species such as suricates *Suricata suricatta*. They eat insects, seeds and vegetable matter.

Breeding

No data for KwaZulu-Natal. In Lesotho (Lynch 1994), pregnant females were collected between October and February and were observed to contain between three and four embryos ($\bar{x} = 3.4$).

Linear measurements (mm) and mass (g)

	$\bar{x}$	s.d.	n	Min	Max
Males					
TL	192.3	19.8	3	170	208
HB	132.0	14.2	3	116	143
T	60.3	5.7	3	54	65
Hf	–	–	1	27	27
E	23.7	1.5	3	22	25
Mass	34.0	–	1	34	34

SPECIMEN RECORDS:

BM: Willbrook Farm;

DM: Sani Pass (near), Mgeni Vlei NR, Willbrook Farm;

SAM: Willbrook;

TM: Dundee (5 km from, on Pomeroy road), Itala GR, Giant's Castle GR (Sani Pass), Mgeni Vlei NR, Vryheid.

ADDITIONAL RECORDS:

Avery (1991): Border Cave, Collingham, eSinhlonhweni, Gehle, Mgede, Nkupe.

> ### Subfamily   CRICETOMYINAE
> ### Genus   *Saccostomus*   Peters 1846

Saccostomus campestris (Peters 1846)   PLATE 36

S. c. campestris (Peters 1846)

Pouched mouse

Wangsakmuis

Taxonomic status

All southern African material is referred to a single subspecies (Meester et al. 1986). Gordon and Rautenbach (1980) described chromosomal variation within the species, but the taxonomic significance of this variation remains unclear. Chromosomal diploid number was constant (2n=46) for 27 karyotyped specimens in the Durban Natural Science Museum and Transvaal Museum collections, identical to that reported for populations in KwaZulu-Natal by Gordon and Rautenbach (1980).

Distribution

The distribution of the pouched mouse in KwaZulu-Natal is closely associated with the limits of the Lowveld and Coast lowlands bioregions in north-eastern KwaZulu-Natal north of 29°S latitude; from the Richards Bay area (29°S) their distribution extends inland along Valley bushveld in the Tugela Basin as far west as Colenso District (Figure 82). The species has been recorded from very late Holocene

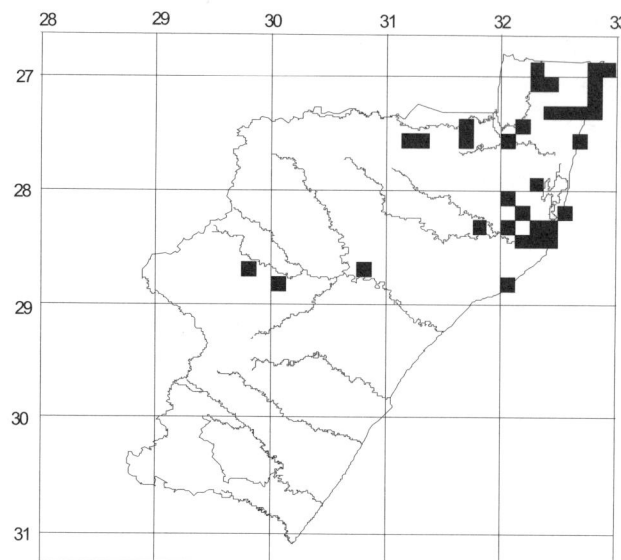

Figure 82

deposits at Mbabane in the Tugela Basin, and in upper Pleistocene and very late Holocene deposits at Border cave in the Lebombo Mountains (see under **Records of occurrence**).

Protected areas

Coastal Forest Reserve (=Kosi Bay), Eastern Shores (Cape Vidal), False Bay Park, Hluhluwe, Itala, Mkuzi, Ndumu, Sodwana Bay, Tembe, Umfolozi, Weenen.

Conservation status

Fairly common.

Habitat

Pouched mice are generally extremely catholic in their habitat requirements; throughout their wide distribution in southern Africa they occupy an extreme range of altitude (sea level to 1 800 m), rainfall (250 to over 1 200 mm) and vegetation type (e.g. grassland fringes of pans, rocky koppies, open *Acacia* bushveld and lowland forest). In KwaZulu-Natal they avoid pure grassland habitats and have been recorded in open *Acacia* bushveld, closed coastal forest (*Acacia karroo*), palm veld, tall grass and on marshy ground. At Umfolozi GR they were shown to increase their numbers following a drought (Bowland 1986).

Habits

Nocturnal and solitary. Their burrows, usually occupied by a single individual, may be excavated themselves in sandy substrates and are characterised by a vertical entrance shaft. On harder ground they use a variety of existing natural holes or crevices including disused burrows of other species. At Pongola in northern Zululand, Swanepoel (1972) found them to use holes in termitaria as well as other naturally ocurring holes. They are known to undergo bouts of torpor and are slow moving animals which often fall prey to carnivores. They feed predominantly on seeds from a wide range of forbs, shrubs and trees (see Skinner and Smithers 1990 and references within) which are carried in their cheek pouches and hoarded in storage chambers in their burrows. In addition, they feed on insects and herbage.

Breeding
In KwaZulu-Natal, young are born between October and February (Swanepoel 1972), compared with between October and April in greater Transvaal (Rautenbach 1982), January and April in Botswana (Smithers 1971), and February and April in Zimbabwe (Smithers and Wilson 1979). Mean number of embryos per pregnant female was 7.4 (range 5–10) in Botswana (Smithers 1971), and 6.7 (range 1–10) in Zimbabwe (Smithers and Wilson 1979). Seven pregnant females (**TM**) were collected during January at Lake Sibayi; the number of embryos per pregnant female ranged from five to six ($\bar{x}$ = 5.6; n = 5). Gestation in captivity is 20–21 days, and females first give birth at 96 days old (Earl 1978).

Linear measurements (mm) and mass (g)

	Males					Females				
	$\bar{x}$	s.d.	n	Min	Max	$\bar{x}$	s.d.	n	Min	Max
TL	165.9	12.1	11	144	190	158.2	16.0	13	135	184
HB	125.9	9.8	11	106	140	117.3	10.8	14	99	140
T	40.0	6.1	11	31	50	41.6	7.5	13	30	55
Hf*cu*	19.6	0.9	8	18	21	19.2	0.9	4	18	20
E	17.0	1.1	8	16	19	15.2	1.8	5	13	17
Mass	–	–	–	–	–	–	–	2	33	40

Males are larger in TL, HB and E than females, but this is not apparent in a larger sample from greater Transvaal (Rautenbach 1982). Pouched mice from KwaZulu-Natal are similar in body size to pouched mice from greater Transvaal (mean TL=166 and 165 mm for males and females respectively; Rautenbach 1982) and larger than pouched mice from Botswana (mean TL=156 and 155 for males and females respectively; Smithers 1971).

Records of occurrence
SPECIMEN RECORDS:
BM: Umfolozi;
DM: Bhanga Nek, Cape Vidal, Lake Eteza, Mfongosi, Richards Bay (ALUSAF), Weenen;
KM: Vergeval Farm;
NM: Boteler Point, Entendweni Bush, Geluk Farm, Hluhluwe GR, Magudu, Mfongosi, Ndumu GR, Sodwana Bay Park, Umfolozi Drift;
SI: Coastal Forest Reserve (=Kosi Bay), Makatini Flats, Maputa Aerodrome, Nkonkoni;
SAM: Lake St Lucia, Mfongosi;
TM: Coastal Forest Reserve (=Kosi Bay), Dukuduku Forest Station, Eastern Shores NR (Mission Rocks), False Bay Park, Hluhluwe GR, Itala GR (Craigadam, Doornpan), Lake Sibayi (Banana Pan, Research Station), Lake St Lucia, Manaba, Manzengwenya Forest Station (Lalanek Inspection Quarters), Maputa, Maputa Store (4 km from, on Ndumu road), Mkuzi Station (=Mkuzi GR), Monzi, Mseleni (6 km from), Ndumu GR (NRC Camp), Ndumu

(Gumede's Kraal), Richards Bay, Sihangwane, St Lucia Forest Station, St Lucia Game Park (Sugarloaf Camp), Tembe Elephant Park, Tete Pan (and 50 km north), Umfolozi GR, Weenen NR.

ADDITIONAL RECORDS:
Avery (1991): Border Cave, Mbabane;
Bourquin et al. (1971): Hluhluwe GR, Umfolozi GR;
Bruton 1978: Banda Banda, Dumila, Manzengwenya, Mbazwane, Mbibi;
Dixon (1966): Mkuzi GR.

Subfamily DENDROMURINAE
Genus *Dendromus* **A. Smith 1829**

KEY TO SPECIES (After Meester et al. 1986)

1. Fifth hind toe with rounded nail; fifth forefinger present, vestigial; ears darker than back; general colour normally grey . . . .
    D. melanotis, p. 91
— Fifth hind toe with pointed nail; fifth forefinger lacking; ears resemble back in colour; general colour normally brown 2

2. Larger, adult skull longer than 22 mm; ventral hair usually slaty-based; black line invariably present on back . . . .
    D. mesomelas, p. 93
— Smaller, adult skull <22 mm; ventral hair white or ochraceous, not slaty . . . .
    D. mystacalis, p. 93

Dendromus melanotis (A. Smith 1834)   PLATE 37
D. m. melanotis (A. Smith 1834)
D. m. vulturnus (Thomas 1916)
Grey climbing mouse
Grysklimmuis

Taxonomic status
The type specimen of this species was recorded from Durban by Andrew Smith in 1834. Five subspecies are recognised for southern Africa, of which two occur in KwaZulu-Natal, *D. m. melanotis* from southern, central and western parts and *D. m. vulturnus* from north-eastern Zululand (Meester et al. 1986).

Distribution
The grey climbing mouse has a scattered distribution in KwaZulu-Natal (Figure 83), occurring from sea level up to 2 700 m at Giant's Castle (Rowe-Rowe and Meester 1982a). There appear to be two foci of distribution (north-eastern and a belt from the Drakensberg to the coast) sug-

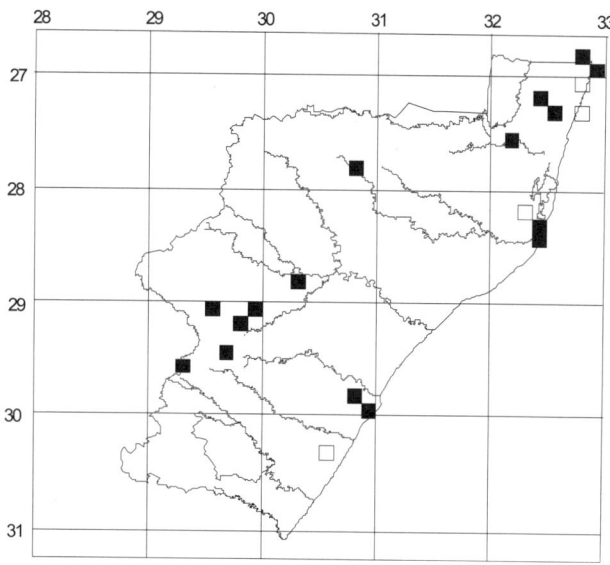

Figure 83

gesting that the subspecies *D. m. melanotis* and *D. m. vulturnus* may be geographically separate in the province. This species was recorded from Holocene deposits at Border Cave in the Lebombo Mountains as well as at two sites in the upper Tugela and Umgeni Basins (Avery 1991; see under **Records of occurrence**).

Protected areas
Bluff, Giant's Castle, Krantzkloof, Mapelane, Mkuzi, Vryheid.

Conservation status
Widespread in KwaZulu-Natal although not commonly collected in traps.

Habitat
Although apparently typically associated with tall stands of grasses such as *Hyparrhenia* and *Merxmuellera spp.* (Lynch 1994), they have been collected in shorter grasslands in the Drakensberg, up to 33 months after fire (Rowe-Rowe and Meester 1982a), and at Vryheid NR, one was trapped in short grassland recently cleared of wattles (present study). Duckworth (1984) collected this species from short, highlying, *Cymbopogon validus-Cyperus natalensis*-dominated grassland at Nyalazi State Forest in Zululand. Dixon (1964) recorded them from reedbeds and aloe plants. According to Rowe-Rowe and Meester (1982a), their avoidance of long grass resulted in habitat segregation between this species and *D. mesomelas* (which preferred longer grass) in the KwaZulu-Natal Drakensberg.

Habits
They are nocturnal and solitary. As the name suggests, grey climbing mice climb tall grass stalks to forage for seeds. The long tail is used for support and balance. During the early summer months, they build ball-shaped nests,

40–60 mm in diameter, about one metre from the ground, using shredded grass and fibres. These are discarded after the breeding season. According to De Graaff (1981), they also make use of burrows up to 500 mm deep and 40–60 mm in diameter, although this was not observed by Rautenbach (1982). Apart from seeds, they also eat a range of insects including termites, grasshoppers, moths and small beetles (Skinner and Smithers 1990). In KwaZulu-Natal Drakensberg, a series of 14 stomachs analysed all contained seed while 24 per cent contained arthropods (Rowe-Rowe 1986).

Breeding
In the Drakensberg, Rowe-Rowe and Meester (1982b) recorded a lactating female in January, perforate females in May and November, males with scrotal testes in April, October and November, and immature animals in April and May, suggesting that the young are born during the summer months in KwaZulu-Natal. Based on limited data elsewhere, two to eight young are born in the summer months between November and April (Skinner and Smithers 1990 and references within; Lynch 1994).

Linear measurements (mm) and mass (g)

	Males					Females				
	$\bar{x}$	s.d.	n	Min	Max	$\bar{x}$	s.d.	n	Min	Max
TL	144	–	–	–	–	145	–	–	–	–
HB	71	1.4	14	63	79	71	1.2	19	64	88
T	73	1.2	14	65	79	74	1.3	19	68	88
Hf	17	0.2	14	16	18	18	0.3	19	15	19
E	16	0.3	13	15	17	16	0.3	19	14	19
Mass	10	0.6	14	6	13	10	0.8	19	6	19

The above data apply to a series collected at Giant's Castle Game Reserve (Rowe-Rowe 1982).

Records of occurrence
SPECIMEN RECORDS:
BM: Willbrook Farm;
DM: Krantzkloof, Mapelane, Vryheid NR;
KM: Mkuzi GR;
SI: Coastal Forest Reserve (=Kosi Bay), Lesotho–KwaZulu-Natal border, Makhake Store (Sani Pass);
TM: Clairwood, Firle Farm, Giant's Castle GR, Lake Sibayi, Maputa (road to), Mpophomeni, St Lucia Forest Station.

ADDITIONAL RECORDS:
Avery (1991): Border Cave, Collingham, Nkupe;
Bourquin and Sowler (1980): Vernon Crookes NR;
Bronner and Meester (1987): Bluff NR;
Dixon (1964): Mkuzi GR;
Duckworth (1984): Nyalazi State Forest;
Rautenbach et al. (1980): Lake Sibayi, Maputa;
Rowe-Rowe and Meester (1982a): Giant's Castle.

Dendromus mesomelas (Brants 1827)
D. m. mesomelas (Brants 1827)
Brant's climbing mouse
Brants-klimmuis

Taxonomic status
Two subspecies occur in southern Africa, of which one is found in KwaZulu-Natal (Meester et al. 1986).

Distribution
Brant's climbing mouse has a wide but scattered distribution in KwaZulu-Natal (Figure 84), occurring from sea level up to 2 400 m at Giant's Castle (Rowe-Rowe and Meester 1982a). Its occurrence in Zululand constitutes an extension to its known range (De Graaff 1981; Skinner and Smithers 1990). It has been recorded from the middle Holocene (7 000–4 000 years BP) onwards in archaeological deposits at four sites in the Tugela and Umgeni Basins (see under **Records of occurrence**).

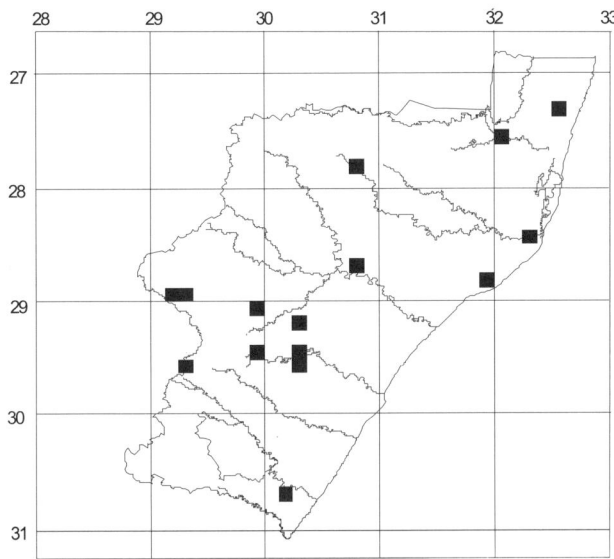

Figure 84

Protected areas
Cathedral Peak, Fort Nottingham, Giant's Castle, Oribi Gorge, Vryheid.

Conservation status
Widespread but not common in KwaZulu-Natal.

Habitat
In KwaZulu-Natal, they have been collected in tall grass clumps in a waterlogged vlei, *Leucosidea*-dominated woodland on a streambank in the Drakensberg, bushveld on a south-facing slope, and in a suburban garden (present study). In the Drakensberg, they have been recorded in forest, scrub, temperate grassland, boulder-bed and tall grassland habitats (Rowe-Rowe and Meester 1982a). Bodbijl (personal communication) collected them in understory *Isoglossa woodii* in the Mfolozi swamp forest (Dukuduku Forest), as well as in adjacent thickets.

Rautenbach (1982) claims that the species requires woody plants, while Smithers and Tello (1976) stated that it prefers tall grass in wetter habitats.

Habits
Similar to the previous species. A series of stomachs collected in the Drakensberg all contained seeds, whereas 75 per cent contained insects (Rowe-Rowe 1986).

Breeding
During the present study, a pregnant female with four foetuses was collected in late November. A lactating female and a male with scrotal testes were collected during November at Cathedral Peak in the Drakensberg (Rowe-Rowe and Meester 1982b). A pregnant female in the Transvaal Museum collection was collected during February at Lake Sibayi. No comparative data are available from elsewhere in the subregion (Skinner and Smithers 1990).

Linear measurements (mm) and mass (g)

	Males					Females				
	$\bar{x}$	s.d.	n	Min	Max	$\bar{x}$	s.d.	n	Min	Max
TL	164.8	4.6	4	159	170	172.8	16.4	4	162	197
HB	68.0	2.9	4	65	71	73.2	6.2	4	68	82
T	96.8	5.6	4	93	105	99.5	10.5	4	92	115
Hf	–	–	2	18	19	19.2	1.0	4	18	20
*E	13.0	–	3	11	16	13.0	1.0	4	12	15
*Mass	11.0	–	3	10	14	14.0	–	1	14	14

*Masses and ear length (for males) obtained from Rowe-Rowe 1982 for a series from Giant's Castle Game Reserve

Records of occurrence
SPECIMEN RECORDS:
DM: Cathedral Peak NR (Catchment 9), Fort Nottingham NR, Vryheid NR, Winterskloof;
NM: Empangeni, Hilton;
SAM: Mfongosi, Rietvlei (Umvoti);
SI: Makhake Store;
TM: Cathedral Peak NR, Lake Sibayi, Oribi Gorge NR.
Bodbijl (private collection): Dukuduku Forest.

ADDITIONAL RECORDS:
Avery 1991: Collingham, Diamond I, Mgede, Nkupe;
Bourquin and Mathias (1984): Oribi Gorge NR;
Rowe-Rowe and Meester (1982a, b): Cathedral Peak, Giant's Castle GR.

Dendromus mystacalis (Heuglin 1863)
D. m. jamesoni (Wroughton 1909)
Chestnut climbing mouse
Roeskleur-klimmuis

Taxonomic status
Two subspecies are recognised from the subregion, of which one is found in KwaZulu-Natal (Meester et al. 1986).

Distribution

As with other species of *Dendromus*, this species has a wide but scattered distribution throughout most of KwaZulu-Natal (Figure 85). It has been recorded from upper Pleistocene (>15 000 years BP) and very late Holocene (<1 000 years BP) deposits at Border Cave in the Lebombo Mountains (Avery 1991).

Protected areas

Chelmsford, Eastern Shores, Fort Nottingham, Giant's Castle NR, Hazelmere Dam, Hluhluwe, Ndumu, Umtamvuna, Umgeni Valley, Vernon Crookes, Weenen.

Conservation status

Widespread but uncommon in KwaZulu-Natal.

Habitat

In KwaZulu-Natal, chestnut climbing mice have been collected in habitats ranging from rank tall coarse grassland near water, reedbeds and dense vegetation in a suburban garden (present study).

Habits

Similar to other species of *Dendromus* found in KwaZulu-Natal, although it is said to make more use of trees, and to occupy the nests of weaver birds (Roberts 1951).

Breeding

No data available for KwaZulu-Natal. Data elsewhere indicate that 3–8 young are born between January and March (Skinner and Smithers 1990).

Linear measurements (mm) and mass (g)

	Males					Females				
	x̄	s.d.	n	Min	Max	x̄	s.d.	n	Min	Max
TL	143.6	8.8	11	132	155	135.5	4.8	6	127	140
HB	62.7	7.2	11	48	70	57.0	4.6	6	52	64
T	80.9	5.9	11	72	86	78.5	3.7	6	74	84
Hf	16.3	1.2	6	15	18	15.7	1.5	3	14	17
E	–	–	2	11	18	9.3	2.5	3	7	12
Mass	–	–	1	8	8	–	–	–	–	–

Records of occurrence

SPECIMEN RECORDS:
DM: Chelmsford NR, Howick, Impendle NR (Scaleby Farm), Mfongosi, Thomas More College, Ubombo District, Westville, Winterskloof (13 Crompton Road);
KM: Lake St Lucia, St Lucia village;
NM: Gwaliweni Forest, Merrivale, Mfongosi, Town Bush, Umgeni Valley;
SI: Giant's Castle NR;
TM: Bluff NR, Hazelmere Dam, Kamberg NR (Stillerust), Karkloof Forest, Kenneth Stainbank, Hluhluwe GR (eGondeni), Magudu, Manaba, Midmar Dam, Ndumu GR (Nyamithi Pan), Pongola River (8 km west of Manaba), Seewater, Tembe Elephant Park, Ubombo, Umtamvuna NR, Vernon Crookes NR.

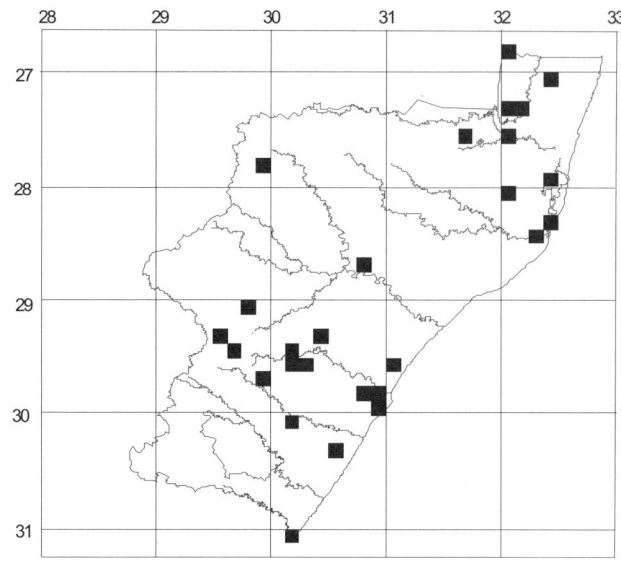

Figure 85

ADDITIONAL RECORDS:
Avery (1991): Border Cave;
Rautenbach et al. (1980): Manaba, Nyamithi Pan, Pongola River, Ubombo (17 km north);
Rautenbach and Bronner (1988): Dukuduku Forest;
Roberts (1931): Pongola River (8 km west of Manaba).

Genus	*Steatomys*	Peters 1846

KEY TO SPECIES (After Meester et al. 1986)

1.   Multimammate, mammae more than
     2 + 2 = 8, usually 12, up to a maximum
     of 16, not necessarily arranged in pairs . . . .
                                    S. pratensis, p. 95
—   Mammae 2 + 2 = 8 2

2.   Smaller, head and body length
     63–80 mm, greatest skull length
     19.5–23.0 mm; ear length sub-equal
     to or greater than hind foot length (*su*);
     tail white; dorsal colour reddish . . . .
                                    S. parvus, p. 95
—   Larger, head and body length
     70–105 mm, greatest skull length
     21.0–27.0 mm; ear length sub-equal
     to or less than hind foot length (*su*);
     tail brownish-grey above, white below;
     dorsal colour brown with a greyish
     tinge . . . . *S. krebsii*, p. 96

Steatomys pratensis (Peters 1846)　　PLATE 38
S. p. natalensis (Roberts 1929)
Fat mouse
Vetmuis

Taxonomic status
The genus *Steatomys* is in need of revision and species listed are regarded as provisional (Coetzee 1977). Three subspecies of *S. pratensis* are recognised from southern Africa, of which only one is found in KwaZulu-Natal (Meester et al. 1986).

Distribution
Sparsely distributed in northern Drakensberg, Zululand and Maputaland (Figure 86).

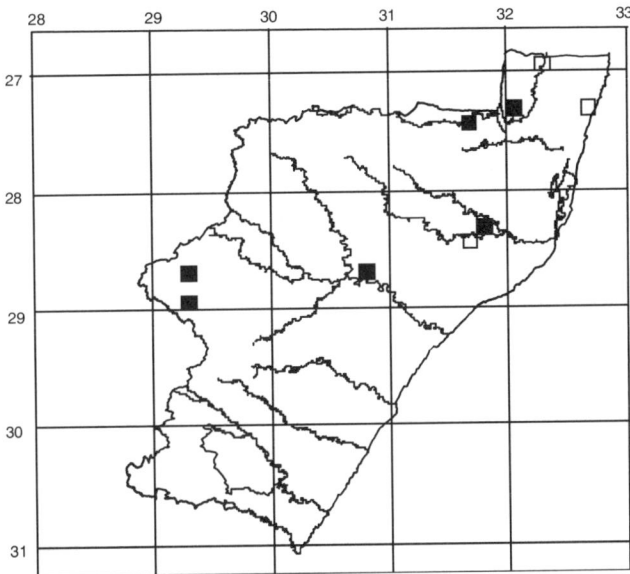

Figure 86

Protected areas
Cathedral Peak, Ndumu, Umfolozi.

Conservation status
Rare in KwaZulu-Natal, although a population flush was recorded during 1984 in the Cathedral Peak region of the Drakensberg (Richardson and Perrin 1992).

Habitat
In KwaZulu-Natal they have been recorded in 'long grass' (Dixon 1966) and 'open grassland' (Bruton 1978). They are generally reported to occur on sandy soils in drier areas, and on the fringes of rivers, pans and cultivated areas.

Habits
They are nocturnal and excavate several tunnels leading down to a central chamber lined with shredded grass (De Graaff 1981). Fat mice undergo torpor. In a captive population collected in the KwaZulu-Natal Drakensberg, the proportion of torpid animals increased in winter and spring, corresponding to the non-breeding season (Richardson and Perrin 1992).

Breeding
Although Richardson and Perrin (1992) recorded some reproductive activity (perforate females and scrotal males) throughout the year in a captive population from the Drakensberg, reproductive activity was depressed between May and December for females, and between September and December for males. This contrasts with a study in the Kruger National Park where no reproductive activity was recorded between April and the end of November (Kern 1977). Between one and nine (mean 3.2) young are born between October and May (Skinner and Smithers 1990). Females have up to 16 mammae, a feature which distinguishes this species from other members of the genus in southern Africa.

Linear measurements (mm)

	Males					Females				
	x̄	s.d.	n	Min	Max	x̄	s.d.	n	Min	Max
TL	126.5	–	2	117	136	–	–	1	137	137
HB	89.0	–	2	80	98	–	–	1	88	88
T	–	–	2	37	38	–	–	1	49	49
Hf	–	–	1	16	16	–	–	–	–	–
E	16.0	–	1	16	16	–	–	–	–	–

Records of occurrence
SPECIMEN RECORDS:
DM: Mfongosi;
KM: Vergeval Farm, Gwaliweni;
NM: Umfolozi Drift;
TM: Bergville, Cathedral Peak.

ADDITIONAL RECORDS:
Bruton (1978): Mbibi;
Dixon (1966): Ndumu GR;
Bourquin et al. (1971): Umfolozi GR.

? Steatomys parvus (Rhoads 1896)
S. p. tongensis (Roberts 1931)
Tiny fat mouse
Dwergvetmuis

De Graaff's (1981) distribution map for this species showed three locality records in northern KwaZulu-Natal. However, no specimens are available to corroborate the occurrence of this species in KwaZulu-Natal. The localities plotted in De Graaff (1981) appear (from their geographical position) to be those mentioned by Roberts (1931) in describing two new taxa from northern KwaZulu-Natal: *S. chiversi chiversi* (=*S. krebsii* according to Coetzee 1977) and *S. chiversi tongensis* (=*S. parvus* according to Coetzee 1977). Based on Roberts (1931) type description of *S. chiversi tongensis* from Manaba in northern Zululand, this form is allied to *S. krebsii* and not to *S. parvus* as supposed by Coetzee 1977) (skull length >23 mm; hair colour brown with grey tinge, not reddish; tail brown above and white below, not white; see **Key**).

Meester et al. (1986) synonymises *chiversii* Roberts 1931 (from Blood River, Buffalo River and Newcastle) with *krebsii*, but still retains *chiversii tongensis* in *parvus*, following Coetzee (1977). I have examined the series of *Steatomys* specimens from northern KwaZulu-Natal examined by Roberts (1931) (**TM** collection): all specimens (including a single specimen of *tongensis* from Manaba) keyed out unequivocally as *S. krebsii*. I suggest that *tongensis* Roberts 1931 should be removed from *S. parvus* and placed in *S. krebsii*, thereby recognising only two species of *Steatomys* from KwaZulu-Natal (*S. pratensis* and *S. krebsii*), pending a taxonomic revision of the genus. This treatment would be consistent with the recent collection of *S. krebsii* but not *S. parvus* from KwaZulu-Natal, and would also explain the anomaly of a disjunct southern KwaZulu-Natal isolate of *S. parvus* (which otherwise occurs from Botswana northwards to the Sudan; Skinner and Smithers 1990).

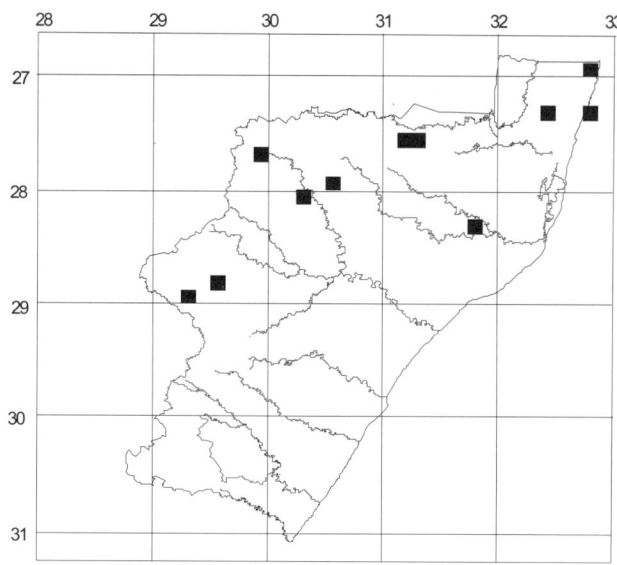

Figure 87

Steatomys krebsii (Peters 1852)
S. k. orangiae (Roberts 1929)
Kreb's fat mouse
Krebs-vetmuis

Taxonomic status
See notes above under *S. parvus*. It is here considered best to refer all material from KwaZulu-Natal described by Roberts (1931), as well as recently collected material other than *S. pratensis*, to *S. krebsii orangiae*. Differences between *chiversi* and *tongensis* on one hand and *orangiae* on the other (Roberts 1931), relate mostly to the smaller size of the former (hence the confusion with the smaller *parvus*). Based on the small size of samples available to Roberts (1931), and the possibility of immature specimens (see Coetzee 1977), these differences may simply indicate geographical variation within one subspecies; however further taxonomic study may reveal that *S. krebsii* from KwaZulu-Natal represents a distinct subspecies or even species.

Distribution
De Graaff (1981) and Smithers (1983) recognised *S. parvus* but not *S. krebsii* from northern KwaZulu-Natal. Meester et al. (1986) and Skinner and Smithers (1990) followed Coetzee (1977) in recognising both of these species in KwaZulu-Natal. However, as discussed above, only *S. krebsii* is regarded here as occurring in KwaZulu-Natal, necessitating a significant northwards contraction in the known range of *S. parvus* and a modest easterly extension in the range of *S. krebsii* from the Free State, compared with published distribution maps (De Graaff 1981; Smithers 1983; Skinner and Smithers 1990). *Steatomys krebsii* is fairly widely distributed from Cathedral Peak, Newcastle and Winterton in western KwaZulu-Natal, to Kosi Bay in north-eastern KwaZulu-Natal (Figure 87).

Protected areas
Cathedral Peak, Coastal Forest Reserve (=Kosi Bay), Itala, Umfolozi.

Conservation status
Rare in KwaZulu-Natal.

Habitat
Specimens from Itala GR were collected in old agricultural lands with red loam soils (Rautenbach et al. 1981).

Habits
Nocturnal and terrestrial, occurring singly or in pairs (Skinner and Smithers 1990).

Breeding
A pregnant female with five embryos was collected at Itala Game Reserve (Rautenbach et al. 1981). No further information is available for the subregion.

Linear measurements (mm)
Roberts (1931) gave the following ranges for measurements obtained from a series of six *S. chiversi* (= *krebsii*) from northern KwaZulu-Natal: HB 70–80; T 36–48; Hf*su* 14–15; E 14–16. These ranges overlap closely the ranges given for *S. parvus* (*n* = 5; HB 64–86; De Graaff 1981) and for *S. krebsii orangiae* from the Free State (*n* = 24; HB 70–80; Roberts 1931), but are slightly lower than *S. krebsii* from Botswana (*n* = 9; mean HB = 79; Smithers 1971).

Records of occurrence
SPECIMEN RECORDS:
BM: Umfolozi;
SI: Coastal Forest Reserve (=Kosi Bay);
TM: Blood River Station, Cathedral Peak, De Jager's Drift (Buffalo River), Itala GR (Craigadam, Doornkraal), Lake Sibayi (Research Station, 3 km north), Manaba, Newcastle, Winterton.

ADDITIONAL RECORDS:
Rautenbach et al. (1980): Lake Sibayi;
Roberts (1931): Blood River Station, Buffalo River, Manaba, Newcastle.

Subfamily MURINAE

KEY TO GENERA (After Meester et al. 1986)

1. Longitudinal lines along mid-back 2
– Back unlined 3

2. One dark line along mid-back . . . .
   Lemniscomys, p. 97
– More than one dark line along mid-back . . . .
   Rhabdomys, p. 98

3. Strong interorbital constriction . . . .
   Dasymys, p. 100
– No strong interorbital constriction 4

4. T7 present on M¹ or at least represented
   by a strong ridge; head and body length
   100–125 mm, tail long and slender,
   145–195 mm . . . . *Grammomys*, p. 101
– T7 lacking on M¹ 5

5. Rows of cusps on M¹ markedly distorted 6
– Rows of cusps on M¹ not markedly
   distorted 7

6. Extreme distortion of cusps; size smaller,
   greatest skull length <24 mm . . . . . *Mus*, p. 103
– Moderate distortion of cusps; size
   larger, greatest skull length in adults
   >25 mm . . . . *Mastomys*, p. 105

7. Dark markings ('spectacles') around eyes . . . . .
   Thallomys, p. 108
– No dark markings around eyes 8

8. Incisive foramina extending between
   molars; molars clearly cuspidate . . . .
   Aethomys, p. 109
– Incisive foramina not, or only just,
   extending between molars; molars
   not clearly cuspidate, with a tendency
   to become laminate . . . . *Rattus*, p. 112

Genus	*Lemniscomys*	Trouessart 1881

Lemniscomys rosalia (Thomas 1904)   PLATE 39
L. r. spinalis (Thomas 1916)
Single-striped mouse
Eenstreepmuis

Taxonomic status
Four subspecies are recognised in southern Africa; of these one (*L. r. spinalis*) occurs in KwaZulu-Natal (Meester et al. 1986).

Distribution
The single-striped mouse is restricted mostly to the drier Valley bushveld and Lowveld bioregions in KwaZulu-Natal (Figure 88). Its distribution extends marginally into Coast lowland and Drier upland bioregions, although it is usually associated with riverine bushveld or forest habitats within these bioregions. The species distribution in KwaZulu-Natal is almost complementary to that of the following species, *Rhabdomys pumilio*, which is almost entirely absent from bushveld habitats (Figure 89).

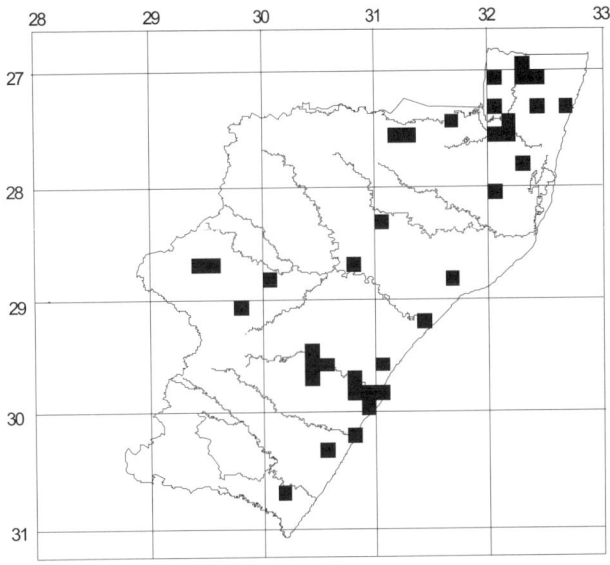

Figure 88

Protected areas
Albert Falls, Harold Johnson, Hazelmere Dam, Hluhluwe, Itala, Kenneth Stainbank, Krantzkloof, Moor Park, Ndumu, Ngoye, Oribi Gorge, Spioenkop, Tembe Elephant Park, Umfolozi, Vernon Crookes, Weenen.

Conservation status
Common in bushveld habitats.

Habitat
In KwaZulu-Natal they appear to prefer open to dense *Acacia spp.* woodland with a dense ground cover of long grass, often in riverine, rocky or hilly situations. Many of the specimens collected during this study were found in ecotones between vlei and bushveld, bushveld and grassland, and coastal forest and grassland habitats. The most important requirement seems to be the presence of tall dense stands of dry grass, although tall stands of herbs or reeds also appear to satisfy their requirements for cover. While they have been collected in grassy and mixed grass-scrub clearings within coastal forest, usually in riverine situations, they are largely absent from coastal forest and grassland habitats within the Coast lowland bioregion, perhaps because of the lack of grassy ground cover in coastal forest. They frequently occupy suburban gardens.

Habits

Although essentially diurnal, they were frequently collected in early morning trap checks during the present study, suggesting either crepuscular or nocturnal activity as well. They excavate their own burrows, use well demarcated runways in grass, and appear to breed in surface grass nests (De Graaff 1981). Males have larger home ranges than do females (Swanepoel 1972). At Itala Game Reserve in northern KwaZulu-Natal, their diet consisted mainly of grass stems (Rautenbach et al. 1981).

Breeding

Two pregnant females collected in November at Albert Falls NR had three and eight embryos (present study). Pregnant females from KwaZulu-Natal in the Transvaal Museum collection were collected in January, April and December; mean number of foetuses was 6.5 (range 6–7; *n*=4). Bourquin et al. (1971) recorded a pregnant female from Hluhluwe GR with seven embryos in April. Between two and eleven young have been recorded elsewhere (Rautenbach 1982; Skinner and Smithers 1990). In northern KwaZulu-Natal, Swanepoel (1972) gave the reproductive season as from September to February for females, and September to May for males.

Linear measurements (mm) and mass (g)

	Males					Females				
	x̄	s.d.	n	Min	Max	x̄	s.d.	n	Min	Max
TL	256.8	29.9	30	198	298	248.0	19.7	23	199	280
HB	125.9	16.4	31	95	161	121.3	9.3	23	104	142
T	131.6	15.2	29	88	160	126.5	14.0	23	84	141
Hf*su*	26.8	1.7	18	24	29	26.5	1.3	16	25	29
E	16.7	2.6	22	12	20	16.4	1.3	19	14	18
Mass	62.8	8.2	18	52	78	55.8	10.2	17	32	70

Records of occurrence

SPECIMEN RECORDS:
BM: Ngoye Hills;
DM: Albert Falls NR, Harold Johnson NR, Hazelmere NR, Itala GR, Kenneth Stainbank NR, Krantzkloof NR, Mfongosi, Mkuzi GR, Moor Park NR, Pietermaritzburg (Jesmond Road), Spioenkop Dam, Tembe Elephant Park, University of Durban-Westville, University of Natal (Durban), Weenen NR, Westville (1 Highgates Avenue);
KM: Vergeval Farm;
NM: Gwaliweni, Langewacht, Mfongosi, Mills Circle (Pietermaritzburg), Nagle Dam, Ndumu GR, Oribi (Pietermaritzburg);
SAM: Mfongosi;
SI: Makatini Flats, Nkonkoni;
TM: Ashburton, Bisley, Darvill Sewage Farm, Hazelmere Dam NR, Heskith Valley, Hluhluwe GR (Mansiya Valley), Ingwavuma River, Itala GR (ranger's house, Craigadam, Doornkraal, Doornpan, Potwe), Manaba, Mkuzi River, Mseleni (6 km from), Ndumu GR, Oribi Gorge NR, Otobotini, Pinetown, Shongwe Store, Sihangwane, Spioenkop NR, Tete Pan Camp, Theunis Bester Game Ranch, Vernon Crookes NR, Weenen.

ADDITIONAL RECORDS:
Avery (1991): Border Cave;
Bourquin et al. (1971): Hluhluwe GR (Hilltop);
Rautenbach et al. (1980): Abercorn Drift, Ubombo.

Genus	*Rhabdomys*	Thomas 1916

Rhabdomys pumilio (Sparrman 1784)   PLATE 40
R. p. dilectus (De Winton 1897)
Striped mouse
Streepmuis

Taxonomic status

Meester et al. (1986) recognised seven subspecies from the subregion, of which one (*R. p. dilectus*) is found in KwaZulu-Natal.

Distribution

The striped mouse occurs widely throughout the higher-lying, moister Montane, Highland, Mistbelt, Coast hinterland and Moist upland bioregions, being absent from or extending marginally into the Drier upland, Coast lowlands, Valley bushveld and Lowveld bioregions (Figure 89). As such, its distribution corresponds very closely to that of the vlei rat (Figure 77), with which it is frequently collected. Avery (1991) recorded *Rhabdomys pumilio* from numerous archaeological sites in KwaZulu-Natal, including Border Cave (up until 1 000 years BP) where the species no longer occurs.

Protected areas

Albert Falls, Cathedral Peak, Chelmsford, Coleford, Dlinza Forest, Fort Nottingham, Garden Castle, Giant's Castle, Highmoor, Impendle, Itala, Kamberg, Karkloof, Loteni, Mount Currie, Midmar, Ncandu NR, Oribi Gorge, Pongola

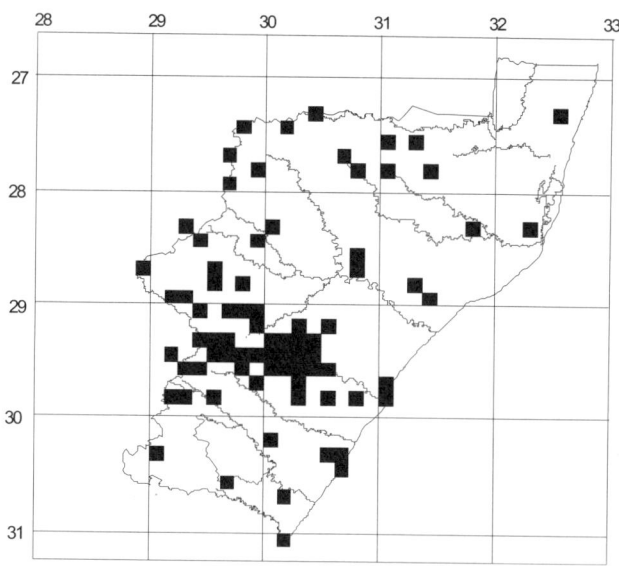

Figure 89

Bush, Queen Elizabeth Park, Royal Natal, Spioenkop, Umgeni Vlei, Umtamvuna, Vernon Crookes, Vryheid.

Conservation status
One of the commonest rodents in KwaZulu-Natal, in mesic grassland habitats.

Habitat
The striped mouse is by far the dominant small mammal species collected in mesic grassland-dominated habitats in KwaZulu-Natal. A maximum trap success of 22 captures per 100 trap-nights was recorded in the Mistbelt bioregion during the present study; Rowe-Rowe (1982) recorded a maximum trap success of 30 captures per 100 trap-nights at Cathedral Peak in the Montane bioregion. Although essentially a grassland species, *Rhabdomys pumilio* has been described as a broad-niche species (Brooks 1974). Based on trapping data for 448 specimens (present study), this characterisation applies to KwaZulu-Natal where they have been recorded from a very wide range of habitats including short to tall, sparse to dense, pure to mixed (shrub, bracken and exotic plant species) grassland and vlei habitats, forest habitats (including exotic pine and wattle plantations, montane *Podocarpus*, mixed mistbelt *Podocarpus*, and, rarely, coastal scarp and dune forest), *Acacia* bushveld, and poorly vegetated rocky ridges and steep slopes. Their presence in atypical habitats such as bushveld and forests is usually dependent on the presence of good grass cover within 1–2 km. For example, in the Drakensberg they have been recorded in montane forest patches bordering on habitats with good grass cover, but not from isolated Afromontane forest patches located high up in the sparsely-vegetated catchments of small tributaries. At Giant's Castle, Rowe-Rowe and Meester (1982a) found that relative abundance of the striped mouse decreased with increasing altitude, particularly from 2 200 m onwards, and Lynch (1994) found this species to occur at low densities in many high-lying areas of Lesotho. Together with *Otomys irroratus*, they are a major pest in commercial forestry plantations where they strip the bark of young pine saplings (Willan 1994).

Habits
They are conspicuously marked, diurnal animals which are frequently observed and therefore better known to the general public than are most other small mammals. In KwaZulu-Natal, they have been observed to build spherical grass nests on the ground, in which the young are born (De Graaff 1981). Unlike species like *Mastomys natalensis*, *Mus minutoides* and *Lemniscomys rosalia*, they are rarely found in association with human dwellings. Although the striped mouse has not been recorded in the Durban Metropolitan area in recent times, early records from Umbilo, Durban (**DM** 200, collected in 1914) and Durban Botanic Gardens (**BM** 86.12.11.5, collected in 1886) suggest that they may once have occurred here, but that they have been displaced by residential development. Their absence from urban and suburban habitats supports this suggestion.

In a series of 89 stomachs of *Rhabdomys pumilio* from the KwaZulu-Natal Drakensberg Park, 94 per cent contained seeds, 28 per cent contained green plant material, and 48 per cent contained arthropods. *Rhabdomys pumilio* remained predominantly granivorous throughout the year, although green plants and insects were consumed more frequently during the wet season compared with the dry season. The most commonly consumed arthropods were beetles (Coleoptera) although crustaceans, arachnids (spiders) and a variety of insects (including Diptera, Hymenoptera, Lepidoptera, Blattoidea, Orthoptera and Hemiptera) were also taken (Rowe-Rowe 1986). Wirminghaus and Perrin (1992) found their diet to be mixed throughout the year at Karkloof Forest, with green plant material comprising the bulk of the diet in all months, followed by fruit, then invertebrates. Flowers were eaten in spring and summer.

Breeding
A mean litter size of 6.1 (*n*=24; range 1–11) was recorded during the present study. The mean number of foetuses recorded in three pregnant specimens (**TM** collection) was 6.0 (range 5–7). These values are somewhat higher than for a population in the KwaZulu-Natal Drakensberg (Rowe-Rowe and Meester 1982b: x̄ = 4.5; range 3–7; *n*=22), but similar to results from the Free State and greater Transvaal (see Skinner and Smithers 1990). Pregnant females were collected between November and February, with the exception of two pregnancies from June and one from September (present study); pregnant females from KwaZulu-Natal in the Transvaal Museum collection were collected in February, March, April and December. These results are similar to those obtained elsewhere in the subregion (Skinner and Smithers 1990), and indicate year-round reproduction with a seasonal peak in the summer months.

Linear measurements (mm) and mass (g)

	Males					Females				
	x̄	s.d.	n	Min	Max	x̄	s.d.	n	Min	Max
TL	192.0	20.5	215	152	268	191.1	19.3	133	151	240
HB	106.1	12.2	208	79	138	106.2	12.5	148	73	140
T	85.5	10.7	213	61	145	83.5	9.6	141	61	105
Hf	20.0	2.3	195	11	33	20.0	1.9	133	11	26
E	12.4	1.7	204	8	19	12.3	1.9	147	8	20
Mass	42.9	10.9	135	16	68	40.9	11.1	74	19	72

Records of occurrence
SPECIMEN RECORDS:

BM: Durban, Durban Botanic Gardens, Estcourt;

CM: Pennington;

DM: Albert Falls NR, Bosch Hoek Farm, Bosch Hoek Military Area (SADF TR6), Cathedral Peak NR (Catchment 9: Mlambonja River), Chelmsford NR, Clearwater Farm, Dlinza Forest NR, Durban, Estcourt, Entumeni NR, Fort Nottingham NR, Garden Castle NR, Goodhope Forest (Mondi), Highmoor NR, Hillcrest, Impendle NR, Karkloof NR, Hlatikulu Vlei, Howick (20 km south), Kamberg, Linwood Forest (Mondi), Loteni NR, Mfongosi, Mgeni Vlei, Midmar Resort, Moor Park NR, Ncandu NR, Ngome, Pongola Bush NR, Royal Natal NP, Umbilo,

Umfolozi GR, Umhlanga, Umtamvuna NR, Vergelegen NR, Vernon Crookes NR, Vryheid NR, Weza State Forest, Willbrook;

KM: Winterton;

NM: Boston, Carter's Nursery, Castle View, Champagne Castle, Donkerhoek Farm, Giant's Castle Hostel, Giant's Castle Rest Huts, Impendle, Kambula, Knoops Farm, Leliefontein, Matatiele District, Mfongosi, Mooi River District, Nagle Dam, Oribi Gorge, Pietermaritzburg, Redlands, Rosslea Estate, Sobantu village (Pietermaritzburg), Spring Grove, The Hoek, Town Bush, Umdoni Park, Umgeni Valley Game Ranch, Umtwalume;

SAM: Mfongosi, Umvoti;

SI: Drakensberg Gardens (3 km west; 2 km north), Groenkloof Farm, Hardingdale, Kilgobbin Farm, Makhake Store, Petchaye;

TM: Avondale, Bisley, Bosch Hoek Farm, Boston Police Station (3 km north-west), Bowel Forest, Bushman's Nek, Carter's Nursery, Cathedral Peak Forest Reserve, Chelmsford Public Resort, Coleford NR, Dargle, Dukuduku Forest Station, Entumeni NR, Estcourt (6.8 km north-west on Ladysmith road), Hilton, Himeville (30 km north), Itala Game Reserve (Craigadam, Potwe), Ixopo village, Izingolweni road, Kamberg NR, Karkloof Forest, Ladysmith (31.5 km from, on Van Reenen road), Lake Sibayi, Loteni NR, Midmar Public Resort, Ngome Forest Reserve, Oribi Gorge NR, Pietermaritzburg, Pinetown, Pongola Bush NR, Port Edward, Pottershill Farm, Qudeni, Queen Elizabeth Park NR, Ringwood Farm, Spioenkop NR, Umtamvuna NR, Umvoti Vlei NR, Underberg, Vernon Crookes NR, Wyford Farm.

ADDITIONAL RECORDS:
Avery (1991): Border Cave, Clarke's, Collingham, Diamond I, Gehle, Mbabane, Mgede, Mhlwazini, Nkupe, Umhlatuzana.

Genus	*Dasymys*	Peters 1875

Dasymys incomtus (Sundevall 1847)   PLATE 41
D. i. incomtus (Sundevall 1847)
Water rat
Waterrot

Taxonomic status
Of the three southern African subspecies described by Meester et al. (1986), one (*D. i. incomtus*) is recorded from KwaZulu-Natal. The type locality of the species is Durban. Gordon (1991) noted differences in karyotype between individuals from Richards Bay in KwaZulu-Natal (2n=46), and individuals from greater Transvaal (2n=38), suggesting the possibility of a distinct chromosomal race or species occurring in KwaZulu-Natal. This urgently warrants further studies involving karyotypes and additional (e.g. morphometric, allozyme and DNA) characters.

Distribution
According to Skinner and Smithers (1990), the water rat occurs throughout the eastern and northern areas of KwaZulu-Natal. The present study indicates a much wider distribution in KwaZulu-Natal (Figure 90), reaching an altitude of 1 500 m in the Drakensberg range in the west. They have been recorded from most if not all of the bioregions. Avery (1991) recorded this species from the upper Pleistocene at Border Cave, and from the Holocene period at Nkupe (in the upper Tugela Basin) and Umhlatuzana.

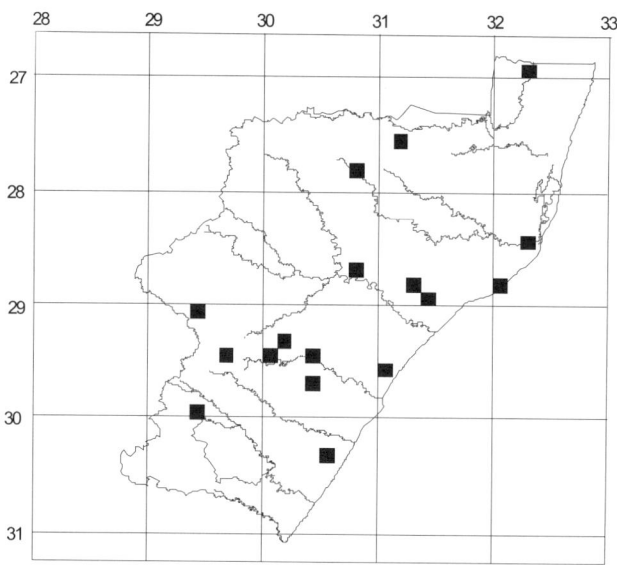

Figure 90

Protected areas
Albert Falls, Coleford, Entumeni, Hazelmere Dam, Itala, Kamberg, Karkloof, Ndumu, Vernon Crookes, Vryheid.

Conservation status
Listed as Indeterminate in the South African Red Data Book (Smithers 1986). Judging from recent trapping surveys, they are not as rare as previously thought, at least in KwaZulu-Natal.

Habitat
In KwaZulu-Natal they have been collected predominantly in lush grass and reedbeds in swampy areas, close to rivers, vleis and dams. However, they have been recorded from scarp forest at Entumeni Forest and from riverine scrub forest at Ndumu. They are frequently collected in the same habitat as *Otomys spp.*

Habits
They were always collected during early morning trap checks in KwaZulu-Natal, suggesting a crepsucular or nocturnal activity cycle. They apparently build nests on the swampy border of rivers with one exit opening to the water, and are excellent swimmers (Skinner and Smithers 1990). Roberts (1951) recorded their use of burrows in stream banks. Like *Otomys*, with which they

often co-exist, they create runways in grass and leave grass and reed cuttings in their feeding area.

Breeding
A single pregnant and lactating female was collected during June, contradicting Skinner and Smithers' (1990) suggestion (based on only four pregant females from Botswana) that breeding is limited to the summer months. The number of foetuses was not recorded; data elsewhere (see Skinner and Smithers 1990) suggest a mean litter size of 5.3 (*n*=4; range 2–9).

Linear measurements (mm) and mass (g)

	Males					Females				
	x̄	s.d.	n	Min	Max	x̄	s.d.	n	Min	Max
TL	315.8	26.5	7	279	345	304.4	21.7	10	256	330
HB	166.9	18.9	8	148	192	164.0	14.6	11	137	184
T	147.3	7.5	7	136	156	141.8	11.0	10	119	160
Hf	34.0	–	2	33	35	32.6	0.8	7	31	33
E	20.3	4.6	3	15	23	21.6	3.7	9	16	29
Mass	176.5	35.8	4	138	218	139.5	18.8	9	103	171

The above measurements exceed those recorded from greater Transvaal (Rautenbach 1982). Mean total lengths of 283 and 265 mm were recorded for males and females respectively in the Transvaal, compared with 316 and 304 mm recorded in the present study. In view of the karyotypic differences mentioned above, these observations support the possibility of taxonomic differentiation between KwaZulu-Natal and Transvaal populations of *Dasymys incomtus*.

Records of occurrence
SPECIMEN RECORDS:
DM: Albert Falls NR, Coleford, Entumeni NR, Eshowe (south of Rutledge Park), Futululu, Hazelmere NR, Kamberg NR, Karkloof NR, Mfongosi, Monk's Cowl, Vryheid NR;
NM: Dukuduku, Mfongosi, Ndumu GR;
TM: Itala GR, Kilgobbin Farm, Poly Shorts Bridge, Richards Bay, Vernon Crookes NR.

ADDITIONAL RECORDS:
Avery (1991): Border Cave, Nkupe, Umhlatuzana.

Genus	***Grammomys***	**Thomas 1915**

KEY TO SPECIES (After Meester et al. 1986)

1. Ears often but not always with subauricular tuft of white hairs; larger and greyer, with greatest skull length usually >31 mm . . . .
   G. cometes, p. 101

– Ears without subauricular tuft; smaller, less grey, with greatest skull length usually <31 mm . . . . *G. dolichurus*, p. 102

Grammomys cometes (Thomas and Wroughton 1908)
G. c. cometes (Thomas and Wroughton 1908)
Mozambique woodland mouse
Mosambiek-woudmuis

Taxonomic status
The nominate *G. c. cometes* is found in KwaZulu-Natal and Mozambique (Meester et al. 1986). Characters separating this species from *G. dolichurus* appear to be unreliable and break down in sympatric situations such as at Ngome Forest in northern KwaZulu-Natal (Bronner, personal communication; Taylor et al. 1994). Specimens having one of two diagnostic characters (subauricular white ear tuft or skull length >31 mm) have been found at Vryheid and Karkloof but are conservatively retained as *G. dolichurus* pending more detailed taxonomic studies of this genus.

Distribution
Previously (De Graaff 1981; Skinner and Smithers 1990) recorded from the north-east of KwaZulu-Natal, as far south as the Ngoye Forest. A single dubious record exists for the Mpumalanga Province (Rautenbach 1982). Recently, two specimens were recorded from the Royal Natal National Park, extending the known southward (28°41′S) and westward (28°56′E) limits of this species (Taylor et al. 1994; Figure 91).

Protected areas
Hluhluwe, Itala, Ngoye Forest, Royal Natal.

Conservation status
Restricted in its distribution and habitat requirements, and occurring only marginally within South Africa. Recent data suggest that the species may be more widespread in KwaZulu-Natal than previously thought. The conservation status of this species cannot be clearly stated until its taxonomic relationship with the commoner *G. dolichurus* is better established.

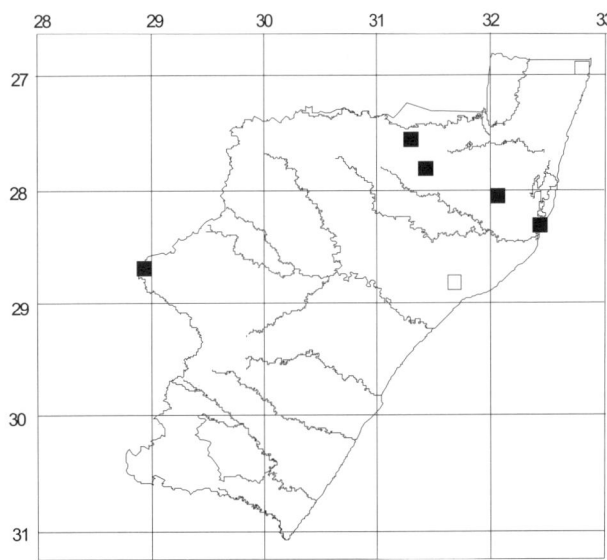

Figure 91

Habitat

The specimen from Royal Natal National Park was collected in mature *Podocarpus* forest along the Tugela River. Specimens have also been taken together with *G. dolichurus* in Afromontane forest at Ngome in northern KwaZulu-Natal (G. Bronner, personal communication), and the species has also been observed from mature scarp forest at Ngoye Forest (Panagis and Nel 1981). *Grammomys cometes* apparently prefers denser, more well developed forest than *G. dolichurus*, and in Malawi is restricted to zones having >1 500 mm rainfall. The latter is clearly not the case in KwaZulu-Natal where the species is also found in drier bushveld or coastal forest associations such as at Hluhluwe and Manguzi Forest. However, until the taxonomic relationships of these two species are better resolved it is perhaps premature to make definitive statements about habitat preferences.

Habits

Assumed to be similar to the following species.

Breeding

A female collected in November was perforate but not pregnant. A pregnant female in the Transvaal Museum collected at Ngome Forest during March had three foetuses. In a captive sample collected from Ngoye Forest, litter size varied from one to four (*n*=8). The young cling to the nipples of the mother for a period of 26 days, although weaning usually takes place at about 19 days (Panagis and Nel 1981).

Linear measurements (mm) and mass (g)

Cat. No.	Sex	TL	HB	T	Hf	E	Mass
DM 2428	M	–	127	–	22	20	–
DM 987	F	310	145	165	–	–	–
DM 2371	F	307	132	175	24	21	52

Records of occurrence

SPECIMEN RECORDS:

DM: Itala Game Reserve, Royal Natal NP, St Lucia village;
TM: Hluhluwe NR (Egodeni), Ngome Forest Reserve.

ADDITIONAL RECORDS:

Panagis and Nel (1981): Ngoye Forest;
Roberts (1951): Manguzi Forest.

Grammomys dolichurus (Smuts 1832)   PLATE 42
G. d. dolichurus (Smuts 1832)
G. d. tongensis (Roberts 1931)
Woodland mouse
Woudmuis

Taxonomic status

This species is distinguished from the previous, less common species on the basis of its smaller size and absence of sub-auricular ear tufts. As discussed above, identifying features of the two species may overlap where they are found together. To further complicate matters, Dippenaar et al. (1983) stated that geographic variation occurs in diploid chromosome number from 2n=44 at Woodbush in the Northern Province to 2n=52 at Ngoye Forest in Zululand. Two subspecies are recognised from KwaZulu-Natal: *G. d. tongensis* was described by Roberts (1931) from Ubombo and Manaba in northern Maputaland, while *G. d. dolichurus* occurs throughout the rest of KwaZulu-Natal (Meester et al. 1986).

Distribution

Skinner and Smithers (1990) described the species as occurring 'coastally' in KwaZulu-Natal. Bourquin (1988) showed the species to be more or less restricted to Zululand, with isolated records from Durban and Cathedral Peak NR. Recent data show the species to be more widespread in KwaZulu-Natal than indicated previously (Figure 92). The woodland mouse has been recorded as far west as Vryheid in the drier northern parts of the province, and is widespread throughout Zululand, the midlands of KwaZulu-Natal and the coastal region as far south as Athlone Park on the South Coast. Inland, it has been recorded as far south as Weza Forest in KwaZulu-Natal, but is also found in the Eastern Cape (De Graaff 1981). Avery (1991) recorded the woodland mouse from Pleistocene and Holocene deposits at Border Cave in the Lebombo range, as well as in the Durban region and the upper Tugela Basin west of Vryheid (see under **Records of occurrence**).

Protected areas

Cathedral Peak, Coastal Forest Reserve (=Kosi Bay), Entumeni, Eastern Shores, False Bay Park, Fort Nottingham, Hluhluwe, Itala, Karkloof, Mapelane, Mkuzi, Monk's Cowl, Ndumu, Umfolozi, Vryheid.

Conservation status

Their frequent collection by cats in suburban areas of Durban suggests that they are commoner than suggested from their low trapping success.

Habitat

Predominantly a forest and woodland species. In KwaZulu-Natal, they were collected most commonly within or on the fringes of Afromontane and other climax forest types, but have also been recorded from open shrubland, open woodland and even grassland. They inhabit modified suburban habitats in the Durban area, and frequently enter houses. These observations suggest that they are less specialised in their requirements for climax evergreen forest than their name suggests.

Habits

They are arboreal, nocturnal and communal, living in family groups (De Graaff 1981; Roberts 1951). They are not strictly arboreal but are often trapped at ground level. At Dududuku Forest in Zululand, their skulls were found in the skats of gaboon adders (*Bitis*) (Bodbijl, personal communication). Woodland mice are known to construct nests

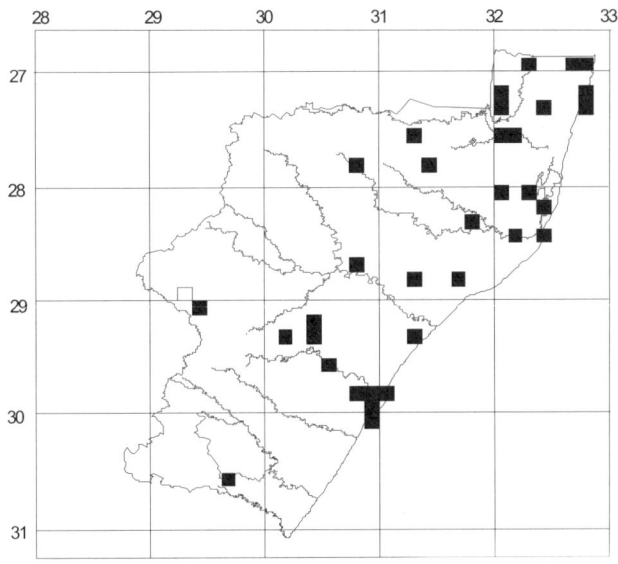

Figure 92

of up to 200 mm in dense bushes and trees (De Graaff 1981; Roberts 1951). This is certainly true in KwaZulu-Natal where nests have been found in dense bushes, between the leaves of Malala palms, the head of a very prickly *Euphorbia* (Roberts 1931), as well as in a *Ficus natalensis* tree at a height of 200 mm (specimen data for **NM** 1366). At Cathedral Peak in the Drakensberg, a single stomach of this species was found to contain 90 per cent white plant matter and 10 per cent insect remains (Rowe-Rowe 1986). Based on stomach contents analysis at Karkloof Forest, Wirminghaus and Perrin (1992) showed that, throughout the year, their diet consisted primarily of fruits and green plant material, with invertebrates being eaten in early summer.

Breeding
During the present study single pregnant females were collected during March (four embryos) and May (five embryos). Juveniles were recorded in May and August. Rautenbach et al. (1980) recorded a pregnant female from Kosi Bay carrying five embryos in April. These limited observations agree with limited data available elsewhere (Rautenbach 1982; Skinner and Smithers 1990) which suggest an extended breeding season.

Linear measurements (mm) and mass (g)

	Males					Females				
	$\bar{x}$	s.d.	n	Min	Max	$\bar{x}$	s.d.	n	Min	Max
TL	265	14.4	6	251	282	273	15.6	5	251	295
HB	102	5.2	6	96	110	110	9.0	5	99	124
T	163	11.3	6	149	175	163	7.2	5	152	171
Hf	22	1.2	4	21	24	23	1.0	3	22	24
E	18	3.3	4	15	23	18	1.5	4	17	20
Mass	43	–	1	43	43	42	–	1	42	42

Records of occurrence

SPECIMEN RECORDS:
DM: Athlone Park, Carrington Heights (5 Kinmont Crescent), Fort Nottingham NR, Hillcrest, Karkloof NR, Kloof, Lake Eteza, Mapelane, Mkuzi GR, Monk's Cowl, Vryheid, Westville, Weza State Forest;
KM: Ndumu GR, False Bay Park (Lister Point);
NM: Gwaliweni Forest, Nagle Dam;
SI: Ndumu GR;
TM: Charter's Creek, Coastal Forest Reserve (=Kosi Bay), Itala GR (Craigadam), Eastern Shores NR, Hluhluwe GR (Egodeni), Ingwavuma, Karkloof Forest, Kosi Lake (Department of Health Camp), Lake Sibayi, Malvern, Manaba, Manguzi Forest, Manzengwenya Forest Station, Mkuzi GR, Ngome Forest Reserve, Ngoye Forest, Stanger Beach, Umfolozi GR.

ADDITIONAL RECORDS:
Avery (1991): Border Cave, Nkupe, Umhlatuzana; Rowe-Rowe (1977): Cathedral Peak.

Genus	*Mus*	Linnaeus 1758
Subgenus	*Mus*	Linnaeus 1758

INTRODUCED SPECIES
Mus (Mus) musculus (Linnaeus 1758)
House mouse
Huismuis

In spite of being a common commensal rodent which is often regarded as a pest of stored foodstuffs, few specimens have been collected from KwaZulu-Natal, and consequently their distribution, although probably widespread, is poorly known.

In KwaZulu-Natal, they appear to be closely associated with human dwellings, barns, food stores, and cultivated fields. In parts of their range in the northern hemisphere, as well as on Marion Island, they are known to take up residence in the field. Specimens collected in KwaZulu-Natal were invariably taken from indoors. They are nocturnal, with pairs constructing nests of paper, cloth or other available material, often under the floors of houses. They live in family groups consisting of a dominant male and several females, young and subordinate males.

In Zimbabwe they are known to breed throughout the year, and to give birth to between two and four young (Smithers 1975). A similar situation probably occurs in KwaZulu-Natal, but, apart from juveniles having been recorded in September and December, no further data are available. A melanistic female collected from Umlalazi Nature Reserve during the present study exceeded the range of total lengths given by Smithers and Wilson (1979) for a population from Zimbabwe.

Subgenus	*Leggada*	Gray 1837

Mus (Leggada) minutoides (A. Smith 1834) PLATE 43
M. m. minutoides
Pygmy mouse
Dwergmuis

Taxonomic status
Three subspecies are recognised from southern Africa, of which only one, *M. m. minutoides,* is found in KwaZulu-Natal (Meester et al. 1986).

Distribution
Widespread in KwaZulu-Natal (Figure 93). Its apparent concentration in the central and north-eastern parts of the province reflects greater sampling effort in these areas.

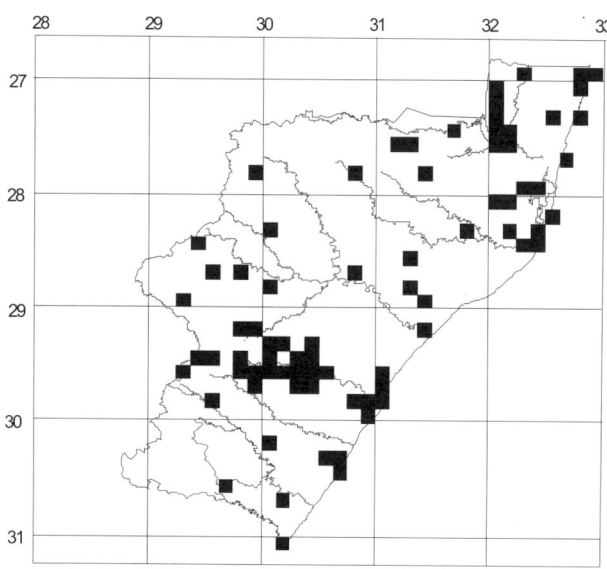

Figure 93

Protected areas
Albert Falls, Cape Vidal, Cathedral Peak, Chelmsford, Eastern Shores, False Bay, Giant's Castle, Hluhluwe, Itala, Karkloof, Loteni, Mapelane, Mgeni Vlei, Midmar, Mkuzi, Mount Currie, Ndumu, Oribi Gorge, Royal Natal, Spioenkop, St Lucia Park, Tembe Elephant Park, Umfolozi, Umtamvuna, Vernon Crookes, Vryheid, Weenen.

Conservation status
Common and widespread in KwaZulu-Natal.

Habitat
They are known to occur in virtually all habitats, and KwaZulu-Natal is no exception. During the present study they were collected in a range of vegetation types, from rank, mesic to recently burnt grasslands, and from bushveld to coastal forest to mistbelt and scarp forests. They have been collected in disturbed areas, in suburban gardens, on fallow lands with secondary growth, and in young pine plantations, but appear to be most common in vlei, riverine or rocky habitats. They have been recorded in all of the bioregions, but are only marginally present in montane habitats in the Drakensberg up to 2 400 m altitude (Rowe-Rowe and Meester 1982a).

Habits
They are nocturnal and terrestrial, and are capable of using any available cover, from sacks and rocks to the burrows of other species (De Graaff 1981). They are known to enter houses where they can live in narrow spaces under floorboards. Pygmy mice construct ball-shaped nests from soft grass and other fibres. Given their wide habitat tolerance, it is not surprising that they have a wide omnivorous diet including seeds, arthropods and green plant matter (Rowe-Rowe 1986).

Breeding
In KwaZulu-Natal, pregnant females have been collected during November ($n=1$), December ($n=2$), February ($n=3$), April ($n=1$) and June ($n=1$). The number of foetuses per pregnant female varied from three to seven, with a mean of 4.5 ($n=4$). Based on captive colonies obtained from the Pietermaritzburg area, Willan and Meester (1978) found them to have a gestation of 19 days or less and a mean litter of 4.0 (range 1–7; $n=27$). Sexual activity was recorded at approximately 42 days of age. Young have a birth weight of 0.8 g and are helpless and blind at birth. From data outside the province, breeding appears to continue throughout the year (De Graaff 1981; Skinner and Smithers 1990).

Linear measurements (mm) and mass (g)

	Males					Females				
	$\bar{x}$	s.d.	n	Min	Max	$\bar{x}$	s.d.	n	Min	Max
TL	103.7	1.4	37	89	120	106.0	2.4	21	82	125
HB	58.8	1.0	37	47	72	61.8	1.7	20	51	75
T	44.9	0.7	37	35	54	45.4	1.2	21	27	53
Hf	12.7	0.2	26	11	15	11.7	0.5	19	7	14
E	9.6	0.2	33	7	12	9.6	0.2	20	8	12
Mass	7.1	0.4	19	4	12	8.7	0.5	14	5	12

Records of occurrence
SPECIMEN RECORDS:
BM: Umfolozi, Willbrook;
DM: Albert Falls NR, Bosch Hoek (SANDF Training Area), Bosch Hoek Farm, Cape Vidal, Chase Valley Heights, Chelmsford NR, Clearwater Farm, Durban, Eastern Shores (Iphiva Campsite), Entumeni NR, Eshowe (Rutledge Park), False Bay Park, Futululu, Goodhope Forest (Mondi), Hilton (12 Forest Lane), Impendle, Karkloof NR, Linwood Forest (Mondi), Loteni NR, Mapelane, Mfongosi, Mgeni Vlei, Nhlosane Forest, Roosfontein NR, Scaleby Farm, Spioenkop NR, Umhlanga (60 Hilken Drive), Umtamvuna NR, Vryheid NR, Weenen NR, Westville (55a Jan Hofmeyr Road), Weza State Forest;
KM: Eastern Shores (Sugar Loaf), Lake Bangazi (north), Mission Rocks, St Lucia, Vergeval Farm;
NM: Castle View, Geluk Farm, Gwaliweni, Mkuzi GR, Mooi River District, Nagle Dam, Oribi Gorge, Pietermaritzburg, Pietermaritzburg (Botanic Gardens,

Carter's Nursery, Drumclog, Leliefontein, Town Bush), Rosslea Estate, Sweetwaters, Umtwalume;
SAM: Durban (Malvern), Mfongosi;
SI: Coastal Forest Reserve (=Kosi Bay), Gardenia, Hardingdale, Kilgobbin, Ladysmith, Makatini Flats, Makhake Store, Nkonkoni, Nyala Statelands;
TM: Ashburton, Cathedral Peak, Chard Farm, Coastal Forest Reserve (=Kosi Bay), Cornhill, Dlinza Forest, Dukuduku Forest, Eastern Shores (Lake St Lucia), False Bay Park (Lister's Point), Hazelmere NR, Hilton, Hluhluwe GR, Ingwavuma, Itala GR (Craigadam, Doornpan), Kloof, Kwaliweni (= Gwaliweni), Lake Bangazi (Eastern Shore), Lake Sibayi (Banana Pan, Eastern Shore, Research Station), Lincolnmead, Loteni NR, Maputa, Melmoth, Midmar Dam, Mission Rocks, Mkuzi GR (Caravan Park), Ndumu GR, Ngome Forest Reserve, Oribi Gorge NR, Otobotini, Pongolapoort River, Ringwood Farm, Ronan Egg Farm, Scottsville, Spioenkop NR, Tete Pan, Umtamvuna NR, Vernon Crookes NR, Weenen, Weenen NR, Wyford Farm.

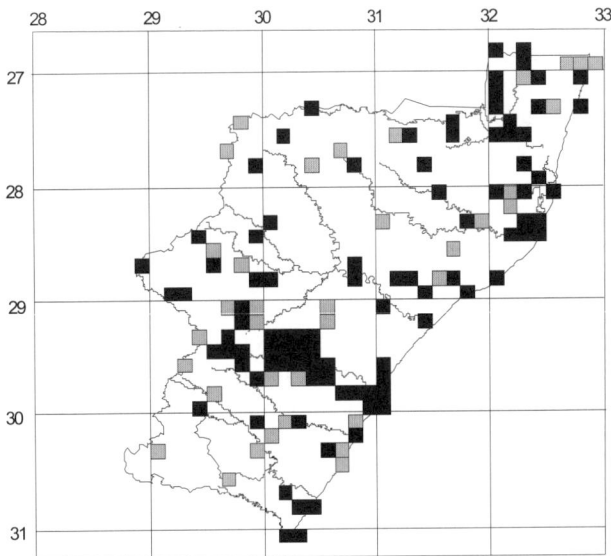

Figure 94

Genus	*Mastomys*	Thomas 1915

Mastomys natalensis (A. Smith 1834)   PLATE 44
M. n. natalensis (A. Smith 1834)
Natal multimammate mouse
Natalse vaalveldmuis

Taxonomic status

This species is difficult to distinguish from the following species, *M. coucha*. Until recently, it could only be distinguished on the basis of chromosome number or electrophoretic mobility of haemoglobin (Green et al. 1980; Meester et al. 1986), making it impossible to determine species identity from museum specimens. However, Dippenaar et al. (1993) and Njobe (1997) demonstrated consistent morphometric differences which allowed the identification of 'unknown' museum specimens based on discriminant analysis of cranial measurements. Morphometrically identified specimens of *Mastomys* from KwaZulu-Natal (Njobe 1997) were used to plot species distributions (Figures 94, 95).

Three subspecies of *M. natalensis* are recognised from southern Africa. Only *M. n. natalensis* is known from KwaZulu-Natal, the type specimen originating from the Durban area (Meester et al. 1986).

Distribution

Widely distributed throughout KwaZulu-Natal from the coast to the lower slopes of the Drakensberg, below 1 500 m (Rowe-Rowe and Meester 1982a). Based on recent morphometric re-classifications of *Mastomys* collections in museums (Njobe 1997), and contrary to earlier results of cytogenetic analyses (Green et al. 1980), it appears that *M. natalensis* is broadly sympatric with its sibling species *M. coucha* throughout much of the province (Figures 94, 95: solid squares represent morphometrically identified

specimens, hatched squares represent unidentified specimens; Njobe 1997), although the latter does not extend into the wetter Highland bioregion (Figure 95). Avery (1991) recorded this species from the upper Pleistocene and very late Holocene at Border Cave, as well as at three sites in central and coastal KwaZulu-Natal (see under **Records of occurrence**).

Protected areas

Albert Falls, Beachwood Mangroves, Cathedral Peak, Chelmsford, Coleford, Dlinza Forest, Eastern Shores, Giant's Castle, Hluhluwe, Ian Ellis, Itala, Karkloof, Kenneth Stainbank, Krantzkloof, Loteni, Mapelane, Midmar, Mkuzi, Ndumu, Oribi Gorge, Queen Elizabeth Park, Royal Natal, Spioenkop, Tembe, Umfolozi, Umgeni Vlei, Umhlanga Lagoon, Umlalazi, Umtamvuna, Vernon Crookes, Vryheid, Weenen.

Conservation status

Probably KwaZulu-Natal's commonest and most widely distributed small mammal.

Habitat

This species has a broad habitat tolerance and is thought to be a pioneer species in the colonisation of disturbed sites (Meester, Lloyd and Rowe-Rowe 1979). It is the first species to colonise early rehabilitation stands of *Acacia karoo* resulting from opencast dune mining by Richards Bay Minerals Co. Ltd on the north coast of KwaZulu-Natal (Ferreira and Van Aarde 1996). At Bluff Nature Reserve, it was the only species caught in burnt grasslands for eight months after the fire (Bronner 1986; Bronner and Meester 1987). Numerous specimen records confirm that this species is often associated with disturbed habitats in KwaZulu-Natal, including young pine stands and established wattle plantations, areas invaded by exotic plants such as *Lantana*, and heavily overgrazed grasslands. The species has also been collected in a wide range of grassland-dominated habitats, particularly in vleis and riverine habitats. Specimens have been taken from Afromontane forest, escarpment forest and coastal forest, although usually close to the forest margin. As elsewhere, multimammate mice have been found to occur in rural homesteads and other human habitations in KwaZulu-Natal.

Habits

They are typically nocturnal, terrestrial and gregarious (Bronner 1986; Skinner and Smithers 1990). During the present study, whenever traps were checked twice a day (morning and afternoon), multimammate mice were invariably collected only during the morning checks, indicating little or no diurnal tendency. They are known to be commensal with man and may construct their nests in secluded spaces in buildings. In the field they can apparently burrow or utilise the burrows of other species (Skinner and Smithers 1990).

Mean home ranges calculated for males and females at Bluff Nature Reserve were 1 750 ± 992 m^2 and 1 607 ± 752 m^2 respectively. Male home ranges overlapped considerably, as expected for a gregarious species, although breeding females tended to be territorial with non-overlapping home ranges (Bronner 1986). In a study of stomachs collected throughout the year at Karkloof Forest, the diet was found to be mixed, with a preference for invertebrates (found in 50 per cent of samples), followed by fruit, then green plant material (Wirminghaus and Perrin 1992). Flowers were eaten in spring and summer.

Breeding

At Bluff Nature Reserve, breeding was observed throughout the year, although with a period of reduced reproduction during the dry winter, from May to July (Bronner 1986). Based on a captive colony collected in the Pietermaritzburg area, Baker and Meester (1977) determined mean litter size to be 11.3 (n=11). The mean number of foetuses observed in five pregnant females collected during the present study was 10.4, with a range of 10–12. The species is prone to episodic population explosions, usually triggered by good rains and abundant food reserves (Skinner and Smithers 1990).

Linear measurements (mm) and mass (g)

	Males					Females				
	$\bar{x}$	s.d.	n	Min	Max	$\bar{x}$	s.d.	n	Min	Max
TL	227.1	25.5	149	170	299	211.3	14.1	96	161	273
HB	119.0	17.0	149	90	169	109.5	10.7	96	90	142
T	108.1	12.2	149	80	137	101.8	10.2	96	71	131
Hf	22.6	4.3	92	17	26	22.0	3.6	65	17	25
E	17.9	2.0	136	14	23	17.5	2.0	83	11	20
Mass	53.2	14.2	97	26	90	38.6	9.8	51	21	70

Mastomys natalensis from KwaZulu-Natal are somewhat larger in body measurements and mass than recorded from greater Transvaal (Rautenbach 1982) and Mozambique (Gliwicz 1985), but marginally smaller than recorded from Zimbabwe (Smithers 1983).

Records of occurrence
SPECIMEN RECORDS:

M. natalensis sensu lato (skulls not measured):
BM: Illovo;
CM: Pennington;
KM: False Bay Park, Vergeval Farm, Zululand;
NM: Boston, Boteler Point, Carter's Nursery, Castle View, Donkerhoek Farm, Drumclog, Hluhluwe GR (Egoteni), Entendweni Bush, Geluk, Giant's Castle GR, Gwaliweni, Hluhluwe GR, Hluhluwe GR (Ngqbateti Corridor, Labour Camp HQ), Ingele Forest, Junction of White and Black Umfolozi Rivers, Kambula, Knoops Farm, Kosi Lake, Langewacht, Leliefontein, Manguzi, Matatiele District, Mfongosi, Mkuzi GR, Mooi River District, Nagle Dam, Ndumu GR, Oribi Gorge, Pietermaritzburg, Redlands, Rosslea Estate, Shalimar Farm, Shemula's Pont, Spring Grove, Sweetwaters, The Hoek, Umdoni Park, Umfolozi Drift, Umgeni Valley Game Ranch, Umtwalume River, Underberg;
SAM: Estcourt, Sibhudeni, Mfongosi.

M. natalensis sensu stricto (identification based on Njobe 1997):
DM: Albert Falls NR, Bluff NR, Bosch Hoek Farm, Bosch Hoek, Carrington Heights (5 Kinmont Crescent), Cathedral Peak (Mlambonja River, Catchment 9), Chelmsford NR, Dlinza Forest NR, Durban, Empisini NR, Eshowe, Eshowe (south of Rutledge Park), Forest Hills, Futululu, Goodhope Plantation (Mondi), Greenwood Park, Harold Johnson NR, Hazelmere NR, Hillcrest, Hilton (21 Walters Road), Hlatikulu Vlei, Ian Ellis NR, Impendle, Inchanga Farm, Itala GR, Kamberg NR, Karkloof NR, Kenneth Stainbank NR, Kloof, Lake Eteza, Linwood Plantation, Loteni NR, Mapelane NR, Mbumbazi NR, Mfongosi, Mgeni Vlei, Midmar Resort, Midmar NR, Mkuzi GR, Moor Park NR, Mount Moreland (Agnes Street, Umhloti River), Ngxwala Hills, Paradise Valley NR, Pietermaritzburg, Pinetown, Pongola Bush NR, Raisethorpe, Richards Bay (ALUSAF Power Station, Bayview Boulevard, Mzingazi Dam, Portnet, Thulazinhleka Pan), Royal Natal NP, Sarnia, Shongweni Dam Resource Reserve, Spioenkop Dam, St Lucia Forest Station, Umbilo (Durban), Umfolozi GR, Umfolozi GR (Makhmisa, Sontuli Loop), Umhlanga NR, University of Natal (Durban), Vernon Crookes NR, Vryheid NR, Weenen NR, Westville (11 Highgates Avenue), Willbrook;
TM: Albert Falls NR, Ashburton, Bisley, Black Umfolozi bridge, Bluff NR (Tara Road), Boschhoek Farm, Boston SAP, Broughton, Bumbeni, Carter's Nursery, Cathedral Peak, Coleford NR, Darvill Sewage Farm, De Hoek Forest Reserve, Dukuduku Forest, Durban, Eastern Shores (Lake St Lucia), Flagstone Spruit, Four Winds Farm, Futululu Research Station, Goedhoop, Greytown, Gumedi's Kraal, Hazelmere Dam NR, Hilltop Farm, Hilton Road, Hluhluwe GR, Hluhluwe GR (Memorial Gate, Research Camp), Ingwavuma, Itala NR (Breda, Craigadam, Doornkraal, Doornpan, ranger's house), Ixopo, Izingolweni road, Karkloof Forest, Kenneth Stainbank NR, Kilgobbin Farm, Klipspruit, Gwaliweni Forest, Lake Sibayi, Loteni,

Magudu (Lot 6), Malvern, Manaba, Mapelane, Maputa, Midmar Dam, Mkuzi, Mkuzi River, Mkuzi River bridge on Candover–Mkuzi road, Ndumu Police Station, Newcastle, Ngome State Forest, Ngoye Forest, NRC camp (Ingwavuma District), Ntambanana, Oribi Gorge NR, Otobotini, Pietermaritzburg, Pinetown, Pongola River, Port Edward, Port Edward Golf Course, Port Shepstone, Pottershill Farm, Queen Elizabeth Park, Ringwood, Santa Suzanne Sanctuary, Sihangwane, Sibhudeni, Spioenkop NR, Tembe Elephant Park, Thabamhlope, The Grange, Theunis Bester Game Ranch, Town Bush Valley, Tugela River, Two Streams Farm, Ubombo, Umfolozi GR, Umfolozi, Umlalazi NR, Umpumulo Post Office, Umtamvuna NR, Umvoti Vlei NR, Umzimkulu, University of Natal (Durban), Vernon Crookes NR, Weenen NR, White Umfolozi, Wyford Farm.

ADDITIONAL RECORDS:

M. natalensis sensu lato:
Avery (1991): Border Cave, eSinhlonhweni, Mbabane, Umhlatazana.

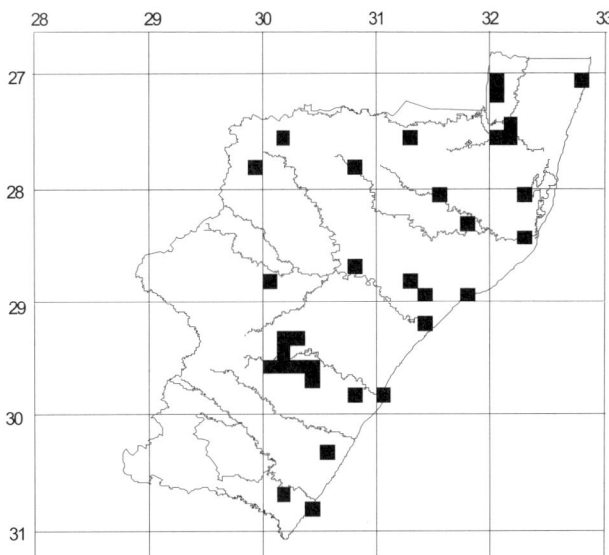

Figure 95

Mastomys coucha (A. Smith 1836)
M. c. coucha (A. Smith 1836)
Multimammate mouse
Vaalveldmuis

Taxonomic status
See under above species account. Meester et al. (1986) listed four subspecies for the subregion. The nominate race, *M. c. coucha*, is the most likely to occur in KwaZulu-Natal.

Distribution
Recorded by Green et al. (1980) from the Eastern Cape and Northern Cape, Zululand, Lesotho, Free State, parts of greater Transvaal, and southern and western Zimbabwe. Based on recent morphometric re-classifications of *Mastomys* collections in museums, and contrary to earlier results of cytogenetic analyses (Green et al. 1980), it appears that *M. coucha* occurs widely throughout the lower-altitude regions of KwaZulu-Natal (Figure 95), and is broadly sympatric with its sibling species *M. natalensis* (Figure 94).

Protected areas
Chelmsford, Dlinza Forest, Entumeni, Harold Johnson, Highmoor, Ian Ellis, Itala, Karkloof, Krantzkloof, Midmar Dam, Mkuzi, Oribi Gorge, Umfolozi, Umlalazi, Vernon Crookes, Vryheid, Weenen.

Conservation status
On the whole, the species appears much less common in KwaZulu-Natal than its sibling species, *M. natalensis*.

Habitat
Similar to the above species (see Skinner and Smithers 1990), although their general distribution in the drier western parts of the subcontinent indicates a preference for drier climatic conditions. The absence of *M. coucha* from the

wetter Highland bioregion of KwaZulu-Natal (annual rainfall >700 mm), and the fact that the two species are equally abundant in the drier Lowveld bioregion (annual rainfall 500–900 mm), while *M. coucha* is much less common at localities where they occur sympatrically in the wetter Moist upland (annual rainfall 700–1 000 mm) and Coast hinterland (annual rainfall 750–1 300 mm) bioregions, reinforces this conclusion.

Habits
They are less aggressive and nervous than *M. natalensis*, being easier to handle and more suitable for domestication as a laboratory species.

Breeding
Probably similar to *M. natalensis*. No data for the province.

Linear measurements (mm) and mass (g)

	Males					Females				
	x̄	s.d.	n	Min	Max	x̄	s.d.	n	Min	Max
TL	192.1	26.4	7	173	237	181.0	17.6	4	159	202
HB	95.0	17.2	7	80	125	95.3	7.0	3	88	102
T	97.1	10.0	7	85	112	87.0	13.3	4	86	100
Hf	21.2	1.5	6	19	23	21.2	2.5	4	18	24
E	16.2	1.0	7	15	18	16.5	1.7	4	14	18
Mass	–	–	–	–	–	–	–	2	26	28

RECORDS OF OCCURRENCE:

Specimen records (identification based on Njobe 1997):
DM: Chelmsford NR, Dlinza Forest NR, Durban, Entumeni NR, Eshowe (5 Poynton Place), Greenwood Park (Durban), Harold Johnson NR, Highmoor, Hillcrest, Ian Ellis NR, Karkloof NR, Krantzkloof NR, Linwood Forest, Mfongosi, Mkuzi GR, Umbilo (Durban), University of Natal (Durban), Vernon Crookes NR, Vryheid NR, Weenen NR;

TM: Ashburton, Bisley, Black Umfolozi Bridge, Broughton (between Broughton and Crammond), Dukuduku Forest, Four Winds Farm, Ingwavuma, Itala GR (Breda, Craigadam), Maputa, Midmar Dam, Mkuzi GR, Mkuzi River Bridge, Njekayi (=Mfkayi), Oribi Gorge, Otobotini, Pietermaritzburg, Pongola River, The Grange, Ubombo, Umfolozi GR, Umlalazi NR, Weenen NR.

Genus	*Thallomys*	Thomas 1920

Thallomys paedulcus (Sundevall 1846)   PLATE 45
Tree rat
Boomrot

Thallomys nigricauda (Thomas 1882)
Black-tailed tree rat
Swartstert boomrot

Taxonomic status

Gordon (1987) found *Thallomys paedulcus sensu lato* to comprise two chromosomal species, *T. paedulcus* (2n=43–47) and *T. nigricauda* (2n=47–50). Based on the distribution of karyotyped specimens, the ranges of the two species appeared parapatric, *T. paedulcus* being associated with the Savanna biome and *T. nigricauda* with the South West Arid biome. The latter species could apparently also be identified on the basis of the black tail and prominent black facial patch.

Taylor et al. (1995) used discriminant analysis to derive craniometric criteria for the two chromosomal species based on known-species samples. These criteria were then used to re-classify available museum collections and plot revised distribution maps. These distribution maps indicated a complicated pattern of widespread sympatry throughout southern Africa. Furthermore, the external characters listed above did not appear to correlate well with the revised species limits. While specimens from Hluhluwe NR were identified by Gordon (1987) on chromosomal grounds as *T. paedulcus*, specimens from other localities in Zululand were identified as *T. nigricauda* on morphometric grounds (Taylor et al. 1995). Specimens from other localities could not be identified morphometrically because of the broken or juvenile status of skulls (see under **Records of occurrence**). The situation is far from clear and requires further karyotypic, morphometric and molecular studies to resolve species limits.

Distribution

Figure 96 shows the distribution of *T. paedulcus sensu lato* in KwaZulu-Natal. Localities represented by karyotypically or morphometrically-identified specimens are listed under **Records of occurrence**. Tree rats appear largely restricted to the Lowveld bioregion of north-eastern KwaZulu-Natal, although isolated populations occur in the Weenen and Mfongosi areas in Valley bushveld associated with the middle reaches of the Tugela River. Further collecting may show the two species to be more widespread throughout

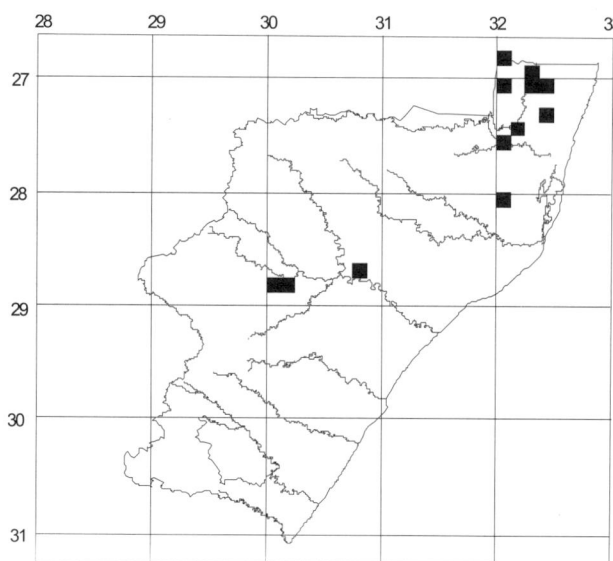

Figure 96

the Valley bushveld region of KwaZulu-Natal. The species was recorded from upper Pleistocene and very late Holocene deposits at Border Cave (Avery 1991).

Protected areas

Hluhluwe, Ndumu, Umlalazi, Weenen.

Conservation status

The species is more restricted in its distribution than was indicated by Bourquin (1988), whose distribution map included a misidentified record from the Pietermaritzburg area.

Habitat

The species is closely associated with *Acacia*-dominated bushveld habitats in KwaZulu-Natal, as shown elsewhere (Skinner and Smithers 1990).

Habits

They are nocturnal and arboreal, living in hollows in trees, or under loose bark (Skinner and Smithers 1990). A specimen obtained from Ubombo (**TM** 6122) was recorded nesting in a large *Aloe sp.* A specimen from Otobothini in the Ingwavuma District (**TM** 7220) was recorded as having its pelage stained with the pollen and nectar of *Aloe sp.* (De Graaff 1981).

Breeding

No data available for the province. From data obtained elsewhere, two to five young are born in the nests during summer (Skinner and Smithers 1990).

Linear measurements (mm) and mass (g)

A female collected at Weenen NR had the following measurments: TL=255, HB=120, T=135, Hf*su*=23, E=15.

Records of occurrence

SPECIMEN RECORDS:
T. paedulcus (identified karyotypically):
TM: Hluhluwe NR;

T. *nigricauda* (identified morphometrically):
TM: Mkuzi River, Ndumu GR, Otobotini, Tete Pan;
Unidentified (not measured or not examined):
DM: Mfongosi, Weenen NR;
KM: Mfongosi;
SI: Mfongosi;
TM: Manaba, Mkuzi River, NRC Camp (Ingwavuma District), Otobotini, Sihangwane, Tete Pan, Tugela Estates, Ubombo, Umlalazi NR, Weenen NR.

ADDITIONAL RECORDS:
Avery (1991): Border Cave.

Genus	*Aethomys*	Thomas 1915

KEY TO SPECIES (After Meester et al. 1986)

1. Size smaller, greatest skull length
   <35 mm; M_1 with three cusps in anterior
   row . . . . *A. namaquensis*, p. 111
- Size larger, greatest skull length
   >35 mm; M_1 with two cusps in anterior
   row . . . . *A. chrysophilus sensu lato*, p. 109

Aethomys chrysophilus (De Winton 1897)
A. c. chrysophilus (De Winton 1897)
Red veld rat
Afrikaanse bosrot

Taxonomic status
Gordon and Rautenbach (1980) and Chimimba (1997) found *A. chrysophilus sensu lato* to consist of two sympatrically-occurring chromosomal species, *A. chrysophilus* (2n=50) and *A. ineptus* (2n=44), which can be separated by discriminant analysis of cranial variables (Chimimba 1997). On craniometric grounds, both species are known to occur in KwaZulu-Natal (Chimimba 1997; Figures 97, 98). While distributions of the two species are discussed separately, details of life history presented below refer to the red veld rat species-complex as a whole (i.e. *A. chrysophilus sensu lato*).

Distribution
Aethomys chrysophilus sensu lato is widespread throughout KwaZulu-Natal, but is absent from grassland-dominated regions in the higher-lying and wetter Montane, Highland, Mistbelt and Moist upland bioregions (Figure 97). Solid squares in Figure 97 indicate positively identified *A. chrysophilus senso stricto* (i.e. specimens subject to craniometric analysis), while barred squares indicate records of *A. crysophilus senso lato* which were not available for morphometric analysis. *Aethomys chrysophilus sensu stricto* is restricted to 11 localities scattered mostly through-

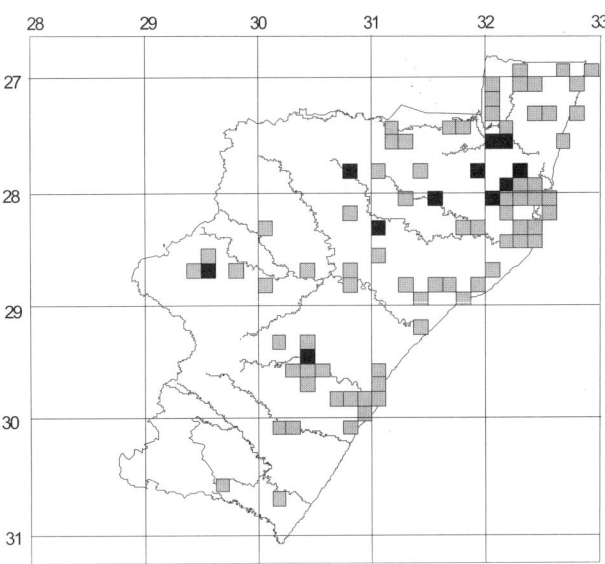

Figure 97

out northern KwaZulu-Natal (Figure 97: solid squares). The ranges of the two sibling species, *A. chrysophilus* (Figure 97) and *A. ineptus* (Figure 98), overlap in KwaZulu-Natal (two species co-occur in seven grid squares), as is the case throughout their ranges (Chimimba 1997). Avery (1991) recorded *A. chrysophilus s.l.* from Pleistocene and Holocene sites at Border Cave and Umhlatuzana (near Durban), as well as from Holocene deposits in the Upper Tugela Basin (see under **Records of occurrence**).

Protected areas
Albert Falls, Coastal Forest Reserve, Eastern Shores, Enseleni, Entumeni, False Bay Park, Hluhluwe, Itala, Krantzkloof, Mkuzi, Ndumu, Oribi Gorge, Queen Elizabeth Park, Sodwana Bay, Sodwana State Forest, Spioenkop, Umfolozi, Umlalazi, Umtamvuna, Vernon Crookes, Vryheid, Weenen.

Conservation status
Widespread and common.

Habitat
The distribution of red veld rats coincides with the availability of bushveld, woodland or forest habitats, indicating their preference for some tree cover. In coastal habitats they are typically collected in ecotones between coastal forest and grassland. They have been collected from a range of forest types including sand forest (e.g. False Bay Park), coastal forest (e.g. Krantzkloof NR), escarpment forest (e.g. Dlinza Forest), and dense riverine *Acacia* woodland (e.g. Shongweni Dam). A high proportion of the specimens collected in the present study were found in close proximity to rivers and dams. The species can tolerate some level of disturbance, and several specimens were collected from alien vegetation, from the borders of agricultural and sports fields and from inside buildings such as pumphouses. Although not usually found in association with man, specimens have been caught by cats in suburban areas of Durban.

Habits
Apparently solitary, nocturnal and terrestrial, excavating their burrows under the cover of bushes or more substantial cover (Skinner and Smithers 1990). During the present study, individuals were collected from the lower branches of bushes and trees, leading to some confusion between this species and *Thallomys paedulcus*, and indicating some arboreal ability.

Breeding
Red veld rats typically breed throughout the year. In the present study, pregnant females were recorded during January (*n*=4), March (*n*=1), September (*n*=2) and November (*n*=4). The mean number of foetuses per pregnant female was 2.7 (range 1–3; *n*=6), which is somewhat lower than recorded elsewhere (range 1–6; mean values ranging from 3.2 to 3.6; Skinner and Smithers 1990).

Linear measurements (mm) and mass (g)

	Males					Females				
	x̄	s.d.	n	Min	Max	x̄	s.d.	n	Min	Max
TL	296.4	25.5	42	216	360	283.3	20.8	37	234	315
HB	138.3	11.8	42	106	160	130.0	12.1	36	103	150
T	158.0	15.9	42	110	200	152.8	13.0	37	107	175
Hf	27.6	2.8	26	21	31	27.0	2.0	18	14	22
E	20.5	1.8	31	16	24	19.7	2.7	23	14	22
Mass	84.1	18.7	27	50	148	70.6	16.3	24	40	102

Records of occurrence
SPECIMEN RECORDS:

A. chrysophilus sensu lato (skulls not examined by Chimimba 1997):

BM: Illovo, Ngoye Hills;

DM: Ashburton, Bosch Hoek SANDF TR6 military area, Cape Vidal (5 km south), Coastal Forest Reserve (Kosi Bay), Dlinza Forest NR, Durban, Enseleni NR, Entumeni NR, False Bay Park, Futululu, Harold Johnson NR, Hluhluwe GR, Karkloof NR, Lake Eteza, Mariannwood NR (Pinetown), Marshall Road (Ashburton), Mount Moreland (12 William Street), Ngome, Roosfontein NR (Westville) Shongweni Resource Reserve, Spioenkop NR, St Lucia village (12 km north), Umbilo (Durban), Umlalazi NR, University of Natal (Durban campus), Weenen NR;

KM: St Lucia Resort, Vergeval Farm;

NM: Boteler Point, Drumclog, Entendweni Bush, Geluk Farm, Gwaliweni, Hluhluwe GR, Hluhluwe GR (Egodeni, Ngqbateti Corridor), Ingwavuma, Mfongosi, Ntombeni, Shemula's Pont, Sodwana Bay Park, Umfolozi Drift, Umfolozi GR (junction of White and Black Umfolozi Rivers);

SAM: Mfongosi;

SI: Buxton, Coastal Forest Reserve (=Kosi Bay), Makatini Flats, Ndumu GR, Nkonkoni, Nyala Statelands, Ladysmith;

TM: Bisley Valley, Ceza Mission Road, False Bay Park, Hilltop Farm, Itala GR, Karkloof Forest, Malvern, Maputa Store, Mposa Hill, Mseleni, Nondweni, Oribi Gorge NR, Otobokim, Queen Elizabeth Park, Richmond Road, Shongwe Store, St Lucia Forest Station, Ubombo, Umfolozi

GR, Umlalazi NR, Umtamvuna NR, Weza (Lower Stinkwood Forest).

A. chrysophilus sensu stricto (identifications from Chimimba 1997):

DM: Albert Falls NR, Vryheid NR;

NM: Langewacht Farm, Mkuzi GR;

TM: Black Umfolozi River (Nongoma–Mhlabatini road), Bumbeni, Hluhluwe (6 km north on Ingweni River bridge), Hluhluwe GR, Mkuzi GR, Mkuzi River, Quixotic Farm, Spioenkop NR.

Aethomys ineptus (Thomas and Wroughton 1908)
Red veld rat
Afrikaanse bosrot

Based on recent craniometric evidence (Chimimba 1997), as discussed under the previous species account, *Aethomys ineptus* is much commoner in KwaZulu-Natal than its sibling species, *A. chrysophilus sensu stricto*, and its distributional limits appear to extend further south (cf. Figures 97, 98).

As the two sibling species have been regarded as a single species in the past, life history details are available only for the species-complex as a whole. It is therefore assumed that the biology of *A. ineptus* is similar to that described under the previous species account.

Records of occurrence
SPECIMEN RECORDS:

A. ineptus sensu stricto (identifications from Chimimba 1997):

DM: Krantzkloof NR, Mapelane NR, Mfongosi, Umfolozi GR, Vryheid NR;

KM: Eastern Shores (Lake St Lucia);

NM: Langewacht Farm, Nagle Dam;

TM: Albert Falls Resort, Ashburton, Black Umfolozi River

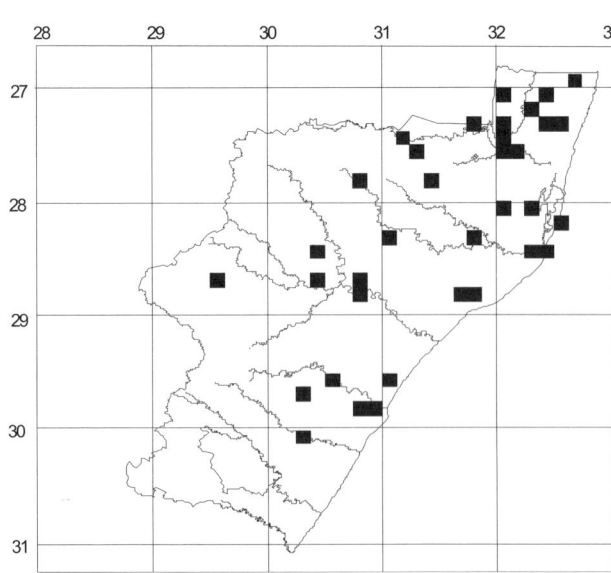

Figure 98

(Nongoma—Mhlabatini road), Candover South African Railways Station (cf. Leeuwspoor 647), Dukuduku Forest Reserve, Eastern Shores (Lake St Lucia), Empangeni, Futululu Research Station, Gwaliweni, Hazelmere, Hluhluwe GR (Mansiya Valley, Research Camp), Ingwavuma Bush, Itala NR (Bivane/Pongola River, Craigadam), Jameson's Drift (Tugela River), Lake Sibayi Research Station, Manaba, Manguzi Forest, Mkusi Bridge, Mkuzi GR (caravan park), Mkusi River, Ngome State Forest, Ngoye Forest, Paradise Valley, Pomeroy (road to Greytown), Pongola River, Quixotic Farm, Richmond Road, Ringwood, Shongwe Store (on road to Ubombo), Sihangwane Store (on road to Maputa), Spioenkop NR, St Lucia Forest Station, Tete Pan, Ubombo, Umfolozi GR, Umkomaas River (Ixopo–Richmond road), Umlalazi NR, Umtamvuma NR, Weza (Lower Stinkwood Forest).

Aethomys namaquensis (A. Smith 1834)   PLATE 46
Namaqua rock mouse
Namakwa-klipmuis

Taxonomic status
All described subspecies are regarded as synonyms of the nominate race (Meester et al. 1986).

Distribution
Occurs mostly throughout the drier northern and western parts of the province, in rocky situations. Less common or absent from the moister Mistbelt, Highlands and Montane bioregions (Figure 99). The species has been recorded from Pleistocene and Holocene deposits at Border Cave, and from Holocene deposits in the Upper Tugela Basin (Avery 1991; see under **Records of occurrence**).

Protected areas
Eastern Shores, Itala, Ndumu, Vryheid, Weenen.

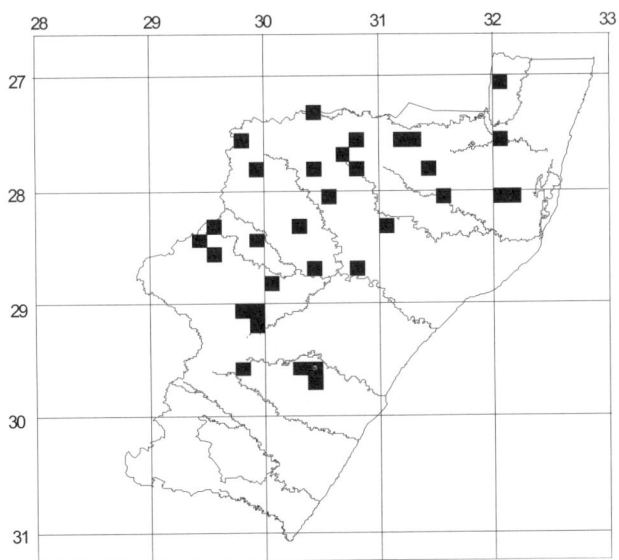

Figure 99

Conservation status
Widespread and common, although restricted to rocky habitats.

Habitat
In KwaZulu-Natal, they have been collected in sparse to fairly dense bushveld habitats in rocky situations (rocky outcrops and ridges as well as piles of rocks). They have also been collected in medium to tall dense grassland within 1 km of rocky terrain. Artificial piles of boulders, such as stone walls used to retain livestock, also provide suitable habitat.

Habits
Nocturnal and communal. They shelter in rocky crevices, hollow trees or in burrows constructed under logs or dense shrubs. Nests of grass and twigs are constructed over the entrances of shelters. Nests have been found in the forks of trees, 1–2 m above ground, suggesting some arboreal ability.

Breeding
No data on reproduction were obtained for this species during the present study. From data obtained elsewhere in southern Africa, pregnant females have been recorded from September to May (i.e. excluding the colder winter months), and pregnant females were found to contain between two and 11 young, with an average of between three and four (Skinner and Smithers 1990).

Linear measurements (mm) and mass (g)

	Males					Females				
	x̄	s.d.	n	Min	Max	x̄	s.d.	n	Min	Max
TL	258.1	18.3	26	224	290	260.2	19.0	18	216	294
HB	114.7	7.9	25	101	130	113.7	7.0	18	99	123
T	144.1	14.3	25	120	170	147.8	14.6	17	116	179
Hf	25.8	0.8	3	25	26	26.0	0.0	3	26	26
E	17.2	0.8	5	16	18	16.6	1.2	4	15	18
Mass	53.5	11.5	4	39	65	51.8	4.2	4	46	56

Records of occurrence
SPECIMEN RECORDS:
DM: Impendle, Moor Park NR, Pongola Bush NR, Vryheid NR, Weenen NR;
NM: Kambula, Langewacht, Louwsburg, Mfongosi, Spring Grove, The Hoek;
SAM: Mfongosi;
SI: Ladysmith (19 km from), Newcastle (16 km south), Nkonkoni (10 km north Mkuze), Petchaye (27 km southeast Estcourt);
TM: Black Umfolozi Bridge, Buffalo River Bridge (Harrismith–Ingogo District), Dundee (9 km from, on Pomeroy road), Goedehoop Farm, Hillside Farm (Ashburton), Hluhluwe GR (Gontshi Guard Camp, Research Camp), Ingwavuma Gorge, Itala GR (Craigadam), Klipspruit, Ladysmith (18 km west, on Van Reenen road), Ladysmith (9 km south, on Estcourt road), Ngome State

Forest, Piet Retief (32 km from, on Paulpietersburg road), Pomeroy (17 km from, on Greytown road), Ubombo, Ubombo (2 km west), Vanodrift (near), Vryheid East, Weenen NR.

ADDITIONAL RECORDS:
Avery (1991): Border Cave, eSinhlonhweni, Mgede, Nkupe.

Genus	*Rattus*	Fischer 1803

INTRODUCED SPECIES
Two species of introduced rats occur in KwaZulu-Natal. The house rat, or black rat (*R. rattus*) occurs in association with human habitations throughout the province, although few specimens have been collected. The much larger brown rat, or Norway rat (*R. norvegicus*) has been recorded only in the vicinity of Durban.

Family	GLIRIDAE	Dormice
Genus	*Graphiurus*	Smuts 1832

?*Graphiurus ocularis* (A. Smith 1829)
Spectacled dormouse
Gemsbokmuis

This species is restricted to the Western Cape, Eastern Cape, Northern Cape and Northwest Province, and is not known from KwaZulu-Natal (Skinner and Smithers 1990; Taylor et al. 1994). However, a skin of *G. ocularis* in the Natal Museum (**NM** 437) was collected on 6 January 1915 from 'Natal, South Africa'. A juvenile skull of this species in the Natal Museum (**NM** 935A) was also apparently collected from the Umtwalume River mouth on the South Coast of KwaZulu-Natal (30°29′S 30°38′E); however, the skull was attached to skin of *Mastomys natalensis* having the same number, casting some doubt on the validity of the provenance data for the skull. Given the dubious nature of both of the above records, the past or present occurrence of the spectacled dormouse, some 400 km outside its known distribution, requires further substantiating evidence.

Graphiurus murinus (Desmarest 1822)   PLATE 47
Woodland dormouse
Boswaaierstertmuis

Taxonomic status
The species is likely to contain more than one chromosomal form (Dippenaar et al. 1983) and is in need of revision (Meester et al. 1986). Taylor et al. (1994) found that specimens from Mkuzi GR (KwaZulu-Natal), Swaziland and Waterpoort (Northern Province of South Africa) differed from typical *G. murinus* (which they resembled in all other respects) in possessing a highly reduced M^1 and

markedly inflated bullae, raising the possibility of an undescribed species occurring in northern KwaZulu-Natal.

Distribution
This species is widely distributed in KwaZulu-Natal from sea level to 2 700 m in Giant's Castle Game Reserve, wherever suitable habitat is present (Figure 100). Avery (1991) recorded the species from Pleistocene and Holocene deposits at Border Cave and from Holocene deposits in the upper Tugela and Mgeni catchments (see under **Records of occurrence**).

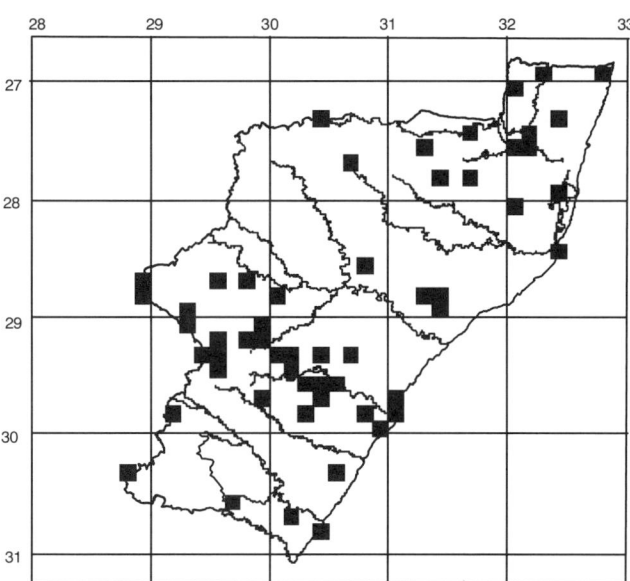

Figure 100

Protected areas
Cathedral Peak, Coastal Forest Reserve (=Kosi Bay), Entumeni, False Bay Park, Giant's Castle, Hluhluwe, Itala, Karkloof, Loteni, Mapelane, Mkuzi, Ndumu, Oribi Gorge, Pongola Bush, Queen Elizabeth Park, Royal Natal, Sodwana Bay, Spioenkop, Vernon Crookes, Vryheid, Weenen.

Conservation status
Fairly common and widespread throughout KwaZulu-Natal.

Habitat
The two main components in the habitat preference of this species are rocks and trees (Rowe-Rowe and Meester 1982a). At Giant's Castle, dormice were collected in forest; grouped-tree woodland (grassland with trees grouped around rocks); *Buddleja-Leucosidea* scrub, temperate grassland boulder-bed (rocky drainage lines on the escarpment); and at the upper limit of temperate grassland (2 700 m altitude). A specimen collected at Mkuzi NR was caught near a rubbish dump in an exotic plantation. At Itala GR they were collected on a slight incline with scattered shrub and grass with a substrate of red loam and scattered small rock rubble (Rautenbach, Root and Nel 1981). At Ndumu GR, Dixon (1964) described this species from thickets and from dry forest east of the Pongola River.

Habits

Woodland dormice are nocturnal and arboreal, but are frequently collected at ground level, suggesting a level of terrestriality. The diet is essentially omnivorous. In the Drakensberg a sample of 11 stomachs all contained insects, while 82 per cent also contained grass seeds (Rowe-Rowe 1986). In a year-long study at Karkloof Forest, Wirminghaus and Perrin (1992) found this species to be predominantly insectivorous (>80 per cent of their diet in any month comprising invertebrates); fruits were also eaten.

Breeding

No pregnant or lactating females were recorded in KwaZulu-Natal during the present study. At Giant's Castle GR, Rowe-Rowe and Meester (1982) noted perforate females during October and November, and from March to May; males with scrotal testes during September to November; and immature animals during March. This suggests that this species breeds mostly during summer. Very little data on reproduction exist for this species in southern Africa (Skinner and Smithers 1990).

Linear measurements (mm) and mass (g)

	Males					Females				
	x̄	s.d.	n	Min	Max	x̄	s.d.	n	Min	Max
TL	173.0	11.1	10	148	183	169.2	14.3	18	132	198
HB	97.2	12.2	10	80	115	92.2	10.2	20	71	113
T	75.8	6.9	10	68	90	76.1	10.5	18	43	87
Hf	20.0	2.6	3	18	23	17.3	1.5	6	15	19
E	14.4	2.5	7	13	19	14.4	1.8	10	11	17
Mass	34.0	6.6	3	28	41	26.6	5.3	7	19	32

Records of occurrence

SPECIMEN RECORDS:

BM: Estcourt, Sentinel Park;

DM: Bellair (Durban), Brentwood Farm, Durban, Entumeni NR, Hillcrest, Karkloof NR, Lute Road?, Loteni NR, Matatiele, Mkuzi GR, Ngome, Nongoma, Phoenix (Durban), Redhill (Durban), Royal Natal NP, St Michael's-on-sea, Weza State Forest;

KM: False Bay Park, Mapelane, Vergeval Farm;

NM: Bishopstowe (Pietermaritzburg), Champagne Castle, Chase Valley (Pietermaritzburg), Dalton, Durban, Estcourt, Geluk Farm, Giant's Castle GR, Howick, Hluhluwe GR, Kambula, Makatini Flats, Mkuzi GR, Mooi River, Pietermaritzburg, Richmond, Weenen, Winterskloof;

SAM: Pietermaritzburg, Zululand;

SI: Drakensberg Gardens (3 km from); Smallhoek Farm;

TM: Ashburton, Cathedral Peak, Coastal Forest Reserve (=Kosi Bay), Itala GR (Craigadam), Dargle, Doornhoek Gold Mine (15 km north-east), Drakensberg, Giant's Castle GR, Hluhluwe GR (Egodeni), Ingwavuma Gorge, Karkloof NR, Loteni NR, Manaba, Mount Edgecombe, Ndumu GR, Ngome Forest Reserve, Oribi Gorge NR, Pietermaritzburg, Pongola Bush NR, Qudeni, Queen Elizabeth Park, Spioenkop NR, Town Hill, Ubombo Bush, Vernon Crookes NR, Winterskloof.

ADDITIONAL RECORDS:

Avery (1991): Border Cave, Collingham (?), Nkupe.

Order	**LAGOMORPHA**	
Family	**LEPORIDAE**	**Hares**
Genus	*Lepus*	**Linnaeus 1758**

KEY TO GENERA (Meester et al. 1986)

1. Tail white below, black above; groove on anterior surface of principal upper incisor invariably filled with cement; chromosomal complement 2n=48; spermatozoa with cap-like acrosome, not pronounced anteriorly . . . . *Lepus*, p. 113
— Tail uniformly reddish-brown, brown or dark brown; groove on anterior surface of principal upper incisor never filled with cement; chromosomal complement never 2n=48; spermatozoa with relatively large acrosome, pronounced anteriorly . . . . *Pronolagus*, p. 115

? Lepus capensis (Linnaeus 1758)
Cape hare
Vlakhaas

Until recently, this species had not been recorded in KwaZulu-Natal, although it had been recorded close to the border with KwaZulu-Natal in the Mpumalanga (Rautenbach 1982) and Free State (Lynch 1983) provinces. Sight records which may represent this species are recorded close to the KwaZulu-Natal border in Lesotho (Lynch 1994). During January 1998, two *Lepus* specimens were shot by KwaZulu-Natal Nature Conservation Service staff at Balele Mountain Lodge on Farm Scurrekopye (27°30´48´´S; 30°12´30´E) in the Newcastle District (**DM** 5569, 5570). One of these specimens (**DM** 5770) was tentatively identified in the field as *L. capensis* on the basis of the presence of a buffy band between the ventral and dorsal coloration (present in *L. capensis* but not *L. saxatilis*). Measurements of the single male specimen were as follows: TL–481, HB–397, T–84, Hf*cu*–120, E–119. However, this specimen was later found to have a skull length of 91.8 mm and a principal upper incisor which was elongated in cross section and broad across its face (width 3.0 mm), these features being diagnostic of *L. saxatilis* (*L. capensis* has a smaller skull, not exceeding 91.1 mm in length, and a narrower principal upper incisor, 2.3–2.8 mm across the face).

Because of the contradictory nature of the above evidence, the presence of *L. capensis* in KwaZulu-Natal remains to be authenticated.

Lepus saxatilis (F. Cuvier 1823)
L. s. zuluensis
Scrub hare
Kolhaas

Taxonomic status
Of 10 southern African subspecies recognised by Meester et al. (1986), only the above is found in KwaZulu-Natal. Because of the difficulty in separating *L. saxatilis* and *L. capensis* on the basis of external appearance, caution should be exercised in accepting some of the sight records listed below under **Records of occurrence**.

Distribution
The species is widely distributed in KwaZulu-Natal; apparent gaps may simply be due to lack of collecting in the former KwaZulu areas (Figure 101).

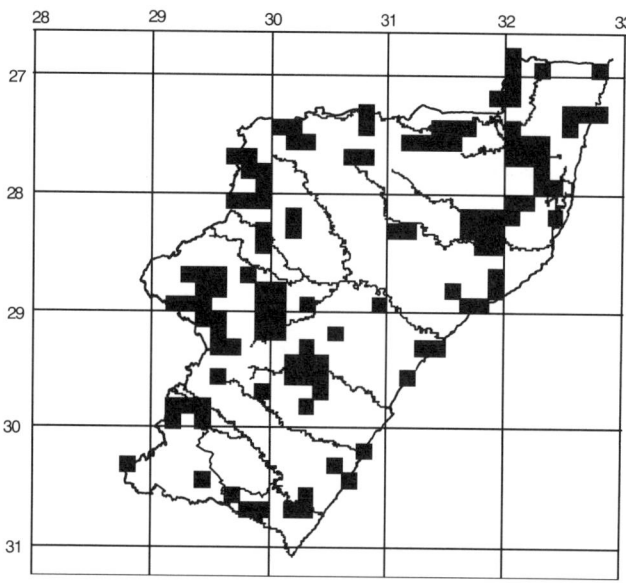

Figure 101

Conservation status
Common in open areas. They apparently adapt well to cultivated or heavily grazed areas (Bourquin 1988), although Lynch (1994) noted that they were rare in Lesotho, possibly at least partially due to human hunting pressures.

Protected areas
Albert Falls, Cathedral Peak, Chelmsford, Coleford, False Bay, Garden Castle, Giant's Castle, Hluhluwe, Itala, Kamberg, Karkloof, Midmar, Mkuzi, Mount Currie, Ndumu, Oribi Gorge, Queen Elizabeth Park, Royal Natal, Sodwana Bay, Spioenkop, Umfolozi, Umlalazi, Vernon Crookes, Weenen.

Habitat
Museum specimens from KwaZulu-Natal were taken in open grassland, on a golf course surrounded by dense coastal forest, in maize fields, open thornveld and roadside verges. Generally speaking, *L. saxatilis* appears to prefer more adequately vegetated habitats than *L. capensis*

(Skinner and Smithers 1990). However, it is clearly able to colonise pure grassland habitats with little or no tree cover, as indicated by its widespread occurrence in the Highland, Moist upland and Drier upland bioregions (Figure 101).

Habits
They are nocturnal, emerging at dusk to graze. During the day they rest in characteristic 'forms' under bushes. Young are also born in these forms and are extremely precocial, being born fully haired with the eyes open, and capable of moving around shortly after birth.

Breeding
Pregnant females in the Transvaal Museum and Natal Museum were collected during August (*n*=7), October (*n*=1), December (*n*=1), June (*n*=4), and July (*n*=2), confirming results from Botswana (Smithers 1971) and Zimbabwe (Smithers and Wilson 1979) where pregnancy occurred throughout the year. The mean number of foetuses per pregnant female in the KwaZulu-Natal sample was 3.5, with a range of one to four (*n*=15). This is somewhat higher than the mean of 1.6 recorded in Zimbabwe and Botswana, where a maximum of three foetuses was recorded in a large sample (*n*=86). The data from KwaZulu-Natal therefore indicate increased breeding success, probably related to higher annual rainfall and the nutritional content of grasses available for grazing.

Linear measurements

	Males					Females				
	$\bar{x}$	s.d.	n	Min	Max	$\bar{x}$	s.d.	n	Min	Max
TL	573.1	40.3	31	468	620	594.4	47.3	25	480	662
HB	480.6	31.4	33	420	520	500.6	40.4	28	405	565
T	96.0	21.2	31	32	125	92.6	18.3	24	40	127
Hfcu	120.6	5.0	13	115	130	122.0	10.7	16	99	140
E	127.1	17.9	13	110	165	125.4	10.9	19	95	140
Mass	2345	414.8	18	1814	2895	2520	438.8	16	1588	3055

Scrub hares from KwaZulu-Natal are similar in their external measurements and mass to those from the Western Cape and greater Transvaal Province, and somewhat larger and heavier than scrub hares from Zimbabwe (Skinner and Smithers 1990).

Records of occurrence
SPECIMEN RECORDS:
BM: Estcourt, Umfolozi;
DM: Brentwood Farm, Chelmsford NR;
KM: False Bay Park (Lister Point), Vergeval Farm;
NM: Ashburton, Babanango, Beginsel, Bergville (18 km from), Bisley (Pietermaritzburg), Broedersrus, Donkerhoek Farm, Duart, Engelbrecht's Drift, Firle Farm, Fricona, Geluk Farm, Groenvlei, Harding (south-east of), Hilton College, Ingwavuma Gorge, Inyala Ranch, Kambula, Kranskop (east of), Langewacht (Babanango), Louwsberg (near), Middlerus, Mkuzi GR, Oribi Gorge NR, Paulpietersburg, Pietermaritzburg, Randfontein, Richmond, Rothswaithe, Seven Oaks, Spring Grove, Stanger area, The

Hoek Farm, Umdoni Park, Umgeni Valley Game Ranch, Umtwalume, Underberg, Weza;
SI: Buxton, Makatini Flats, Papama;
TM: Babanango, Itala GR (Craigadam), Droogdedal, Hilton, Hluhluwe GR, Ingwavuma River, Lake Sibayi Research Station, Lebombo, Leeuwspoor, Magut (Jozini Dam), Mkuzi River, Ndumu GR, Spioenkop NR, Ubombo, Umfolozi GR, Vernon Crookes NR, Weenen NR, Winterton.
Sight records:
KwaZulu-Natal Nature Conservation Service: numerous un-named grid squares.

Genus	*Pronolagus*	Lyon 1904

KEY TO SPECIES (After Meester et al. 1986)

1. Total length of adult skull 77.4–87.5 mm; upper principal incisor 2.2–2.8 mm across face; bulla robust, breadth 6.4–9.2 mm, i.e., 7.4–11.4 per cent of total skull length, 16.5–25.1 per cent of mandibular height . . . . . .
    P. rupestris, p. 115
–   Total length of adult skull 85.5–96.3 mm; upper principal incisor 2.6–3.7 mm across face; bulla narrow, breadth 5.1–7.2 mm, i.e., 5.6–8.0 per cent of total skull length, 11.6–16.7 per cent of mandibular height . . . .
    P. crassicaudatus, p. 116

Pronolagus rupestris (A. Smith 1834)
P. r. barretti (Roberts 1949)
Smith's red hare
Smith rooi-haas

Taxonomic status
Eight subspecies were listed by Meester et al. (1986), of which one occurs in KwaZulu-Natal.

Distribution
This species occupies the escarpment and top of the Drakensberg along most of its length from the districts of Underberg to Newcastle, and has also penetrated some distance into the midlands where its range overlaps with that of *P. crassicaudatus* (Pringle 1974; Figure 102). They also occur in isolated areas where suitable rocky habitat occurs, such as the Babanango area.

Protected areas
None recorded.

Conservation status
Because of their specialised habitat and sparse distribution, and the fact that they occur in no protected areas, they are fairly rare in this province.

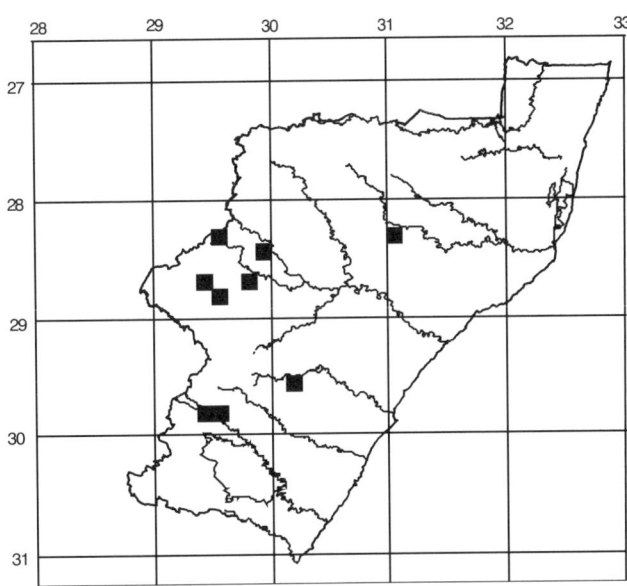

Figure 102

Habitat
Smith's red hare is associated with open grassland in rocky and mountainous areas on slopes or summits within the Highland, Mistbelt, Moist upland and Drier upland bioregions (Figure 102).

Habits
They are nocturnal, and hide up during the day in shelters among rocks. Nests are excavated at the bases of bushes. They create runs of up to 300 m in the vegetation. Diet consists entirely of grasses.

Breeding
Based on museum specimens, one pregnant female having a single embryo was recorded in October. According to Skinner and Smithers (1990) the species breeds from September to February, during which time three to four litters (of one to two young) are born.

Linear measurements (mm) and mass (g)

	Males					Females				
	x̄	s.d.	n	Min	Max	x̄	s.d.	n	Min	Max
TL	484.2	25.9	4	465	522	475.0	11.4	3	467	488
HB	415.0	26.8	4	400	455	407.3	6.4	3	400	412
T	69.2	8.3	4	60	80	67.7	11.7	3	55	78
Hfcu	87.7	4.6	3	85	93	82.5	–	2	80	85
E	99.7	9.9	3	95	111	77.5	–	2	75	80
Mass	1512	346.3	3	1134	1814	1500	–	2	1500	1500
Skull	83.6	2.7	4	81.2	87.2	–	–	1	67	67

Records of occurrence
SPECIMEN RECORDS:
NM: Beginsel, Bergville, Colenso, Dartford, Langewacht, Underberg;
TM: Babanango, Inhlosane, Matiwane, Winterton.

Pronolagus crassicaudatus (I. Geoffroy 1832)
P. c. crassicaudatus (I. Geoffroy 1832)
P. c. ruddi (Thomas and Schwann 1905)
P. c. lebombo (Roberts 1936)
Natal red hare
Natalse rooihaas

Taxonomic status

Meester et al. (1986) recognised three subspecies in KwaZulu-Natal, with the following type localities: Durban (*P. c. crassicaudatus*), Sibhudeni in the Nkandhla District of Zululand (*P. c. ruddi*), and Ubombo in north-eastern Zululand (*P. c. lebombo*). Pringle (1974) separated specimens in the Natal Museum into two subspecies based on the colour of the nape patch, with *P. c. crassicaudatus* (rufous-coloured nape patch) occurring to the south of the Tugela River and *P. c. ruddi* (greyish-coloured nape patch) occurring to the north. No specimens from the range of the third subspecies, *P. c. lebombo*, were available for study by Pringle (1974). Apart from the type series from Ubombo (collected in 1928), and dated reports of the species occurrence at Mkuzi and Ndumu Game Reserves (Dixon 1964, 1966), there are no modern records of the species' occurrence in the Maputaland region, and Pringle (1974) considered that the species' distribution had decreased in this and other areas of the province where it once occurred.

Distribution

The Natal red hare occurs from the coast throughout the midlands to an altitude of 1 550 m in the foothills of the Drakensberg (Pringle 1974; Figure 103).

Protected areas

Coleford, Giant's Castle, Hluhluwe, Itala, Loteni, Mkuzi, Ndumu, Oribi Gorge, Royal Natal, Spioenkop, Umfolozi, Umtamvuna, Vernon Crookes.

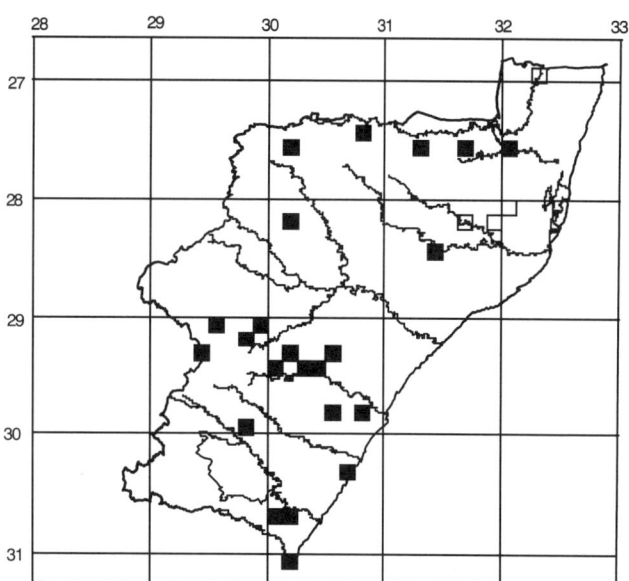

Figure 103

Conservation status

More common and widespread, and better protected in the province, than *P. rupestris*.

Habitat

Similar to that described for Smith's red hare. A number of specimens in the Natal Museum collection were collected from grassy slopes along the edges of cliff faces. Smith's red hares require the presence of rocks and boulders for sheltering during the day time. Pringle (1974) described a large colony near Magudu which had taken up residence in an artificial rock pile formed from boulders excavated during the construction of a farm road.

Habits

Similar to Smith's red hare. Although relying mostly on rocks for shelter, they have also been found resting in densely grassed areas where they are extremely difficult to flush.

Breeding

Pregnant females were collected during June (*n=2*) and August (*n=2*). The number of foetuses per pregnant female varied from one (*n=1*) to two (*n=3*). Lactating females were collected in August (*n=1*), October (*n=1*) and February (*n=2*), and a juvenile was collected in July. These data suggest that breeding occurs throughout the year. The present data represent all that is currently known on the reproduction of this species (cf. Skinner and Smithers 1990).

Linear measurements (mm) and mass (g)

	Males					Females				
	x̄	s.d.	n	Min	Max	x̄	s.d.	n	Min	Max
TL	541.9	59.0	10	467	640	541.0	32.6	10	480	595
HB	488.1	45.0	12	425	560	481.6	30.2	10	425	540
T	56.6	20.2	10	420	110	59.9	11.1	10	40	75
Hfcu	96.0	4.7	5	905	100	98.8	8.5	6	88	110
E	78.6	8.1	5	67	90	82.7	8.9	6	70	95
Mass	1959	395	7	1361	2601	2219	748	4	1240	2895
Skull	88.9	4.0	10	81.0	93.0	89.1	6.2	6	78	94

Records of occurrence

SPECIMEN RECORDS:
CM: Kokstad (5 km west);
DM: Kranzkloof;
NM: Broedersrus, Faraway, Giant's Castle GR, Harding (south-east of), Hilton College, Karkloof, Knoop's Farm, Magudu, Meadow Farm, Melmoth (19 km north), New Hanover, Ntabamhlope, Oribi Gorge NR, Otto's Bluff, Pongola River, Randfontein, Rosslea Estate;
TM: Albert Falls (near), Gillitts, Hilton, Itala GR (Craigadam), Itala GR (10 km north-west), Kilgobbin Farm, Magut (=Magudu), Port Edward, Ubombo.

ADDITIONAL RECORDS:
Bourquin et al. (1971): Hluhluwe GR, Umfolozi GR;
Dixon (1964): Mkuzi GR;
Dixon (1966): Ndumu GR.

Order	MACROSCELIDEA Elephant shrews
Family	MACROSCELIDIDAE

KEY TO GENERA (After Meester et al. 1986)

1. Size larger, length of head and body (HB) of adults >160 mm, condylobasal skull length (CB) >45 mm, upper toothrow (UTR) >25 mm; hallux absent; two pairs of mammae . . . . *Petrodromus*, p. 117

- Size smaller, HB <160 mm, CB <40 mm, UTR <22 mm; hallux present; three pairs of mammae . . . . *Elephantulus*, p. 118

Genus	*Petrodromus*	Peters 1846

Petrodromus tetradactylus (Peters 1946)   PLATE 48
P. t. warreni (Thomas 1918)
Four-toed elephant shrew
Bosklaasneus

Taxonomic status
Five southern African subspecies were listed by Meester et al. (1986), one of which occurs in KwaZulu-Natal.

Distribution
This species is restricted to the Lowveld and Coast lowland bioregions in the extreme north-east of KwaZulu-Natal, with isolated populations extending as far south as Lake St Lucia (Figure 104).

Protected areas
False Bay Park, Hluhluwe, Mkuzi, Ndumu, Sodwana Bay, Sodwana State Forest, Ndumu.

Conservation status
Listed as Rare in the South African Red Data Book (Smithers 1986).

Habitat
In KwaZulu-Natal, they occur in mature riparian and evergreen forests having dense undergrowth, and in coastal forest and scrub (Bourquin 1988). Habitat of specimens collected at Ndumu GR was recorded as 'dry thorn bush' and 'thicket forest'.

Habits
They are either solitary or live in pairs, and are active during night and day. They rest under thick cover formed at the base of shrubs, under logs, or under the roots of trees. Conspicuous runs can be seen radiating from their shelters, these being marked by evenly-spaced bare patches

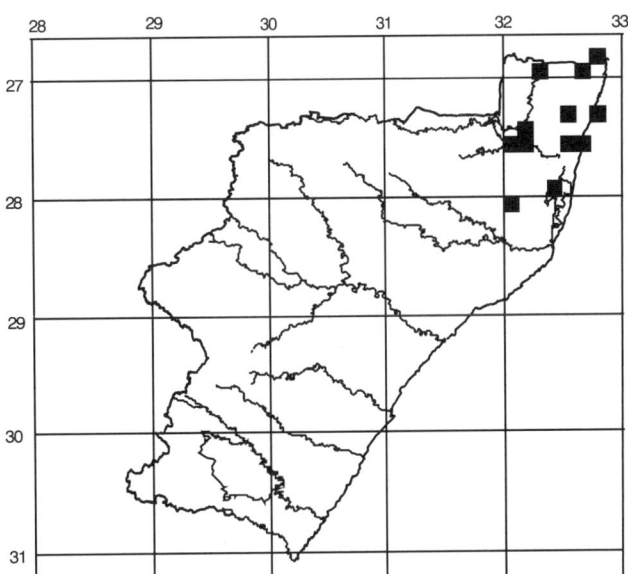

Figure 104

where the animals take off and land as they proceed in jumps along the runs. Four-toed elephant shrews are insectivorous and ants have been found to comprise an important component of their diet (Skinner and Smithers 1990).

Breeding
No data available for the province. Meagre data available elsewhere suggests that the species may have an extended breeding season throughout the year (Skinner and Smithers 1990).

Linear measurements (mm)

	Males					Females				
	$\bar{x}$	s.d.	n	Min	Max	$\bar{x}$	s.d.	n	Min	Max
TL	–	–	2	319	335	341.8	15.8	5	323	359
HB	–	–	2	172	178	186.8	9.6	5	171	194
T	–	–	2	147	157	155.0	9.8	5	140	165
Hfcu	–	–	2	52	53	53.5	2.9	4	50	57
E	–	–	2	32	34	35.0	1.2	5	33	36

Records of occurrence
SPECIMEN RECORDS:
DM: Mkuze, Umziki Pan;
NM: Manguzi, Mseleni, Ndumu GR;
SI: Makatini Flats;
TM: Hluhluwe, Lake Sibayi Research Station, Makatini Research Station, Mkuzi River, Mosi Swamp, road to Maputa, Sodwana Bay.

ADDITIONAL RECORDS:
Bruton (1978): Lake Sibayi (South Basin), Mseleni Stream;
Rautenbach et al. (1980): Mkuzi Swamp.
Sight records (KwaZulu-Natal Nature Conservation Service): Mkuzi GR, False Bay NR.

Genus	*Elephantulus*	Thomas and Schwann 1906

Elephantulus myurus (Thomas and Schwann 1906)
Rock elephant shrew
Bosveldklaasneus

Taxonomic status
Meester et al. (1986) did not recognise any subspecies.

Distribution
Their distribution in KwaZulu-Natal is restricted to the Drier upland bioregion in the north-west (Figure 105).

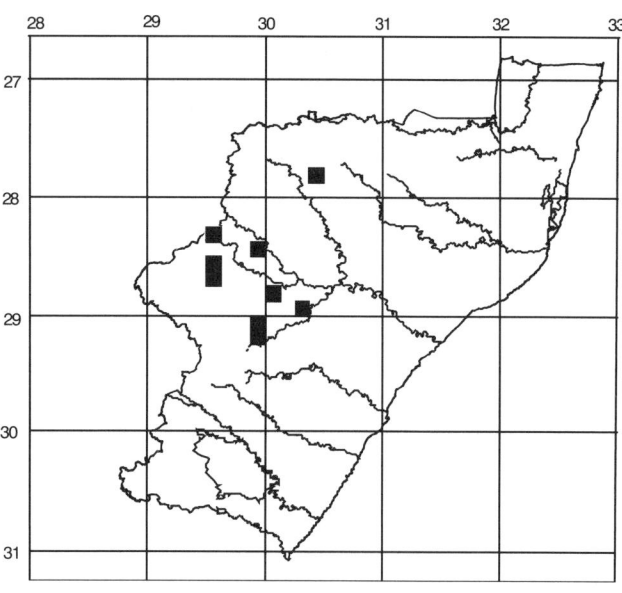

Figure 105

Protected areas
Spioenkop, Weenen.

Conservation status
They are locally common in areas of suitable rocky habitat in the Drier upland bioregion of KwaZulu-Natal.

Habitat
Specimens from Weenen NR were collected from rock piles in open *Acacia* woodland as well as dense riverine *Acacia* woodland in a steep-sided valley.

Habits
Rock elephant shrews are predominantly diurnal. They remain camouflaged in concealing vegetation or rocky overhangs during the day, making rapid sorties to seize their prey. They vocalise with a series of high-pitched squeaks, during which the head is held high with the trunk curled back over the top of the muzzle and the mouth held wide open. They also 'purr' when alarmed, by vibrating the hind legs against the substrate. Their insectivorous diet includes large numbers of ants and termites which are often foraged from dassie middens (Skinner and Smithers 1990).

Breeding
Two pregnant females, each with two foetuses, were collected in November during this study.

Linear measurements (mm) and mass (g)

	Males					Females				
	x̄	s.d.	n	Min	Max	x̄	s.d.	n	Min	Max
TL	259.7	11.2	6	242	272	258.0	14.1	8	230	280
HB	122.3	7.1	6	113	134	123.4	4.1	8	116	129
T	137.3	10.8	6	117	145	134.6	14.6	8	103	151
Hf*cu*	36.7	0.6	3	36	37	–	–	2	37	37
E	22.0	6.2	3	15	27	–	–	2	25	26
Mass	–	–	2	54	62	–	–	–	–	–

Records of occurrence
SPECIMEN RECORDS:
DM: Weenen NR;
NM: Muden (10 km west), Spring Grove, The Hoek;
SI: Estcourt, Ladysmith;
TM: Klipspruit, Ladysmith (19 km west), Spioenkop, Weenen NR.

Sight records: Renlee Farm, Frischgewaagd Farm.

GAZETTEER OF COLLECTING LOCALITIES

Latitude and **longitude** are expressed as degrees (first two digits) and minutes (third and fourth digits). **Gridref** refers to the code used to describe 1/64th-degree grid squares.

LOCALITY	LATITUDE	LONGITUDE	GRIDREF
ADDINGTON (DURBAN)	2950	3100	2931CCA
ALBERT FALLS (NEAR)	2926	3025	2930ADD
ALBERT FALLS NATURE RESERVE	2928	3023	2930ADD
ALLER RIVER VALLEY (WESTVILLE)	2950	3056	2930DDB
AMANZIMTOTI	3003	3053	3030BBB
ASHBURTON	2938	3028	2930CBD
ASSEGAI	2951	3043	2930DCB
ATHLONE PARK (AMANZIMTOTI)	3001	3055	3030BBB
AVONDALE	2952	3037	2930DCA
BABANANGO	2822	3105	2831ACA
BALGOWAN	2924	3003	2930ACC
BALLITO BAY	2932	3113	2931CAB
BAYNESFIELD	2937	3023	2930CBB
BEACHWOOD COUNTRY CLUB	2946	3102	2931CCA
BEACHWOOD NATURE RESERVE	2948	3103	2931CCA
BEACON HILL (IFAFA DISTRICT)	3025	3036	3030BCC
BEAUMONT WATTLE ESTATE (=SUGAR ESTATE, ADIT)	2953	3034	2930DCC
BEGINSEL FARM	2804	2945	2829BBA
BELFORT	2933	3022	2930CBA
BELLAIR	2953	3057	2930DDD
BELLEVUE FARM (NO. 25)	2752	3052	2730DDA
BENEVA MINE TUNNEL (=SEZELA MINE)	3021	3038	3030BCB
BEREA	2950	3100	2931CCA
BERGVILLE	2844	2921	2829CBC
BERGVILLE (18 KM FROM)	2842	2923	2829CBD
BHANGA NEK	2701	3251	2732BBA
BISHOPSTOWE	2937	3023	2930CBB
BISHOPSTOWE CAVE	2936	3023	2930CBB
BISI RIVER	2911	3017	2930ABC
BISLEY	2936	3023	2930CBB
BLACK UMFOLOZI RIVER (NONGOMA–MHLABATINI RD)	2807	3130	2831BAA
BLACKRIDGE	2936	3023	2930CBB

LOCALITY	LATITUDE	LONGITUDE	GRIDREF
BLINKWATER NATURE RESERVE	2914	3027	2930ABD
BLOOD RIVER STATION	2754	3035	2730DCC
BLUFF	2956	3101	2931CCC
BLUFF NATURE RESERVE	2956	3059	2930DDD
BONAMANZI GAME RESERVE	2806	3218	2832ABA
BONCRAY (30 KM NORTH OF HIMEVILLE)	2952	2937	2929DCA
BONGAZI NORTH (=LAKE BHANGAZI)	2739	3238	2732DAD
BOPOMENI	2708	3227	2732ABD
BOSCH HOEK FARM (20 KM NORTH-WEST OF HOWICK)	2921	3006	2930ACA
BOSCH HOEK SADF AREA (50 KM NORTH-EAST OF LADYSMITH)	2917	3006	2930ACA
BOSTON	2938	3001	2930CAC
BOSTON (13 KM NORTH-WEST)	2936	2952	2929DBA
BOSTON POLICE STATION (3 KM NORTH-WEST)	2937	2952	2929DBA
BOTELER POINT	2701	3252	2732BBA
BOTHA'S HILL (ROB ROY HOTEL)	2945	3045	2930DDA
BOWEL FOREST	2752	3107	2731CCA
BOWER FOREST (=WEZA STATE FOREST)	3036	2943	3029DAB
BRACCO	2917	3023	2930ADB
BRENTWOOD FARM	2940	2953	2929DBD
BRIGHTON BEACH	2956	3101	2931CCC
BROEDERSRUS FARM	2737	3122	2731CBA
BROUGHTON (BETWEEN BROUGHTON AND CRAMOND)	2922	3022	2930ADA
BUCKLAND DOWNS	2822	2907	2829ACA
BUFFELS RIVER BRIDGE (INGOGO–WAKKERSTROOM RD)	2737	2952	2729DBA
BUFFELS RIVER BRIDGE (UTRECHT–NEWCASTLE RD)	2737	3007	2730CAA
BULWER	2948	2945	2929DDA
BULWER (23 KM FROM)	2952	2937	2929DCA
BUMBENI	2752	3222	2732CDA
BURMAN BUSH (DURBAN)	2950	3100	2931CCA
BUSHMEN'S NEK	2952	2937	2929DCA
BUXTON (10 KM EAST OF ESHOWE)	2850	3131	2831DCA
CAMPERDOWN	2945	3045	2930DDA
CANDOVER (7.5 KM NORTH, ON COLLEL RD)	2722	3152	2731BDA
CAPE VIDAL	2808	3233	2832BAC
CAPE VIDAL (5 KM SOUTH)	2808	3233	2832BAC
CARRINGTON HEIGHTS	2950	3100	2931CCA

LOCALITY	LATITUDE	LONGITUDE	GRIDREF
CARTER'S NURSERY	2937	3023	2930CBB
CASTLE VIEW	2950	2925	2929CDB
CATHEDRAL PEAK NATURE RESERVE	2856	2915	2829CDC
CATHEDRAL PEAK (CATCHMENT 9)	2859	2915	2829CDC
CATHEDRAL PEAK (MLAMBONJA RIVER)	2856	2914	2829CCD
CATO RIDGE (3 KM NORTH-EAST)	2937	3037	2930DAA
CAVENDISH	2953	3059	2930DDD
CEZA MISSION RD	2807	3122	2831ABA
CHAMPAGNE CASTLE	2903	2915	2929ABA
CHARD FARM (KARKLOOF FOREST)	2917	3023	2930ADB
CHARTER'S CREEK	2812	3225	2832ABD
CHASE VALLEY HEIGHTS	2937	3023	2930CBB
CHATSWORTH	2947	3047	2930DDA
CHELMSFORD NATURE RESERVE	2749	2955	2729DDB
CLAIRWOOD PARK	2956	3059	2930DDD
CLARENDON	2936	3023	2930CBB
CLARIDGE	2937	3023	2930CBB
CLEARWATER FARM	3102	3010	3130AAB
CLIFTON PREP SCHOOL (NOTTINGHAM ROAD)	2921	3000	2930ACA
CLIFTON SCHOOL (MORNINGSIDE)	2950	3100	2931CCA
COASTAL FOREST RESERVE (=KOSI BAY)	2655	3250	2632DDC
COASTLANDS	?	?	?
COLBOURNE	2918	3013	2930ACB
COLEFORD NATURE RESERVE	2957	2927	2929CDD
COLENSO	2844	2949	2829DBC
CORNHILL FARM	3012	3000	3030AAC
COWIES HILL	2949	3053	2930DDB
CRAIGIEBURN	3012	3046	3030BBC
CRAMMOND	2925	3026	2930ADD
CROOKE'S FARM	3012	3046	3030BBC
CURRY'S POST	2929	3013	2930ACD
DALTON	2920	3038	2930BCB
DARGLE PLANTATION (BOSCH HOEK FARM, NO. 973)	2937	3007	2930CAA
DARGLE STATE FOREST	2928	3006	2930ACC
DARTFORD FARM	2952	2928	2929CDB
DARVILL SEWERAGE FARM	2936	3023	2930CBB

LOCALITY	LATITUDE	LONGITUDE	GRIDREF
DE HOEK FOREST RESERVE	2350	3002	2330CCA
DE JAGERS DRIFT (BUFFALO RIVER)	2807	3022	2830ABA
DENNY DALTON MINE	2817	3114	2831ACB
DLINZA FOREST NATURE RESERVE	2853	3127	2831CDD
DOMLEO'S CAVE (=LAAGER CAVE)	2924	3044	2930BCD
DONKERHOEK FARM	2738	2943	2729DAD
DONNYBROOK	2956	2952	2929DDC
DOORNHOEK FARM	2822	2922	2829ADA
DOORNHOEK MINE (PIETERMARITZBURG)	2936	3031	2930DAA
DOREEN CLARK NATURE RESERVE	2937	3023	2930CBB
DRAKENSBURG GARDENS HOTEL	2946	2915	2929CDA
DRAKENSBERG GARDENS TROUT HATCHERY	2947	2930	2929DCA
DROOGDEDAL	2807	2952	2829BBA
DRUMCLOG	2937	3023	2930CBB
DUART	2930	2931	2929DAA
DUKUDUKU FOREST (6 KM NORTH NORTH-EAST MTUBATUBA)	2822	3221	2832ADC
DUMISA GOLD MINE	3022	3022	3030ADA
DUNDEE	2810	3015	2830ABC
DUNDEE (8 KM FROM, ON POMEROY RD)	2815	3020	2830ADA
DUNDEE (18 KM FROM, ON POMEROY RD)	2817	3020	2830ADA
DURBAN	2952	3100	2931CCA
DURBAN BOTANICAL GARDENS	2950	3100	2931CCA
DURBAN CITY HALL	2950	3100	2931CCA
DURBAN COUNTRY CLUB	2950	3102	2931CCA
DURBAN NORTH	2948	3102	2931CCA
EASTERN SHORES NATURE RESERVE (CAPE VIDAL)	2809	3232	2832BAC
EASTERN SHORES NATURE RESERVE (GENERAL)	2810	3230	2832BAC
EASTERN SHORES NATURE RESERVE (IPHIVA CAMP SITE)	2821	3225	2832ADB
EASTERN SHORES NATURE RESERVE (MISSION ROCKS)	2816	3229	2832ADB
EASTERN SHORES NATURE RESERVE (ST LUCIA GAME PARK)	2821	3225	2832ADB
EERSTELING FARM	3012	3000	3030AAC
ELANDSLAAGTE	2825	2958	2829BDD
ELLIOT FARM	2916	3013	2930ACB
EMPANGENI	2846	3154	2831DDB
EMPISINI NATURE RESERVE	3012	3046	3030BBC
ENGELBRECHT'S DRIFT	2737	3020	2730CBA

LOCALITY	LATITUDE	LONGITUDE	GRIDREF
ENSELENI NATURE RESERVE	2841	3200	2832CAC
ENTENDWENI BUSH	2811	3210	2832AAD
ENTUMENI NATURE RESERVE	2850	3122	2831CDA
ESCOMBE	2952	3054	2930DDB
ESCOMBE SCHOOL (NORTHDENE)	2952	3054	2930DDB
ESHOWE	2854	3128	2831CDD
ESPERANZA	3020	3039	3030BCB
ESTCOURT	2900	2953	2929BBB
ESTCOURT (16 KM FROM, ON NOTTINGHAM ROAD RD)	2907	2952	2929BBA
ESTON	2952	3030	2930DCA
EVERTON	2952	3100	2931CCA
FALSE BAY PARK	2801	3221	2832ABA
FALSE BAY PARK (DUGUNDHLOVU CAMP SITE)	2801	3221	2832ABA
FALSE BAY PARK (LISTER POINT)	2758	3223	2732CDD
FARAWAY	2952	3031	2930DCA
FERNCLIFF	2937	3023	2930CBB
FIELD'S HILL (=PINETOWN)	2949	3051	2930DDA
FIRLE FARM	2913	2959	2929BBD
FLAGSTONE SPRUIT (9 KM SOUTH OF LADYSMITH)	2837	2937	2829DAA
FOREST HILLS	2949	3050	2930DDA
FOREST INN	?	?	?
FORT NOTTINGHAM NATURE RESERVE	2924	2954	2929BDD
FORT YOLLAND	2851	3112	2831CCB
FOUR WINDS FARM	2941	3026	2930CBD
FOXTAIL FARM (GIANT'S CASTLE NATURE RESERVE)	2912	2931	2929BAC
FRICONA FARM	2725	3050	2730BDC
FUTULULU RESEARCH STATION	2825	3216	2832ADC
FYNNLANDS	2955	3102	2931CCC
GARDEN CASTLE NATURE RESERVE	2945	2915	2929CDA
GARDENIA (17 KM SOUTH OF NEWCASTLE)	2750	2956	2729DDB
GELUK FARM	2844	2949	2829DBC
GHOST MOUNTAIN (NEAR MKUZE)	2737	3204	2732CAA
GIANT'S CASTLE GAME RESERVE	2920	2929	2929ADB
GIANT'S CASTLE GAME RESERVE (HILLSIDE CAMP)	2914	2930	2929BAC
GIANT'S CASTLE GAME RESERVE (INJASUTI OUTPOST)	2910	2925	2929ABD
GILLITTS	2948	3048	2930DDA

LOCALITY	LATITUDE	LONGITUDE	GRIDREF
GLEN HILLS	2946	3102	2931CCA
GLENASHLEY	2948	3102	2931CCA
GLENBELLA FARM	2905	2955	2929BBB
GLENHILLS	2950	3100	2931CCA
GLENMORE	2950	3100	2931CCA
GLENWOOD	2950	3100	2931CCA
GLENWOOD HIGH SCHOOL	2950	3100	2931CCA
GOEDEHOOP FARM	2752	3055	2730DDB
GOEDHOEK FARM	2823	3105	2831ACC
GOEDVERSWACHT FARM (NO. 455)	2807	3107	2831AAA
GOOD HOPE FARM	2925	3004	2930ACC
GOOD HOPE FOREST (MONDI)	2939	2958	2929DBD
GOUDHOEK FARM	2823	3105	2831ACC
GREENWOOD PARK	2947	3101	2931CCA
GREYTOWN	2907	3037	2930BAA
GROENKLOOF FARM (5 KM SOUTH OF WAKKERSTROOM)	2724	3010	2730ACD
GROENVLEI FARM	2727	3015	2730ADC
GUMEDI'S KRAAL (USUTU RIVER)	2652	3222	2632CDA
GWALIWENI (=KWALIWENI)	2720	3205	2732ACA
HAPPY HILLS FARM	2927	3007	2930ACC
HARDING	3040	2958	3029DBD
HARDINGDALE (11 KM NORTH-EAST OF PIETERMARITZBURG)	2932	3024	2930CBB
HAROLD JOHNSON NATURE RESERVE	2920	3125	2931ADB
HAWAAN FOREST (UMHLANGA ROCKS)	2948	3102	2931CCA
HAZELDENE FARM (14 KM FROM VANTS DRIFT)	2807	3022	2830ABA
HAZELMERE NATURE RESERVE	2935	3102	2931CAA
HELLA HELLA GAME RANCH	2954	3005	2930CCC
HESKITH VALLEY	2936	3023	2930CBB
HIBBERDENE	3034	3035	3030DAA
HIGHMOOR NATURE RESERVE	2919	2937	2929BCA
HILLARY	2953	3057	2930DDD
HILLCREST	2946	3046	2930DDA
HILLSIDE FARM	2937	3022	2930CBA
HILTON	2929	3018	2930ADC
HILTON COLLEGE	2929	3018	2930ADC
HILTON RAILWAY TUNNEL	2929	3018	2930ADC

LOCALITY	LATITUDE	LONGITUDE	GRIDREF
HILLTOP FARM	3003	3018	3030ABA
HIMEVILLE	2945	2931	2929DCA
HLABENI FOREST	2957	2944	2929DCD
HLABISA	2808	3153	2831BBD
HLAMBANYATI–MKUZE	2738	3214	2732CAD
HLATIKULU VLEI	2917	2941	2929BCB
HLATINI ROSE FARM	3054	3018	3030CDC
HLATWA DISTRICT	?	?	?
HLUHLUWE (MKUZE RD)	2807	3207	2832AAA
HLUHLUWE FLATS	2807	3207	2832AAA
HLUHLUWE GAME RESERVE	2802	3207	2832AAA
HLUHLUWE GAME RESERVE (EGODENI)	2805	3202	2832AAA
HLUHLUWE GAME RESERVE (EZUNQUIMENE CORRIDOR)	2802	3208	2832AAB
HLUHLUWE GAME RESERVE (FIELD STAFF HQ)	2805	3204	2832AAA
HLUHLUWE GAME RESERVE (GONTSHI GUARD CAMP)	2803	3208	2832AAB
HLUHLUWE GAME RESERVE (MAGANDA)	2805	3202	2832AAA
HLUHLUWE GAME RESERVE (MANSIYA VALLEY)	2804	3202	2832AAA
HLUHLUWE GAME RESERVE (MANZIBOMVU RIVER)	2804	3207	2832AAA
HLUHLUWE GAME RESERVE (MEMORIAL GATE)	2805	3202	2832AAA
HLUHLUWE GAME RESERVE (RESEARCH CAMP)	2804	3202	2832AAA
HLUHLUWE GAME RESERVE (STAFF QUARTERS)	2805	3202	2832AAA
HLUHLUWE RIVER (6 KM EAST MAIN RD)	2807	3222	2832ABA
HLUHLUWE VILLAGE (3 KM NORTH, ON INGWENI R BRIDGE)	2758	3214	2732CCD
HLUHLUWE–UMFOLOZI CORRIDOR RESERVE (MASIMBA CAMP)	2816	3157	2831DDB
HOWICK	2930	3014	2930CAB
IAN ELLIS NATURE RESERVE	3046	3025	3030CDB
IFAFA	3028	3039	3030BCD
ILLOVO	3006	3050	3030BBA
IMPENDLE NATURE RESERVE	2940	2953	2929DBD
IMPENDLE VILLAGE	2936	2952	2929DBA
INANDA	2941	3056	2930DBD
INANDA GAME PARK	2941	3056	2930DBD
INCHANGA FARM	2944	3036	2930DAC
INGELE FOREST	3037	2943	3029DAB
INGOGO	2734	2954	2729DBB
INGWAVUMA BUSH	2707	3207	2732AAA

LOCALITY	LATITUDE	LONGITUDE	GRIDREF
INGWAVUMA GORGE	2707	3207	2732AAA
INGWAVUMA VILLAGE	2708	3201	2732AAC
INHLOSANE FARM (EAST OF HOWICK)	2930	3014	2930CAB
INHLUZANI MOUNTAIN	2932	2954	2929DBB
INJASUTI NATURE RESERVE	2907	2922	2929ABA
INSUZI VALLEY	2853	3103	2831CCC
INYALA RANCH	2845	3154	2831DDB
IRON WATCH	3037	3022	3030CBA
ISIPINGO	2958	3055	2930DDD
ITALA GAME RESERVE	2731	3122	2731CBA
ITALA GAME RESERVE (BREDA)	2730	3112	2731CAB
ITALA GAME RESERVE (CONFLUENCE OF BIVANE & PONGOLA)	2727	3113	2731ACD
ITALA GR (CRAIGADAM)	2731	3121	2731CBA
ITALA GAME RESERVE (DOORNKRAAL)	2730	3113	2731CAB
ITALA GAME RESERVE (DOORNPAN)	2731	3113	2731CAB
IXOPO	3009	3004	3030AAC
IXOPO (25.6 KM NORTH)	3000	2955	3029BBB
IXOPO (RICHMOND RD)	3001	3015	3030ABA
IXOPO (40 KM SOUTH-EAST)	3017	3027	3030ADB
IZINGOLWENI RD	3103	3012	3130AAB
JAMESONS DRIFT (TUGELA RD, KRANSKOP–VRYHEID RD)	2852	3052	2830DDA
JOZINI DAM (=JOZINI VILLAGE)	2725	3204	2732ACC
KAMBERG NATURE RESERVE	2924	2940	2929BCD
KAMBULA	2741	3043	2730DAD
KARKLOOF (MERENSKY TRUST FOREST)	2920	3011	2930ACB
KARKLOOF FOREST (SHAFTON HOUSE)	2924	3017	2930ADC
KARKLOOF NATURE RESERVE	2918	3013	2930ACB
KEARSNEY COLLEGE	2919	3116	2931ADA
KENNETH STAINBANK NATURE RESERVE	2955	3056	2930DDD
KETELFONTEIN	2934	3021	2930CBA
KHARWASTAN	2955	3054	2930DDD
KILGOBBIN FARM	2928	3003	2930ACC
KINGSBURGH	3006	3050	3030BBA
KINGTHORPE FARM	2939	3027	2930CBD
KLIPSPRUIT	2748	3025	2730CDB
KLOOF	2949	3050	2930DDA

LOCALITY	LATITUDE	LONGITUDE	GRIDREF
KNOOPS FARM	3042	3014	3030CAD
KOK'S FARM (OTTO'S BLUFF)	2929	3023	2930ADD
KOKSTAD	3032	2925	3029CBB
KOSI BAY NATURE RESERVE (=COASTAL FOREST RESERVE)	2655	3250	2632DDC
KOSI BAY (LAKE AMANZAMYAMA)	2702	3250	2732BBA
KOSI LAKE (DEPARTMENT OF HEALTH CAMP)	2657	3250	2632DDC
KRANSKOP	2745	2948	2729DDA
KRANSKOP (13 KM EAST)	2907	3104	2931AAA
KRANTZKLOOF NATURE RESERVE	2946	3050	2930DDA
KULU	2838	3106	2831CAC
KWALIWENI (=GWALIWENI)	2720	3205	2732ACA
KWAMAGWAZA MISSION	2835	3125	2831CBB
KWAMASHU	2945	3058	2930DDB
KWASHELENI	?	?	?
LAAGER CAVE (=DOMLEO'S CAVE)	2924	3044	2930BCD
LADYSMITH	2833	2947	2829DBA
LADYSMITH (50 KM FROM, ON VAN REENEN RD)	2822	2922	2829ADA
LADYSMITH (18 KM WEST, ON VAN REENEN MAIN RD)	2822	2937	2829BCA
LAKE BANGAZI	2808	3233	2832BAC
LAKE BHANGAZI NORTH	2739	3239	2732DAD
LAKE ETEZA	2829	3209	2832ACD
LAKE SIBAYI	2720	3245	2732BDA
LAKE SIBAYI (BANANA PAN)	2722	3237	2732BCA
LAKE SIBAYI (BANDABANDA BAY)	2721	3243	2732BCB
LAKE SIBAYI (STAMCO)	2721	3243	2732BCB
LAKE ST LUCIA	2800	3232	2832BAA
LANGEWACHT	2822	3105	2831ACA
LEBOMBO MOUNTAINS	2707	3207	2732AAA
LEEUWSPOOR FARM (NO. 647)	2725	3152	2731BDC
LELIEFONTEIN (PIETERMARITZBURG)	2937	3023	2930CBB
LIDGETTON	2927	3007	2930ACC
LINCOLNMEAD	2936	3023	2930CBB
LINWOOD FOREST	2934	3004	2930CAA
LONGSTEAD FARM	2930	3022	2930CBA
LOSKOP	2856	2937	2829DCC
LOTENI NATURE RESERVE	2927	2932	2929BCC

LOCALITY	LATITUDE	LONGITUDE	GRIDREF
LOUWSBURG	2735	3117	2731CBA
LOWER UMGENI RIVER	2902	3008	2930AAB
MADLANGULA	2657	3250	2632DDC
MAGUDU	2732	3139	2731DAB
MAKATINI FLATS (=OTOBOTINI)	2725	3208	2732ACD
MAKHAKE STORE (=SANI PASS)	2936	2919	2929CBA
MALVERN	2953	3056	2930DDD
MANABA	2715	3225	2732ADB
MANDERSTON	2944	3026	2930CBD
MANDINI	2909	3125	2931ABD
MANGUZI FOREST	2658	3244	2632DCD
MANOR GARDENS (DURBAN)	2952	3100	2931CCA
MANZENGWENYA FOREST STATION (LALANEK)	2713	3246	2732BBC
MAPELANE	2825	3225	2832ADD
MAPUTA (=MANGUZI VILLAGE)	2700	3245	2732BBA
MARIANNHILL	2952	3050	2930DDA
MARIANNWOOD NATURE RESERVE (PINETOWN)	2949	3051	2930DDA
MARITZDAAL	2928	3003	2930ACC
MASON'S MILL (PIETERMARITZBURG)	2937	3023	2930CBB
MATATIELE	3020	2848	3028BDA
MATIWANE	2826	2954	2829BDD
MAURANN (UMFOLOZI FLATS)	2828	3215	2832ADC
MBUMBAZI NATURE RESERVE (PORT SHEPSTONE)	3048	3016	3030CDA
MEADOW FARM	2956	2952	2929DDC
MELMOTH	2834	3119	2831CBA
MELMOTH (19 KM FROM, ON NONGOMA RD)	2823	3125	2831ADD
MERRIVALE	2930	3014	2930CAB
MFKAYI	2807	3222	2832ABA
MFONGOSI	2843	3048	2830DBC
MGENI VLEI NATURE RESERVE	2929	2948	2929BDC
MHLATUZE RIVER (BRIDGE ON ESHOWE–MELMOTH RD)	2946	3130	2831CCA
MHLOPENI RANCH	2901	3025	2930ABB
MICHAELHOUSE GOLF COURSE	2924	3003	2930ACC
MID-ILLOVO	2952	3037	2930DCA
MIDDLERUS	2858	3019	2830CDC
MIDMAR DAM PUBLIC RESORT/NATURE RESERVE	2932	3012	2930CAB

LOCALITY	LATITUDE	LONGITUDE	GRIDREF
MILLS CIRCLE (PIETERMARITZBURG)	2937	3023	2930CBB
MISSION ROCKS – SEE EASTERN SHORES			
MITCHELL DAM (=KARKLOOF FOREST)	2920	3011	2930ACB
MITCHELL PARK (DURBAN)	2950	3100	2931CCA
MKONDENI	2936	3023	2930CBB
MKUZI GAME RESERVE	2737	3214	2732CAB
MKUZI GAME RESERVE (SAND FOREST)	2740	3212	2732CAD
MKUZI RIVER	2737	3207	2732CAA
MKUZI VILLAGE	2737	3202	2732CAA
MKYOLA (40 KM SOUTH-EAST MKUZI)	2800	3212	2832AAB
MONKS COWL FOREST STATION	2902	2924	2929ABB
MONZI	2826	3218	2832ADC
MONZI (BETWEEN MONZI AND MTUBATUBA)	2822	3207	2832ACA
MOOI RIVER	2913	3000	2930AAC
MOOIPLAAS PLANTATION	2833	3109	2831CAB
MOOR PARK NATURE RESERVE	2905	2948	2929BBA
MOORFIELD FARM	2753	2943	2729DCD
MORNINGSIDE (DURBAN)	2950	3100	2931CCA
MOSELEY	2951	3053	2930DDB
MOSIE SWAMP	2735	3235	2732DAA
MOUNT CURRIE NATURE RESERVE	3025	2930	3029BCC
MOUNTAIN LAKES ADVENTURE CAMPSITE (=JOZINI DAM)	2725	3204	2732ACC
MPOSA (=MPOSA HILL)	2839	320S	2832CAC
MSELENI	2722	3231	2732BCA
MSINYINI PAN	2822	3152	2831BDA
MOUNT EDGECOMBE	2942	3102	2931CAC
MOUNT MORELAND	2938	3105	2931CAC
MTOUGON	?	?	?
MTUBATUBA	2825	3210	2832ACD
MTUBATUBA/ST LUCIA RD	2823	3222	2832ADC
MTUNZINI	2857	3146	2831DDC
MUDEN	2858	3019	2830CDC
NAGLE DAM	2935	3037	2930DAA
NCANDU NATURE RESERVE	2752	2942	2729DCB
NDUMU GAME RESERVE	2654	3220	2632CDC
NDUMU GAME RESERVE (BANZI PAN)	2653	3215	2632CDC

LOCALITY	LATITUDE	LONGITUDE	GRIDREF
NDUMU GAME RESERVE (SITE 5)	2654	3215	2632CDC
NDUMU (NYANITI)	2653	3215	2632CDC
NDUMU POLICE STATION	2655	3216	2632CDC
NDUMU STORE	2655	3216	2632CDC
NEW GERMANY	2948	3053	2930DDB
NEW HANOVER	2921	3032	2930BCA
NEWCASTLE	2744	2955	2729DBD
NEWCASTLE (5 KM FROM, ON DUNDEE RD)	2746	2955	2729DDB
NEWCASTLE (10 KM FROM, ON UTRECHT RD)	2739	2957	2729DBD
NEWCASTLE (16 KM SOUTH)	2755	2955	2729DDD
NGELI FOREST	3037	2943	3029DAB
NGOME FOREST RESERVE	2748	3126	2731CDB
NGOME POLICE STATION	2750	3124	2731CDB
NGOYE FOREST RESERVE (=NGOYE HILLS, UMGOYE FOREST)	2849	3139	2831DCB
NGUBEVU MINE	2844	3038	2830DAD
NGUTSHANE DIP	2714	3235	2732BAC
NGXWALA HILL	2735	3205	2732CAA
NHLOSANE FOREST	2935	2958	2929DBB
NJEKAYI	2807	3222	2832ABA
NKANDHLA	2837	3105	2831CAA
NKONJENI HOSPITAL (ON ULUNDI RD)	2813	3125	2831ABD
NKONKONI (10 KM NORTH MKUZE)	2733	3201	2732CAA
NONDWANA FALLS	2952	3102	2931CCA
NONDWENI	2810	3048	2830BBC
NONGOMA	2750	3139	2731DCB
NORTH PARK	2952	3053	2930DDB
NORTHDENE	2952	3054	2930DDB
NORTHERN SEWAGE WORKS (=SEACOW LAKE)	2952	3100	2931CCA
NOTTINGHAM ROAD	2921	3000	2930ACA
NTABAMHLOPE (NEAR)	2905	2935	2929BAA
NTAMBANANA	2836	3144	2831DAB
NTOMBENI	?	?	?
NYALAZI FOREST	2812	3222	2832ABC
NYAMITI PAN (=NDUMU?)	2652	3207	2632CCA
ORIBI (PIETERMARITZBURG)	2937	3023	2930CBB
ORIBI GORGE NATURE RESERVE	3042	3014	3030CAD

LOCALITY	LATITUDE	LONGITUDE	GRIDREF
ORIBI–LIND VALLEY GAME RANCH	3042	3014	3030CAD
OTHOBOTHINI	2707	3207	2732AAA
OTTO'S BLUFF	2929	3023	2930ADD
PAPAMA (24 KM EAST OF ESHOWE)	2853	3143	2831DCD
PARADISE VALLEY NATURE RESERVE	2950	3054	2930DDB
PARK RYNIE	3019	3044	3030BCB
PAULPIETERSBURG	2725	3050	2730BDC
PENNINGTON	3023	3041	3030BCD
PERNA PERNA RESORT (ST LUCIA)	2817	3225	2832ADB
PETCHAYE (27 KM SOUTH-EAST OF ESTCOURT)	2905	2957	2929BBB
PHOENIX	2943	3101	2931CAC
PIET RETIEF (32 KM FROM, ON PAULPIETERSBURG RD)	2718	3052	2730BDA
PIETERMARITZBURG	2937	3023	2930CBB
PIETERMARITZBURG BOTANIC GARDENS	2936	3020	2930CBA
PIETERMARITZBURG GOLF COURSE	2937	3023	2930CBB
PIGEON VALLEY PARK (DURBAN)	2950	3100	2931CCA
PINETOWN	2949	3051	2930DDA
POLY SHORTS BRIDGE	2937	3022	2930CBA
POMEROY	2830	3025	2830CBB
POMEROY (17 KM FROM, ON GREYTOWN RD)	2840	3027	2830CBD
PONGOLA BUSH NATURE RESERVE	2719	3029	2730ADB
PONGOLA NATURE RESERVE	2725	3155	2731ADD
PONGOLA RIVER (=JOZINI DAM)	2725	3204	2732ACC
PONGOLAPOORT DAM (=JOZINI DAM)	2725	3204	2732ACC
PONT CARAVAN PARK	3104	3010	3130AAB
PORT EDWARD	3103	3012	3130AAB
PORT SHEPSTONE	3046	3024	3030CDB
POTTERSHILL FARM (NO. 4085)	2729	2947	2729BDC
PRESTBURY (PIETERMARITZBURG)	2937	3023	2930CBB
PUNTANS HILL	2950	3100	2931CCA
QUDENI	2837	3047	2830DBA
QUEEN ELIZABETH PARK	2937	3023	2930CBB
QUIXOTE FARM (23 KM SOUTH MKUZE)	2745	3158	2731DDB
RAISETHORPE	2932	3024	2930CBB
RAMSGATE	3055	3020	3030CDC
RANDFONTEIN	2810	3014	2830AAD

LOCALITY	LATITUDE	LONGITUDE	GRIDREF
REDCLIFF FARM	2913	3036	2930BAC
REDHILL	2947	3102	2931CCA
REDLANDS	3036	2943	3029DAB
RENISHAW	3016	3045	3030BDA
RICHARDS BAY	2847	3205	2832CCA
RICHARDS BAY (11 KM WEST, ON EMPANGENI RD)	2847	3200	2832CCA
RICHARDS BAY HOTEL (7 KM FROM, ON EMPANGENI RD)	2847	3200	2832CCA
RICHMOND	2952	3016	2930CDA
RINGWOOD FARM	2937	3037	2930DAA
ROCKCLIFF FARM	2837	2901	2829CAA
RONAN EGG FARM	2944	3023	2930CBD
ROOSFONTEIN NATURE RESERVE (WESTVILLE)	2951	3055	2930DDB
ROSEGLEN	2950	3100	2931CCA
ROSELEIGH FARM	2920	3035	2930BCA
ROSETTA	2920	2958	2929BDB
ROSSLEA ESTATE	3019	3040	3030BCB
ROSTHWAITE	2741	3043	2730DAD
ROYAL NATAL NATIONAL PARK	2840	2855	2828DBD
ROYAL NATAL NATIONAL PARK (SENTINEL PEAK)	2845	2854	2828DDB
SANI PASS	2936	2918	2929CBA
SANTA SUZANNE SANCTUARY	2804	3218	2832ABA
SARNIA	2950	3053	2930DDB
SAXONY CAVE (PIETERMARITZBURG)	2928	3020	2930ADC
SCALEBY FARM	2940	2959	2929DBD
SCOTTBURGH	3017	3045	3030BDA
SCOTTSVILLE (PIETERMARITZBURG)	2936	3022	2930CBB
SEA VIEW	2954	3058	2930DDD
SEACOW LAKE (=NORTHERN SEWAGE WORKS)	2952	3100	2931CCA
SEAFORTH	2952	3016	2930CDA
SEEWATER	?	?	?
SEVEN OAKS	2913	3036	2930BAC
SEZELA (=BENEVA TUNNEL)	3021	3038	3030BCB
SHALIMAR FARM (NEAR ORIBI GORGE)	3042	3014	3030CAD
SHALLCROSS	2953	3052	2930DDC
SHEMULA (=SHEMULA'S PONT)	2706	3215	2732ABA
SHEPSTONE RESERVE (=UNIVERSITY OF NATAL, DURBAN)	2952	3059	2930DDB

LOCALITY	LATITUDE	LONGITUDE	GRIDREF
SHIHANGWANA (6 KM FROM, ON ROAD TO MAPUTA)	2703	3225	2732ABB
SHONGWE STORE	2722	3207	2732ACA
SHONGWENI POLO CLUB	2950	3044	2930DCB
SHONGWENI RESOURCE RESERVE (=DAM)	2951	3043	2930DCB
SIBHUDENI	2846	3110	2831CCB
SIHANGWANE (=SHIHANGWANA, SIHANGWANA)	2703	3225	2732ABB
SMALLHOEK FARM (19 KM WEST OF LADYSMITH)	2833	2935	2829DAA
SOBANTU VILLAGE (PIETERMARITZBURG)	2937	3023	2930CBB
SODWANA BAY	2733	3242	2732DAB
SOUTHBROOM	3054	3019	3030CDC
SOUTHPORT	3041	3030	3030DAC
SPEEDWELL MINES	2840	3059	2830DBD
SPIOENKOP NATURE RESERVE	2841	2931	2829DAC
SPRING GROVE FARM	2825	2958	2829BDD
ST BARNABAS CHURCH	2956	3101	2931CCC
ST LUCIA FOREST STATION	2817	3225	2832ADB
ST LUCIA VILLAGE	2823	3225	2832ADD
ST LUCIA VILLAGE (12 KM NORTH)	2816	3228	2832ADB
ST MICHAEL'S-ON-SEA	3050	3025	3030CDB
STANGER (17 KM FROM, ON GLEN DALE RD)	2922	3107	2931ACA
STANGER AREA	2920	3118	2931ADA
STANGER BEACH	2922	3122	2931ADA
STELLAWOOD (DURBAN)	2950	3100	2931CCA
STRATHFIELDSAY FARM	2952	3028	2930CDB
SUGAR LOAF CAMP (ST LUCIA)	2823	3225	2832ADD
SWEETWATERS (PIETERMARITZBURG)	2937	3016	2930CBA
SYDENHAM (DURBAN)	2952	3107	2931CCA
TABAMHLOPE	2907	2939	2929BAB
TABLE MOUNTAIN	2936	3036	2930DAA
TANGLEWOOD FARM (PINETOWN)	2949	3052	2930DDA
TAYSIDE	2804	3023	2830ABB
TEMBE ELEPHANT PARK	2702	3225	2732ABB
TETE PAN	2709	3216	2732ABC
TETE PAN CAMP/SCHOOL (59 KM NORTH OF UBOMBO)	2709	3216	2732ABC
THE GRANGE FARM	2928	3010	2930ACD
THE HOEK	2913	2959	2929BBD

LOCALITY	LATITUDE	LONGITUDE	GRIDREF
THEUNIS BESTER GAME FARM (LOWER MKUZI)	2750	3215	2732CDA
THOMAS MORE COLLEGE (KLOOF)	2947	3051	2930DDA
TOWN BUSH (PIETERMARITZBURG)	2933	3021	2930CBA
TOWN HILL	2936	3023	2930CBB
TRETOWER FARM	2945	2925	2929CDB
TREVERTON	2921	3000	2930ACA
TUGELA ESTATES	2845	3009	2830CCB
TUMBLE INN (PIETERMARITZBURG)	2937	3023	2930CBB
TWEEDIE	2929	3010	2930ACD
TWIN STREAMS FARM	2859	3142	2831DCD
TWO STREAMS FARM (KARKLOOF FOREST)	2923	3017	2930ADC
UBOMBO VILLAGE/BUSH/MISSION	2734	3204	2732CAA
UITSICAT FARM (NO. 501)	2752	3107	2731CCA
ULUNDI (10 KM FROM)	2813	3125	2831ABD
UMBILO (DURBAN)	2950	3100	2931CCA
UMBUMBULU	2959	3043	2930DCD
UMDONI PARK	3024	3041	3030BCD
UMFOLOZI GAME RESERVE	2820	3152	2831BDA
UMFOLOZI GAME RESERVE (MASINDA CAMP)	2817	3152	2831BDA
UMFOLOZI GAME RESERVE (WHITE/BLACK UMFOLOZI RIVERS)	2821	3159	2831BDB
UMFOLOZI STATION	2814	3114	2831AAD
UMGENI HATCHERY	2935	3037	2930DAA
UMGENI PARK (DURBAN)	2950	3100	2931CCA
UMGENI VALLEY GAME RANCH	2929	3020	2930ADC
UMHLANGA ROCKS (=UMHLANGA NATURE RESERVE)	2944	3105	2931CAC
UMHLOTI RIVER	2938	3103	2931CAC
UMKOMAAS RIVER (IXOPO–RICHMOND RD)	3001	3015	3030ABA
UMKOMAAS TOWN/RIVER	3013	3048	3030BBC
UMLALAZI NATURE RESERVE (=UMLALAZI RIVER)	2856	3146	2831DDC
UMPAMBINYONI RIVER MOUTH (=SCOTTBURGH)	3017	3045	3030BDA
UMSUNDUZI RIVER	?	?	?
UMTAMVUMA NATURE RESERVE	3102	3012	3130AAB
UMTWALUME RIVER MOUTH	3029	3038	3030BCD
UMVOTI	2911	3019	2930ABC
UMVOTI VLEI NATURE RESERVE	2908	3034	2930BAC
UMZIKI PAN	?	?	?

LOCALITY	LATITUDE	LONGITUDE	GRIDREF
UMZIMKULU	3016	2956	3029BDB
UNDERBERG (TOWN, TROUT HATCHERY)	2947	2930	2929DCA
UNIVERSITY OF NATAL (DURBAN CAMPUS)	2950	3100	2931CCA
UNIVERSITY OF NATAL (PIETERMARITZBURG CAMPUS)	2936	3023	2930CBB
USUTU RIVER (ABERCORN PONT)	2652	3207	2632CCA
UTRECHT (33 KM FROM, ON BLOOD RIVER STATION RD)	2752	3032	2730DCA
VERGELEGEN NATURE RESERVE	2932	2927	2929CBB
VERGEVAL FARM	2729	3142	2731BCD
VERNON CROOKES NATURE RESERVE	3017	3035	3030BCA
VIRGINIA FARM	2940	2957	2929DBD
VRYHEID	2745	3047	2730DDA
VRYHEID (11 KM FROM, ON LOUWSBURG RD)	2745	3055	2730DDB
VRYHEID (11 KM FROM, ON UTRECHT RD)	2750	3040	2730DCB
VRYHEID NATURE RESERVE	2745	3047	2730DDA
WARBURTUN (?)	?	?	?
WARNER BEACH	3004	3052	3030BBA
WARTBURG	2926	3035	2930BCC
WATERFALL	2945	3050	2930DDA
WATERFALL FARM	2945	3050	2930DDA
WEENEN NATURE RESERVE	2851	3000	2830CCA
WEMBLY (PIETERMARITZBURG)	2937	3023	2930CBB
WESTON AGRICULTURAL COLLEGE	2912	3002	2930AAC
WESTVILLE	2950	3056	2930DDB
WEZA	3036	2943	3029DAB
WEZA STATE FOREST	3034	2943	3029DAB
WEZA STATE FOREST (LOWER STINKWOOD FOREST)	3034	2943	3029DAB
WILLBROOK (=WILLBROEK)	2908	2952	2929BBC
WINSLOW FARM	?	?	?
WINTERSKLOOF	2935	3016	2930CBA
WINTERTON	2849	2932	2829DCA
WOODHOUSE	2936	3023	2930CBB
WYFORD FARM	2825	2926	2829ADD
WYLDE HOLME FARM	2947	2940	2929DCB
YARM? (ESTON)	2953	3031	2930DCC
YELLOWWOOD PARK	2955	3056	2930DDD
ZIQHUMENE	2811	3157	2831BBD
ZWARTKOP VALLEY (PIETERMARITZBURG)	2937	3023	2930CBB

BIBLIOGRAPHY

Acocks, J.P.H. 1988. Veld types of South Africa. Third Edition. *Memoirs of the Botanical Survey of South Africa* 57: 1–146.

Avery, D.M. 1991. Late Quaternary incidence of some micromammalian species in Natal. *Durban Museum Novitates* 16: 1–11.

Baker, C.M. & Meester, J. 1977. Postnatal physical and behavioural development of *Praomys (Mastomys) natalensis* (A. Smith 1836). *Zeitschrift für Säugetierkunde* 42: 295–306.

Baker, C.M. & Meester, J.A.J. 1991. A new record of the greater dwarf shrew (*Suncus lixus*) from Natal. *Durban Museum Novitates* 16: 36.

Barrett, L. & Henzi, S.P. 1997. An inter-population comparison of body weight in chacma baboons. *South African Journal of Science* 93: 436–8.

Bearder, S.K. 1997. Thick-tailed bushbaby. In *The complete book of southern African mammals*, edited by G. Mills & L. Hes. Cape Town: Struik Winchester.

Bennett, N.C. 1989. The social structure and reproductive biology of the common mole-rat, *Cryptomys h. hottentotus* and remarks on the trends in reproduction and sociality in the family Bathyergidae. *Journal of Zoology (London)* 219: 45–59.

Bergmans, W. 1994. Taxonomy and biogeography of African fruit bats (Mammalia, Megachiroptera). The genus *Rousettus* Gray. *Beaufortia* 44(4): 79–126.

Bernard, R.T.F. 1980. Reproductive cycles of *Miniopterus schreibersii natalensis* (Kuhl 1819) and *Miniopterus fraterculus* (Thomas & Schwann 1906). *Annals of the Transvaal Museum* 32(3): 55–64.

Bernard, R.T.F. 1982a. Female reproductive cycle of *Nycteris thebaica* (Microchiroptera) from Natal, South Africa. *Zeitschrift für Säugetierkunde* 47(1): 12–18.

Bernard, R.T.F. 1982b. Monthly changes in the female reproductive organs and the reproductive cycle of *Myotis tricolor* (Vespertilionidae: Chiroptera). *South African Journal of Zoology* 17: 79–84.

Bernard, R.T.F. 1983. Reproduction of *Rhinolophus clivosus* (Microchiroptera) in Natal, South Africa. *Zeitschrift für Säugetierkunde* 48: 321–9.

Bernard, R.T.F. & Meester, J.A.J. 1982. Female reproduction and the female reproductive cycle of *Hipposideros caffer caffer* (Sundevall 1846) in Natal, South Africa. *Annals of the Transvaal Museum* 33(8): 131–44.

Bernard, R.T.F., Bronner, G.N., Taylor, P.J., Bojarski, C. & Tsita, J.N. 1994. A seasonal reproduction in the Hottentot golden mole, *Amblysomus hottentotus*, from the summer rainfall region of South Africa. *South African Journal of Science* 90: 547–9.

Bernard, R.T.F., Happold, D.C.D. & Happold, M. 1997. Sperm storage in a seasonally reproducing African vespertilionid, the banana bat (*Pipistrellus nanus*) from Malawi. *Journal of Zoology (London)* 241: 161–74.

Bourquin, O. 1988. Insectivora, Chiroptera, Primates, Pholidota, Lagomorpha, Rodentia and Hyracoidea. Distribution, importance and management/research requirements in Natal. KwaZulu-Natal Nature Conservation Service Internal Report.

Bourquin, O. & Mathias, I. 1984. The vertebrates of Oribi Gorge Nature Reserve: 1. *Lammergeyer* 33: 35–44.

Bourquin, O., Vincent, J. & Hitchins, P.M. 1971. The vertebrates of the Hluhluwe Game Reserve – Corridor (State land) – Umfolozi Game Reserve complex. *Lammergeyer* 14: 5–63.

Bourquin, O. & Sowler, S.G. 1980. The vertebrates of Vernon Crookes Nature Reserve. *Lammergeyer* 28: 20–32.

Bothma, J. du P. 1967. Recent Hyracoidea of southern Africa. *Annals of the Transvaal Museum* 25: 117–52.

Bothma, J. du P. 1971. Order Hyracoidea, Part 12. In *The Mammals of Africa: an identification manual*, edited by J. Meester & H.W. Setzer. Washington D.C.: Smithsonian Institution Press.

Bowie, R.C.K., Jacobs, D.S. & Taylor, P. Resource utilisation by two morphologically similar insectivorous bats (*Nycteris thebaica* and *Hipposideros caffer*) sharing the same night roost. *South African Journal of Zoology* (in press).

Brooks, P.M. 1974. The ecology of the four-striped field mouse *Rhabdomys pumilio* (Sparrman 1984) with particular reference to a population on the Van Riebeeck Nature Reserve. DSc Thesis, University of Pretoria.

Bronner, G.N. 1986. Demography of the multimammate mouse *Mastomys natalensis* (A. Smith 1834) on Bluff Nature Reserve, Durban. MSc Thesis, University of Natal.

Bronner, G.N. 1990. New distribution records for four mammal species, with notes on their taxonomy and ecology. *Koedoe* 33(2): 1–7.

Bronner, G.N. 1995. Systematic revision of the golden mole genera *Amblysomus, Chlorotalpa* and *Calcochloris* (Insectivora: Chrysochloromorpha; Chrysochloridae). PhD Thesis, University of Natal.

Bronner, G.N. & Meester, J.A.J. 1987. Mammals of Bluff Nature Reserve. *Lammergeyer* 38: 1–7.

Bronner, G.N. & Meester, J.A.J. 1988. *Otomys angoniensis. Mammalian Species* 306: 1–6.

Brown, C.R. & Bernard, R.T.F. 1994. Thermal preference of Schreiber's long-fingered (*Miniopterus schreibersii*) and Cape horseshoe (*Rhinolophus capensis*) bats. *Comparative Biochemistry and Physiology* 107A(3): 439–49.

Bruton, M.N. 1978. Recent mammal records from Eastern Tongaland in KwaZulu, with notes on hippopotamus in Lake Sibayi. *Lammergeyer* 24: 19–27.

Camp, K.G.T. 1995. The bioresource units of KwaZulu-Natal. Cedara Report N/A/95/32. KwaZulu-Natal Department of Agriculture, Pietermaritzburg.

Chimimba, C. 1997. Systematic revision of southern African *Aethomys*, Thomas 1915 (Rodentia: Muridae). PhD Thesis, University of Pretoria.

Coetzee, C.G. 1977. Genus *Steatomys*, Part 6.8, pp. 1–4. In *The mammals of Africa: an identification manual*, edited by J. Meester & H.W. Setzer. Washington D.C.: Smithsonian Institution Press.

Contrafatto, G., Meester, J.A., Willan, K., Taylor, P.J., Roberts, M.A. & Baker, C.M. 1992a. Genetic variation in the

African rodent subfamily Otomyinae (Muridae). II. Chromosomal changes in some populations of *Otomys irroratus*. *Cytogenetics and Cell Genetics* 59: 293–9.

Contrafatto, G., Meester, J., Bronner, G., Taylor, P.J. & Willan, K. 1992b.   Genetic variation in the African rodent subfamily Otomyinae (Muridae). IV: Chromosome G-banding anlysis of *Otomys irroratus* and *O. angoniensis*. *Israel Journal of Zoology* 38: 277–91.

Crozier, M. 1994.   Clues from the baboons. *University of Natal Focus* 5(4): 18–19.

Davis, D.H.S. 1975.   Genera *Tatera* and *Gerbillurus*. Part 6.4, pp. 1–7. In *The mammals of Africa: an identification manual*, edited by J. Meester & H.W. Setzer. Washington D.C.: Smithsonian Institution Press.

Davis, R.M. 1973.   The ecology and life history of the vlei rat, *Otomys irroratus* (Brants 1827) on the Van Riebeeck Nature Reserve. DSc Thesis, University of Pretoria.

Davis, R.M. & Meester, J. 1981.   Reproduction and postnatal development in the vlei rat, *Otomys irroratus*, on the Van Riebeeck Nature Reserve. *Mammalia* 45: 99–116.

De Graaff, G. 1981.   *The rodents of southern Africa*. Pretoria: Butterworths.

De Moor, P.P. 1969.   Seasonal variation in local distribution, age classes and population density of the gerbil, *Tatera brantsii*, on the South African highveld. *Journal of Zoology (London)* 157: 399–411.

Dean, W.R.J. 1978.   Conservation of the white-tailed rat in South Africa. *Biological Conservation* 13(2): 133–40.

Dippenaar, N.J. 1979.   Variation in *Crocidura mariquensis* (A. Smith 1844), Part 2 (Mammalia: Soricidae). *Annals of the Transvaal Museum* 32: 1–34.

Dippenaar, N.J., Swanepoel, P. & Gordon, D.H. 1993.   Diagnostic morphometrics of two medically important southern African rodents, *Mastomys natalensis* and *M. coucha* (Rodentia: Muridae). *South African Journal of Science* 89: 300–3.

Dixon, J.E.W. 1964.   Preliminary notes on the mammal fauna of the Mkuzi Game Reserve. *Lammergeyer* 3: 40–58.

Dixon, J.E.W. 1966.   Notes on the mammals of Ndumu Game Reserve. *Lammergeyer* 6: 24–40.

Duckworth, A. 1984.   Aspects of small habitat preference at Nyalazi State Forest, Zululand. Directorate of Forestry, Pietermaritzburg. Unpublished report.

Earl, Z. 1978.   Postnatal development of *Saccostomus campestris*. *African Small Mammal Newsletter* 2: 10–12.

Ellerman, J.R., Morrison-Scott, T.C.S. & Hayman, R.W. 1953.   *Southern African mammals: 1758–1951: a reclassification*. London: Trustees, British Museum (Natural History).

Fenton, M.B. 1985.   The feeding behaviour of insectivorous bats: echolocation, foraging strategies, and resource partitioning. *Transvaal Museum Bulletin* 21: 5–19.

Ferreira, S.M. & Van Aarde, R.J. 1996.   Changes in community characteristics of small mammals in rehabilitating coastal dune forests in northern KwaZulu/Natal. *African Journal of Ecology* 34: 113–30.

Fotheringham, P.J. 1981.   *Agriquest*. Pietermaritzburg: Department of Agriculture (Natal Region).

Fourie, P.B. 1977.   Accoustic communication in the rock hyrax, *Procavia capensis*. *Z. Tierpsychol* 44: 194–219.

Gordon, D.H. 1987.   Discovery of another species of tree rat. *Transvaal Museum Bulletin* 22: 30–2.

Gordon, D.H. 1991.   Chromosomal variation in the water rat *Dasymys incomtus* (Rodentia: Muridae). *Journal of Mammalogy* 72(2): 411–14.

Gordon, D.H. & Rautenbach, I.L. 1980.   Species complexes in medically important rodents: chromosomal studies of *Aethomys, Tatera* and *Saccostomus* (Rodentia: Muridae, Cricetidae). *South African Journal of Science* 76: 559–61.

Goulden, E.A. & Meester, J. 1978.   Notes on the behaviour of *Crocidura* and *Myosorex* (Mammalia: Soricidae) in captivity. *Mammalia* 42: 197–207.

Green, C.A., Keogh, H., Gordon, D.H., Pinto, M. & Hartwig, E.K. 1980.   The distribution, identification, and naming of the *Mastomys natalensis* species complex in southern Africa (Rodentia: Muridae). *Journal of Zoology (London)* 192: 17–23.

Groves, C.P. 1993.   Order Primates. In *Mammal species of the world*, edited by D.E. Wilson & D. Reeder. Washington D.C.: Smithsonian Institution Press.

Happold, D.C.D. & Happold, M. 1996.   The social organisation and population dynamics of leaf-roosting banana-bats, *Pipistrellus nanus* (Chiroptera, Vespertilionidae), in Malawi, east-central Africa. *Mammalia* 60(4): 517–44.

Hayman, R.W. & Hill, J.E. 1971.   Order Chiroptera, Part 2, pp. 1–73. In *The mammals of Africa: an identification manual*, edited by J. Meester & H.W. Setzer. Washington D.C.: Smithsonian Institution Press.

Heim de Balsac, H. & Meester, J. 1977.   Order Insectivora, pp. 1–129. In *The mammals of Africa: an identification manual*, edited by J. Meester & H.W. Setzer. Washington D.C.: Smithsonian Institution Press.

Henzi, S.P. & Lawes, M.J. 1987.   Breeding season influxes and the behaviour of adult male samango monkeys (*Cercopithecus mitis albogularis*). *Folia primatologica* 48: 125–36.

Henzi, S.P. & Lycett, J.E. 1995.   Population structure, demography, and dynamics of mountain baboons: an interim report. *American Journal of Primatology* 35: 155–63.

Herselman, J.C. & Norton, P.M. 1985.   The distribution and status of bats (Mammalia: Chiroptera) in the Cape Province. *Annals of the Cape Provincial Museums* (Natural History) 16(4): 73–126.

Hickman, G.C. 1979.   Burrow system construction of the bathyergid *Cryptomys hottentotus* in Natal, South Africa. *Zeitschrift für Säugetierkunde* 44: 153–62.

Hickman, G.C. 1982.   Copulation of *Cryptomys hottentotus* (Bathyergidae), a fossorial rodent. *Mammalia* 46: 293–8.

Hill, J.E. 1974.   A review of *Scotoecus* Thomas 1901 (Chiroptera: Vespertilionidae). *Bulletin of the British Museum of Natural History* 27: 169–88.

Honeycutt, R.L., Edwards, S.V., Nelson, K. & Nevo, E. 1987.   Mitochondrial DNA variation and the phylogeny of African mole rats (Rodentia: Bathyergidae). *Systematic Zoology* 36: 280–92.

Kearney, T.C. 1993. A craniometric analysis of three taxa of *Myosorex* from Natal and Transkei. MSc Thesis, University of Natal, Pietermaritzburg.

Kearney, T. & Taylor, P.J. 1997.   New distribution records of bats in KwaZulu-Natal. *Durban Museum Novitates* 22: 53–6.

Kern, N.G. 1977.   The influence of fire on populations of small mammals of the Kruger National Park. MSc Thesis, University of Pretoria.

Killick, D.J.B. 1978.   The Afro-Alpine region, pp. 515–60. In *Biogeography and ecology of Southern Africa*, Vol. 1. The Hague: Dr W. Junk bv Publishers.

Kingdon, J. 1971.   *East African mammals*, Vol. 1, (Primates, Hyraxes, Pangolins, Protoungulates, Sirenians). London: Academic Press.

Koopman, K.F. 1993.   Chiroptera. In *Mammal species of the world*, edited by D.E. Wilson & D. Reeder. Washington D.C.: Smithsonian Institution Press.

Kyle, R. 1996.   Sight records of *Galago moholi*, the lesser bushbaby, in KwaZulu-Natal. *Lammergeyer* 44: 39–40.

LaVal, R.K. & LaVal, M.L. 1977. Reproduction and behaviour of the African banana bat, *Pipistrellus nanus*. *Journal of Mammalogy* 58: 403–10.

LaVal, K. & LaVal, M.L. 1980. Prey selection by the slit-faced bat *Nycteris thebaica* (Chiroptera: Nycteridae) in Natal, South Africa. *Biotropica* 12(4): 241–6.

Lawes, M.J. 1990a. The distribution of the samango monkey (*Cercopithecus mitis erythrarchus* Peters 1852 and *Cercopithecus mitis labiatus* I. Geoffroy 1843) and forest history in southern Africa. *Journal of Biogeography* 17: 669–80.

Lawes, M.J. 1990b. The socioecology and conservation of the samango monkey (*Cercopithecus mitis erythrarchus* in Natal). PhD Thesis, University of Natal, Pietermaritzburg.

Lawes, M.J. 1992. Estimates of population density and correlates of the status of the samango monkey *Cercopithecus mitis* in Natal, South Africa. *Biological Conservation* 60: 197–210.

Lawes, M.J. 1997. Samango monkey. In *The complete book of southern African mammals*, edited by G. Mills & L. Hes. Cape Town: Struik Winchester.

Lawes, M.J., Henzi, S.P. & Perrin, M.R.P. 1990. Diet and feeding behaviour of samango monkeys (*Cercopithecus mitis labiatus*) in Ngoye Forest, South Africa. *Folia Primatalogica* 54: 57–69.

Laycock, P.A. 1976. A study of cave-dwelling microchiroptera in the Natal midlands. MSc Thesis, University of Natal, Pietermaritzburg.

Lee, A.K. 1976. Report on small mammals at Royal Natal National Park. KwaZulu-Natal Nature Conservation Service. Unpublished report.

Lynch, C.D. 1983. The mammals of the Orange Free State. *Memoirs van die Nasionale Museum Bloemfontein* 18: 1–218.

Lynch, C.D. 1989. The mammals of the North-Eastern Cape Province. *Memoirs van die Nasionale Museum Bloemfontein* 25: 1–116.

Lynch, C.D. 1994. The mammals of Lesotho. *Navorsinge van die Nasionale Museum Bloemfontein* 10(4): 177–241.

Lynch, C.D. & Watson, J.P. 1992. The distribution and ecology of *Otomys sloggetti* (Mammalia: Rodentia) with notes on its taxonomy. *Navorsinge van die Nasionale Museum Bloemfontein* 8(3): 141–58.

Maddalena, T., Mehmeti, A.M., Bronner, G. & Vogel, P. 1987. The karyotype of *Crocidura flavescens* (Mammalia, Insectivora) in South Africa. *Zeitschrift für Säugetierkunde* 52: 129–32.

Measroch, V. 1954. Growth and reproduction in the females of two species of gerbils, *Tatera brantsii* (A. Smith) and *Tatera afra* (Gray). *Proceedings of the Zoological Society of London* 124: 631–58.

Meester, J. 1963. A systematic revision of the shrew genus *Crocidura* in southern Africa. *Transvaal Museum Memoir* 13: 1–126.

Meester, J. & Lambrechts, A. Von W. 1971. The southern African species of *Suncus* Ehrenberg (Mammalia: Soricidae). *Annals of the Transvaal Museum* 27: 1–14.

Meester, J.A.J., Rautenbach, I.L., Dippenaar, N.J. & Baker, C.M. 1986. Classification of southern African mammals. *Transvaal Museum Monograph* 5: 1–359.

Melton, D.A. 1976. The biology of the aardvark (Tubulidentata Orycteropodidae). *Mammal Review* 6: 75–88.

Melton, D.A. & Daniels, C. 1986. A note on the ecology of the aardvark, *Orycteropus afer*. *South African Journal of Wildlife Research* 16: 112–14.

Mentis, M.T. 1974. Distribution of some wild animals in Natal. *Lammergeyer* 20: 1–68.

Mutere, F.A. 1973. A comparative study of reproduction in two populations of the insectivorous bats, *Otomops martiensseni*, at latitudes 1°5′S and 2°30′S. *Journal of Zoology* (London) 171: 79–92.

Nevo, E., Capanna, E., Corti, M., Jarvis, J.U.M. & Hickman, G.C. 1986. Karyotype differentiation in the endemic subterranean mole rats of South Africa (Rodentia, Bathyergidae). *Zeitschrift für Säugetierkunde* 51: 36–49.

Nevo, E., Ben-Shlomo, R., Beiles, A., Jarvis, J.U.M. & Hickman, G.C. 1987. Allozyme differentiation and systematics of the endemic subterranean mole rats of South Africa. *Biochemical Systematics and Ecology* 15: 489–502.

Njobe, K. 1997. The distribution and biogeography of the cryptic multimammate mice, *Mastomys natalensis* (A. Smith 1834) and *Mastomys coucha* (A. Smith 1836) in southern Africa. MSc Thesis, University of Pretoria.

Nowak, R.M. & Paradiso, J.L. 1983. *Walker's mammals of the world*, Vol. 2. Fourth Edition. Baltimore: John Hopkins University Press.

Panagis, K. & Nel, J.A.J. 1981. Growth and behavioural development in *Thamnomys dolichurus* (Rodentia: Muridae). *Acta Theriologica* 26: 381–92.

Perrin, M.R. 1980. The breeding stategies of two coexisting rodents, *Rhabdomys pumilio* (Sparrman 1784) and *Otomys irroratus* (Brants 1827). *Acta Oecologia Generalis* 1: 383–410.

Pillay, N. 1990. The breeding and reproductive biology of the vlei rat *Otomys irroratus*. MSc Thesis, University of Natal, Durban.

Pillay, N. 1993. The evolution and socio-ecology of two populations of vlei rat, *Otomys irroratus*. PhD Thesis, University of Natal, Durban.

Pillay, N., Willan, K. & Meester, J. 1995. Post-zygotic reproductive isolation in two populations of the African vlei rat *Otomys irroratus*. *Acta Theriologica* 40(1): 69–76.

Pringle, J.A. 1974. The distribution of mammals in Natal. Part 1. Primates, Hyracoidea, Lagomorpha (except *Lepus*), Pholidota and Tubulidentata. *Annals of the Natal Museum* 22: 173–86.

Pringle, J.A. 1977. The distribution of mammals in Natal. Part 2. Carnivora. *Annals of the Natal Museum* 23: 93–115.

Rautenbach, I.L. 1982. The mammals of the Transvaal. *Ecoplan Monograph* 1: 1–211.

Rautenbach, I.L. & Bronner, G.N. 1988. The mammals of Dukuduku Forest. The results of a survey conducted 1–7 October 1988. Unpublished report.

Rautenbach, I.L., Bronner, G.N. & Schlitter, D.A. 1993. Karyotypic data and attendant systematic implications for the bats of southern Africa. *Koedoe* 36(2): 87–104.

Rautenbach, I.L., Nel, J.A.J. & Root, G.A. 1981. Mammals of Itala Nature Reserve, Natal. *Lammergeyer* 31: 21–37.

Rautenbach, I.L., Skinner, J.D. & Nel, J.A.J. 1980. The past and present status of the mammals of Maputaland. In *Studies on the ecology of Maputaland*, edited by M.N. Bruton & K.H. Cooper. Grahamstown: Rhodes University.

Richardson, P.R.K. 1987. Aardwolf: the most specialized myrmecophagous mammal? *South African Journal of Science* 83: 643–6.

Richardson, E.J. & Perrin, M.R. 1992. Seasonal changes in body mass, torpidity, and reproductive activity of captive fat mice, *Steatomys pratensis*. *Israel Journal of Zoology* 38: 315–22.

Richardson, E.J. & Taylor, P.J. 1995. New observations on the large-eared free-tailed bat *Otomops martiensseni* in Durban, South Africa. *Durban Museum Novitates* 20: 72–4.

Roberts, A. 1931. New forms of South African mammals. *Annals of the Transvaal Museum* 13(3): 221–36.

Roberts, A. 1936. Report on a survey of the higher vertebrates of north-eastern Zululand. *Annals of the Transvaal Museum* 18: 163–251.

Roberts, A. 1951. *The mammals of South Africa.* Cape Town: Central News Agency.

Robbins, C.B. 1978. Taxonomic identification and history of *Scotophilus nigrata* (Schreber) (Chiroptera: Vespertilionidae). *Journal of Mammalogy* 59: 212–13.

Rowe-Rowe, D.T. 1977. Mammal survey of the Cathedral Peak area of the Natal Drakensberg. KwaZulu-Natal Nature Conservation Service. Unpublished report.

Rowe-Rowe, D.T. 1986. Ecology of some mammals in relation to conservation management in Giant's Castle Game Reserve. PhD Thesis, University of Natal, Durban.

Rowe-Rowe, D.T. 1986. Stomach contents of small mammals from the Drakensberg, South Africa. *South African Journal of Wildlife Research* 16: 32–5.

Rowe-Rowe, D.T. 1992. *The carnivores of Natal.* Pietermaritzburg: KwaZulu-Natal Nature Conservation Service.

Rowe-Rowe, D.T. 1994. *The ungulates of Natal.* Second Edition. Pietermaritzburg: KwaZulu-Natal Nature Conservation Service.

Rowe-Rowe, D.T. 1995. Small-mammal recolonisation of a fire-exclusion catchment after unscheduled burning. *South African Journal of Wildlife Research* 25(4): 133–7.

Rowe-Rowe, D.T. & Lowry, P.B. 1982. Influence of fire on small mammal populations in the Natal Drakensberg. *South African Journal of Wildlife Research* 12: 130–9.

Rowe-Rowe, D.T. & Meester J. 1982a. Habitat preferences and abundance relations of small mammals in the Natal Drakensberg. *South African Journal of Zoology* 17: 202–9.

Rowe-Rowe, D.T. & Meester, J. 1982b. Population dynamics of small mammals in the Drakensberg of Natal, South Africa. *Zeitschrift für Säugetierkunde* 47: 347–56.

Rowe-Rowe, D.T. & Meester, J. 1985. Altitudinal variation in external measurements of two small-mammal species in the Natal Drakensberg. *Annals of the Transvaal Museum* 34(3): 49–53.

Rydell, J. & Yalden, D.W. 1997. The diets of two high-flying bats from Africa. *Journal of Zoology* (London) 242: 69–76.

Schröder, W. & Mensah, G.A. 1987. Reproductive biology of *Thryonomys swinderianus* (Temminck). *Zeitschrift für Säugetierkunde* 52: 164–8.

Skinner, J.D. & Smithers, R.H.N. 1990. *The mammals of the southern African subregion.* Second Edition. Pretoria: University of Pretoria.

Smithers, R.H.N. 1971. The mammals of Botswana. *Museum Memoirs of the National Museums and Monuments of Rhodesia* 4: 1–340.

Smithers, R.H.N. 1983. *The mammals of the southern African subregion.* First Edition. Pretoria: University of Pretoria.

Smithers, R.H.N. 1986. South African Red Data Book: terrestrial mammals. *South African National Scientific Programmes Report* 125: 1–216.

Smithers, R.H.N. & Tello, J.L.P. Lobao. 1976. Checklist and atlas of the mammals of Mozambique. *Museum Memoirs of the National Museums and Monuments of Rhodesia* 8: 1–184.

Smithers, R.H.N. & Wilson, V.J. 1979. Checklist and atlas of the mammals of Zimbabwe Rhodesia. *Museum Memoirs of the National Museums and Monuments of Rhodesia* 9: 1–147.

Swanepoel, P. 1972. The population dynamics of rodents at Pongola, northern Zululand, exposed to dieldrin cover spraying. MSc Thesis, University of Pretoria.

Stuart, C.T. 1981. Notes on the mammalian carnivores of the Cape Province, South Africa. *Bontebok* 1: 1–58.

Swart, J. & Lawes, M.J. 1996. The effect of habit patch connectivity on samango monkey (*Cercopithecus mitis*) metapopulation persistence. *Ecological modelling* 93: 57–74.

Tainton, N.M., Bransby, D.I. & Booysen, P. De V. 1976. *Common veld and pasture grasses of Natal.* Pietermaritzburg: Shuter & Shooter.

Taylor, P.J. 1991. First record of Welwitsch's hairy bat from Natal. *Durban Museum Novitates* 16: 35–6.

Taylor, P.J. 1994. To catch a mole. *Farmer's Weekly.* 11 February: 33–4.

Taylor, P.J., Contrafatto, G. & Willan, K. 1993. Climatic correlates of chromosomal variation in the African vlei rat, *Otomys irroratus. Mammalia* 58(4): 623–34.

Taylor, P.J., Jarvis, J.U.M., Crowe, T.M. & Davies, K.C. 1985. Age determination in the Cape molerat *Georychus capensis. South African Journal of Zoology* 20: 261–7.

Taylor, P.J., Rautenbach, I.L., Gordon, D., Sink, K. & Lotter, P. 1995. Diagnostic morphometrics and southern African distribution of two sibling species of tree rat, *Thallomys paedulcus* and *Thallomys nigricauda* (Rodentia: Muridae). *Durban Museum Novitates* 20: 49–62.

Taylor, P.J., Richardson, E.J., Meester, J. & Wingate, L. 1994. New distribution records for six small mammal species in Natal, with notes on their taxonomy and ecology. *Durban Museum Novitates* 19: 59–66.

Tudge, C. 1994. Going bats over conservation. *New Scientist.* 12 March, 27–31.

Van der Merwe, M. 1975. Preliminary study on the annual movements of the Natal clinging bat. *South African Journal of Science* 71: 237–41.

Van der Merwe, M. & Rautenbach, I.L. 1986. Multiple births in Schlieffen's bat, *Nycticeius schlieffenii* (Peters 1859) (Chiroptera: Vespertilionidae) from the Southern African Subregion. *South African Journal of Zoology* 21: 48–50.

Van der Merwe, M. & Rautenbach, I.L. 1987. Reproduction in Schlieffen's bat, *Nycticeus schlieffenii*, in the eastern Transvaal lowveld, South Africa. *Journal of Reproductive Fertility* 81: 41–50.

Viljoen, S. 1980. A comparative study on the biology of two subspecies of tree squirrels, *Paraxerus palliatus tongensis* Roberts 1931 and *Paraxerus palliatus ornatus* (Gray 1864) in Zululand. DSc Thesis, University of Pretoria.

Viljoen, S. 1989. Taxonomy and historical zoogeography of the red squirrel *Paraxerus palliatus* (Peters 1852) in the Southern African subregion (Rodentia: Sciuridae). *Annals of the Transvaal Museum* 35: 49–60.

Vivier, L. & Van der Merwe, M. 1996. Reproductive pattern in the male Angolan free-tailed bat, *Tadarida (Mops) condylura* (Microchiroptera: Molossidae) in the Eastern Transvaal, South Africa. *Journal of Zoology (London)* 239: 465–76.

Vivier, L. & Van der Merwe, M. 1997. Reproduction in the female Angolan free-tailed bat, *Tadarida (Mops) condylura* (Microchiroptera: Molossidae), in the Eastern Transvaal, South Africa. *Journal of Zoology (London)* 243: 507–21.

Watson, J.P. 1990. Westward range extension of Temminck's hairy bat in South Africa and Lesotho. *South African Journal of Wildlife Research* 20: 119–21.

Whiten, A., Byrne, R.W. & Henzi, S.P. 1987. The behavioral ecology of mountain baboons. *International Journal of Primatology* 8: 367–88.

Willan, K. 1992. *Problem rodents and their control: A handbook for southern African farmers, foresters and smallholders.* Pietermaritzburg: The Natal Witness.

Wingate, L. 1983. The population status of five species of Microchiroptera in Natal. MSc Thesis, University of Natal.

Wirminghaus, J.O. & Nanni, R.F. 1989. Some additional *Suncus lixus* records for Natal. *Lammergeyer* 40: 4–5.

Wirminghaus, J.O. & Perrin, M.R. 1992. Diets of small mammals in a southern African temperate forest. *Israel Journal of Zoology* 38: 353–61.

INDEX OF SCIENTIFIC NAMES

Aethomys chrysophilus   109
Aethomys ineptus   110
Aethomys namaquensis   111
Amblysomus hottentotus   24
Amblysomus marleyi   26
Calcochloris obtusirostris   23
Cercopithecus mitis   69
Chaerephon ansorgei   62
Chaerephon pumila   62
Chalinolobus variegatus   49
Chlorocebus aethiops   68
Chlorotalpa sclateri   23
Chrysospalax villosus   22
Cloeotis percivali   41
Crocidura cyanea   15
Crocidura flavescens   11
Crocidura fuscomurina   17
Crocidura hirta   13
Crocidura maquassiensis   17
Crocidura mariquensis   13
Crocidura silacea   16
Cryptomys hottentotus   78
Dasymys incomtus   100
Dendrohyrax arboreus   72
Dendromus melanotis   91
Dendromus mesomelas   93
Dendromus mystacalis   93
Eidolon helvum   30
Elephantulus myurus   118
Epomophorus gambianus (EXCLUDED)   30
Epomophorus wahlbergi   28
Eptesicus capensis   52
Eptesicus hottentotus   51
Eptesicus rendalli   50
Eptesicus zuluensis   52
Galago moholi   66
Georychus capensis   80
Grammomys cometes   101
Grammomys dolichurus   102
Graphiurus murinus   112
Graphiurus ocularis (EXCLUDED)   112
Hipposideros caffer   40
Hystrix africaeaustralis   76
Kerivoula argentata   57
Kerivoula lanosa   58
Laephotis wintoni   49
Lemniscomys rosalia   97
Lepus capensis (POSSIBLE)   113
Lepus saxatilis   114
Manis temminckii   73
Mastomys coucha   107
Mastomys natalensis   105
Miniopterus fraterculus   42
Miniopterus schreibersii   42
Mops condylurus   64
Mormopterus acetabulosus (EXTINCT)   60

Mus minutoides   104
Mus musculus (INTRODUCED)   103
Myosorex cafer   9
Myosorex sclateri   10
Myosorex varius   7
Myotis tricolor   44
Myotis welwitschii   44
Mystromys albicaudatus   89
Nycteris hispida   33
Nycteris thebaica   33
Nycticeius schlieffenii   56
Orycteropus afer   72
Otolemur crassicaudatus   65
Otomops martiensseni   58
Otomys angoniensis   83
Otomys irroratus   84
Otomys laminatus   82
Otomys sloggetti   85
Papio hamadras   66
Paraxerus palliatus   75
Pedetes capensis   76
Petrodromus tetradactylus   117
Pipistrellus anchietai   47
Pipistrellus kuhlii   46
Pipistrellus nanus   48
Procavia capensis   70
Pronolagus crassicaudatus   116
Pronolagus rupestris   115
Rattus norvegicus (INTRODUCED)   112
Rattus rattus (INTRODUCED)   112
Rhabdomys pumilio   98
Rhinolophus blasii   37
Rhinolophus clivosus   35
Rhinolophus darlingi   36
Rhinolophus hildebrandtii (POSSIBLE)   35
Rhinolophus landeri   37
Rhinolophus simulator   38
Rhinolophus swinnyi   39
Rousettus aegyptiacus   31
Saccostomus campestris   90
Scotoecus albofuscus   56
Scotophilus dinganii   53
Scotophilus viridis   55
Steatomys krebsii   96
Steatomys parvus (EXCLUDED)   95
Steatomys pratensis   95
Suncus infinitesimus   20
Suncus lixus   18
Suncus varilla   19
Tadarida aegyptiaca   60
Taphozous mauritianus   32
Tatera brantsii   88
Tatera leucogaster   87
Thallomys nigricauda   108
Thallomys paedulcus   108
Thryonomys swinderianus   77

INDEX OF COMMON NAMES

Aardvark   72
Baboon, Chacma   66
Banana bat   48
Brown rat (INTRODUCED)   112
Bushbaby, lesser   66
Bushbaby, thick-tailed   65
Butterfly bat   49
Cane-rat, greater   77
Climbing mouse, Brant's   93
Climbing mouse, chestnut   93
Climbing mouse, grey   91
Dassie, rock   70
Dassie, tree   72
Dormouse, spectacled (EXCLUDED)   112
Dormouse, woodland   112
Dwarf shrew, greater   18
Dwarf shrew, least   20
Dwarf shrew, lesser   19
Elephant shrew, four-toed   117
Elephant shrew, rock   118
Fat mouse   95
Fat mouse, Kreb's   96
Fat mouse, tiny (EXCLUDED)   95
Forest shrew   7
Forest shrew, dark-footed   9
Forest shrew, Sclater's   10
Free-tailed bat, Angola   64
Free-tailed bat, Ansorge's   62
Free-tailed bat, Egyptian   60
Free-tailed bat, large-eared   58
Free-tailed bat, little   62
Free-tailed bat, Natal (EXTINCT)   60
Fruit bat, Egyptian   31
Fruit bat, Peters's epauletted (EXCLUDED)   30
Fruit bat, straw-coloured   30
Fruit bat, Wahlberg's epauletted   28
Gerbil, bushveld   87
Gerbil, highveld   88
Golden mole, Hottentot   24
Golden mole, Marley's   26
Golden mole, rough-haired   22
Golden mole, Sclater's   23
Golden mole, yellow   23
Hairy bat, Temminck's   44
Hairy bat, Welwitsch's   44
Hare, Cape (POSSIBLE)   113
Hare, Natal red   116
Hare, Scrub   114
Hare, Smith's red   115
Horseshoe bat, Bushveld   38
Horseshoe bat, Darling's   36
Horseshoe bat, Geoffroy's   35
Horseshoe bat, Hildebrandt's (POSSIBLE)   35
Horseshoe bat, Lander's   37
Horseshoe bat, Peak-saddle   37
Horseshoe bat, Swinny's   39

House bat, lesser yellow   55
House bat, Thomas's   56
House bat, yellow   53
House rat (INTRODUCED)   112
House mouse (INTRODUCED)   103
Leaf-nosed bat, Sundevall's   40
Long-eared bat, Winton's   49
Long-fingered bat, lesser   42
Long-fingered bat, Schreiber's   42
Molerat, Cape   80
Molerat, common   78
Monkey, samango   69
Monkey, vervet   68
Multimammate mouse   107
Multimammate mouse, Natal   105
Musk shrew, greater   11
Musk shrew, lesser grey-brown   16
Musk shrew, lesser red   13
Musk shrew, Makwassie   17
Musk shrew, reddish-grey   15
Musk shrew, swamp   13
Musk shrew, tiny   17
Namaqua rock mouse   111
Pangolin   73
Pipistrelle, Anchieta's   47
Pipistrelle, Kuhl's   46
Porcupine   76
Pouched mouse   90
Pygmy mouse   104
Red veld rat (two species)   109, 110
Schlieffen's bat   56
Serotine bat, Cape   52
Serotine bat, long-tailed   51
Serotine bat, Rendall's   50
Serotine bat, Zulu   52
Single-striped mouse   97
Slit-faced bat, Egyptian   33
Slit-faced bat, hairy   33
Sloggett's rat   85
Springhare   76
Squirrel, red   75
Striped mouse   98
Tomb bat, Mauritian   32
Tree rat   108
Tree rat, black-tailed   108
Trident bat, short-eared   41
Vlei rat   84
Vlei rat, Angoni   83
Vlei rat, laminate   82
Water rat   100
White-tailed rat   89
Woodland mouse   102
Woodland mouse, Mozambique   101
Woolly bat, Damara   57
Woolly bat, lesser   58